Applied Engineering Sciences

Applied Engineering Sciences

Edited by **Edgar Miller**

C WILLFORD PRESS

New York

Published by Willford Press,
118-35 Queens Blvd., Suite 400,
Forest Hills, NY 11375, USA
www.willfordpress.com

Applied Engineering Sciences
Edited by Edgar Miller

International Standard Book Number: 978-1-68285-042-8 (Hardback)

Printed in the United States of America.

Contents

Permissions

List of Contributors

Preface

Applied engineering is a field which focuses on the practical application of engineering principles for the design and implementation of new techniques for production. This book explores all the important aspects of applied engineering in the present day scenario. It includes some of the vital pieces of work being conducted across the world, on various topics such as laboratory-specific custom instrumentation, diagnostics, experimental techniques, etc. This text aims to serve as a resource guide for students and experts alike and contribute to the growth of the discipline.

The information shared in this book is based on empirical researches made by veterans in this field of study. The elaborative information provided in this book will help the readers further their scope of knowledge leading to advancements in this field.

Finally, I would like to thank my fellow researchers who gave constructive feedback and my family members who supported me at every step of my research.

Editor

Negative-resistance models for parametrically flux-pumped superconducting quantum interference devices

Kyle M Sundqvist[1][*] and Per Delsing[2]

*Correspondence:
kyle.sundqvist@gmail.com
[1] Electrical and Computer
Engineering, Texas A&M University,
College Station, TX 77843, USA
Full list of author information is
available at the end of the article

Abstract

A Superconducting QUantum Interference Device (SQUID) modulated by a fast oscillating magnetic flux can be used as a parametric amplifier, providing gain with very little added noise. Here, we develop linearized models to describe the parametrically flux-pumped SQUID in terms of an impedance. An unpumped SQUID acts as an inductance, the Josephson inductance, whereas a flux-pumped SQUID develops an additional, parallel element which we have coined the "pumpistor." Parametric gain can be understood as a result of a negative resistance of the pumpistor. In the degenerate case, the gain is sensitive to the relative phase between the pump and signal. In the nondegenerate case, gain is independent of this phase.

We develop our models first for degenerate parametric pumping in the three-wave and four-wave cases, where the pump frequency is either twice or equal to the signal frequency, respectively. We then derive expressions for the nondegenerate case where the pump frequency is not a multiple of the signal frequency, for which it becomes necessary to consider idler tones that occur. For the nondegenerate three-wave case, we present an intuitive picture for a parametric amplifier containing a flux-pumped SQUID where current at the signal frequency depends upon the load impedance at an idler frequency. This understanding provides insight and readily testable predictions of circuits containing flux-pumped SQUIDs.

Keywords: parametric amplifiers; SQUIDs; Josephson devices

1 Introduction

Parametric amplifiers are attractive in that they can in principle amplify a signal while only adding a minimum of noise [1]. From this point of view, parametric amplifiers may be divided into two groups; *phase sensitive* amplifiers which amplify only one of the incoming quadratures, and *phase insensitive* amplifiers which amplify both quadratures, thereby preserving the phase of the signal. A phase sensitive amplifier can in principle amplify the signal without adding any noise. The minimum noise added by a phase insensitive amplifier corresponds to half a quantum of the amplified frequency, $\hbar\omega/2$.

In a parametric amplifier, some parameter of the system must be varied in time. By pumping the system, *i.e.* modulating that parameter at one frequency, it is possible to amplify a signal at a different frequency. Power is transferred from the pump frequency to the signal frequency.

Parametric amplifiers can be realized in a large number of systems, both in optics and in electronics. A typical example in optics is a fiber-based amplifier where the refractive index of the fiber material is modulated by the pump. In other systems utilizing varactor diodes, it is the nonlinear diode capacitance which is modulated. Varactor diodes are typically used at frequencies ranging from radio to THz frequencies.

Superconducting circuits can also be used to build parametric amplifiers in the microwave domain. The use of parametric amplifiers with Josephson junctions was pioneered by several researchers in the 1970s [2–6], as well as Bernard Yurke in the 1980s [7, 8]. Josephson junctions are used as parametric inductances, and may be pumped either by a time varying current through the junction [9–11], or in a SQUID geometry by a time-varying magnetic flux [11–14]. Alternatively the kinetic inductance of a thin superconductor can be used as the parametric component [15, 16].

Parametric amplifiers based on superconducting devices have recently regained interest because of the need for better amplifiers for qubit readout and microwave quantum optics. The utility of these amplifiers have been demonstrated in a number of experiments showing single shot qubit readout [17], quantum feedback [18], vacuum squeezing [19], and in determining the statistics of nonclassical photon states [20]. There are two major advantages of superconducting parametric amplifiers: (i) they have very low dissipation, and (ii) they have well characterized and engineer-able properties. This makes it possible to design well functioning parametric amplifiers with good gain and little added noise [9, 21].

To understand and implement a parametric amplifier, one often needs to solve a system of coupled equations where it may be difficult to fully appreciate the amplifier's overall properties. Along with the resurgent use of parametric amplifiers as applied to quantum systems, a quantum optics formalism is also typically adopted to explain the amplifier.

By contrast, we recently presented [14] a linearized impedance model for a flux-pumped SQUID following the engineering formalism [22–24] developed for (classical) varactor diodes in the 1960s and 1970s. While a similar formalism had also been utilized for early treatments of Josephson junction parametric amplifiers [4], this had not been applied to the flux-pumped SQUID. The flux-pumped SQUID can be represented as a parallel combination of a Josephson inductance and an additional circuit element which we named the "pumpistor." The pumpistor has the frequency dependence of an inductance, but it is an inductance which is *complex*. The phase of this complex inductance (or impedance) depends on the phase angle of the pump relative to the signal. By properly adjusting the pump, the pumpistor can act as a negative resistance. Thus, it can provide gain in the circuit. In this recent paper, we treated only the three-wave degenerate case, *i.e.* where the pump is applied at exactly twice the signal frequency.

In this work, we extend this pumpistor model. We revisit the three-wave degenerate case to include higher-order saturation effects. We also explore the four-wave degenerate case, which couples to the pump at higher order. Perhaps most importantly, we also treat the *nondegenerate* case, where the pump frequency is not a multiple of the signal frequency. Here a matrix formalism provides for the exploration of many types of nondegenerate frequency mixing, which, in addition to gain as a negative resistance, also describes up- and down-conversion of a signal.

2 The current response of a simple dc SQUID

In this section, we briefly review the relations between external magnetic flux, effective junction phase, and series current in a dc SQUID. In this work, we refer to a dc SQUID simply as a "SQUID," and we consider it free of parasitic internal impedances. To begin, we first consider a single Josephson junction in order to introduce the Josephson relations due to the *dc* and *ac* Josephson effects [25].

2.1 Current and voltage in a simple Josephson junction

In a Josephson junction, the *dc Josephson effect* denotes the relation between the phase difference ϕ_J, *i.e.*, the difference in phase between the superconducting order parameters on either side of the junction, and the current I which flows through the junction. This is given by

$$I = I_c \sin(\phi) \tag{1}$$

Here, I_c is the critical current for this single Josephson junction, which is its maximum allowed super-current. The *ac Josephson effect* relates the *time derivative* of the phase difference to the voltage, V, across the junction.

$$V = \left(\frac{\Phi_0}{2\pi}\right)\frac{d\phi}{dt} \tag{2}$$

where, $\Phi_0 \equiv h/(2e)$ is the *superconducting flux quantum*. By taking the time derivative of Eq. (1) and combining it with Eq. (2), we see that the Josephson junction acts like an inductor, $dI/dt = V/L_J$, with the *Josephson inductance*

$$L_J = \frac{\Phi_0}{2\pi I_c \cos\phi} \tag{3}$$

2.2 Extending the Josephson relations to a SQUID

Placing two Josephson junctions ("1" and "2") in parallel, we form a SQUID, where the currents combine as a sum. We adopt the sign conventions suggested in Ref. [26].

$$I = I_{c1} \sin(\phi_1) - I_{c2} \sin(\phi_2) \tag{4}$$

Going around the loop and returning to the same point, the phase can only subtend multiples of 2π. We therefore find a quantization condition for the superconducting loop flux. We regard the phase differences to occur only at the two Josephson junctions, *i.e.*, neglecting the inductance of the loop. Furthermore we assume that the two junctions are equal, $I_{c1} = I_{c2} = I_c/2$, and we define the SQUID phase to be $\phi = (\phi_1 - \phi_2)/2$. Then we arrive at the SQUID current,

$$I = I_c \cos\left(\pi\frac{\Phi_{ext}}{\Phi_0}\right)\sin(\phi) \tag{5}$$

We see that the SQUID acts just like a Josephson junction, but with a critical current tunable by the external magnetic flux Φ_{ext}. Note that our choice of sign convention following Ref. [26] eliminates the need for taking the *absolute value* of the quantity $\cos(\pi\Phi_{ext}/\Phi_0)$

in Eq. (5). This is not the case in the definition commonly used in other very good and popular references (*e.g.*, [27, 28]). In any case, for this work we consider only the situation where $|\Phi_{\text{ext}}/\Phi_0| < |1/2|$. Here, the quantity corresponding to $\cos(\pi\,\Phi_{\text{ext}}/\Phi_0)$ is always positive regardless of convention.

Thus, we recover a device phenomenology similar to the single Josephson junction depicted in Eqs. (1) and (2). Specifically, the SQUID acts as a tunable inductance such that

$$L_J = \frac{\Phi_0}{2\pi I_c \cos(\pi \frac{\Phi_{\text{ext}}}{\Phi_0})\cos\phi} \tag{6}$$

In this section, we have defined the system of a SQUID by current and voltage relations similar to a single Josephson junction. We found the SQUID to be tunable by an externally applied magnetic flux. Using this framework, in the next section we examine the SQUID circuit response to a magnetic flux, *applied dynamically*.

3 The signal impedance of a SQUID, subject to a dynamically pumped external magnetic flux

We investigate how a SQUID responds as an impedance due to the presence of a periodic perturbation of the external magnetic flux. To this end, we assume the external flux is of the following form.

$$\Phi_{\text{ext}} = \Phi_{\text{dc}} + \delta\Phi \tag{7}$$

Here Φ_{dc} is a static ("quiescent") magnetic flux, and we use a time-dependent perturbation of the form $\delta\Phi = \Phi_{\text{ac}}\cos(\omega_3 t + \theta_3)$.

For convenience of notation, we define these following normalized flux amplitudes.

$$F = \pi\frac{\Phi_{\text{dc}}}{\Phi_0} \tag{8}$$

$$\delta f = \pi\frac{\Phi_{\text{ac}}}{\Phi_0} \tag{9}$$

3.1 An aside regarding labels and conventions

For clarity, we take the opportunity to introduce a handful of electromagnetic disturbances necessary to understand our system. These small-signal disturbances occur at different frequencies. We follow the nomenclature for frequency terms as presented by Blackwell and Kotzebue [22].

Regarding frequencies and how we label them, in this work we consider at most six frequencies due to possible mixing effects. Foremost, we consider a "signal" which exists at a frequency assigned to index 1. For a parametric amplifier, the signal frequency serves as the frequency of both the input and output of the device. In this case, we determine both the small-signal current and voltage components at this same signal frequency. This gives us a "signal impedance" upon which we base our subsequent reasoning. Some driven "pump" disturbance occurs at a frequency of index 3. This pump frequency corresponds to the frequency at which the SQUID is driven externally. The pumping of the SQUID provides for a nonlinear interaction to occur. Another frequency we consider is the "idler" frequency. An idler response comes about due to the nonlinear mixing between signal

Table 1 Our convention for the frequencies involved in mixing effects

(Angular) frequency	Designation	Relation
ω_1	"signal"	ω_1
ω_2	"idler" (three-wave difference)	$\omega_3 - \omega_1$
ω_3	"pump"	ω_3
ω_4	"idler" (three-wave sum)	$\omega_3 + \omega_1$
ω_5	"idler" (four-wave difference)	$2\omega_3 - \omega_1$
ω_6	"idler" (four-wave sum)	$2\omega_3 + \omega_1$

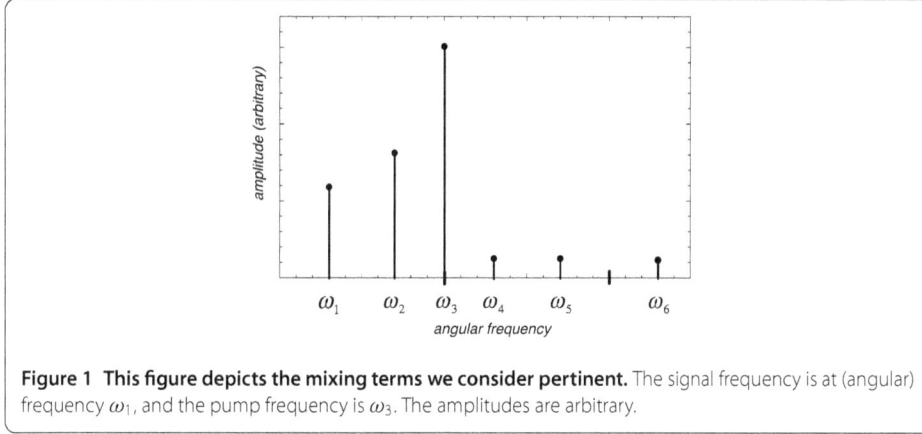

Figure 1 This figure depicts the mixing terms we consider pertinent. The signal frequency is at (angular) frequency ω_1, and the pump frequency is ω_3. The amplitudes are arbitrary.

and pump. In the general case, we need to provide for the possibility for the idler response to exist, even if it remains as an internal state variable (serving neither as an externally accessible input or output to the circuit). Among the various topologies which allow frequency mixing, an idler tone occurs at a frequency that is some linear combination of the signal and pump frequencies. In this work we delineate an idler as either a *sum* or a *difference* between signal and pump frequencies, for either the *three-wave* or *four-wave* case. An underlying principle of the parametric amplifier is that (some portion of) the power absorbed at the pump frequency is transferred to signal and idler frequencies, allowing for an amplified response.

We list all considered mixing frequencies in Table 1, and provide a depiction in Figure 1.

3.2 Small-signal disturbances and the modulated SQUID current

We consider different types of electromagnetic disturbances in the SQUID. Generally, we may account for *voltage*, *current*, *junction phase*, and *magnetic flux*. We have accounted for magnetic flux by Eq. (7). Consider also a general, small-signal response of the voltage, current, and junction phase at any of the six frequencies of Table 1. We assume ideal, sinusoidal tones.

$$v_n(t) = \frac{1}{2}V_n e^{j\omega_n t} + \frac{1}{2}V_n^* e^{-j\omega_n t} \quad n \in \{1,2,\ldots,6\} \tag{10}$$

$$i_n(t) = \frac{1}{2}I_n e^{j\omega_n t} + \frac{1}{2}I_n^* e^{-j\omega_n t} \quad n \in \{1,2,\ldots,6\} \tag{11}$$

$$\phi_n(t) = \frac{1}{2}\tilde{\phi}_n e^{j\omega_n t} + \frac{1}{2}\tilde{\phi}_n^* e^{-j\omega_n t} \quad n \in \{1,2,\ldots,6\} \tag{12}$$

The amplitudes V_n, I_n, and $\tilde{\phi}_n$ are complex. Eqs. (10)-(12) also demonstrate that we have adopted the electrical engineering convention for complex number, j, rather than the physics convention, i, leading to a sign convention opposite of what one would find in the quantum optics literature.

The SQUID current is directly related to the junction phase by the dc Josephson effect as in Eq. (5). Note that we specify the current as two multiplicative terms; the "flux" term, and the "phase" term.

$$i(t) = \underbrace{I_c \cos(\pi\Phi_{ext}(t)/\Phi_0)}_{\text{"flux" term}} \underbrace{\sin(\phi(t))}_{\text{"phase" term}} \tag{13}$$

When treating these dynamics involving sinusoids, a common approximation is to implement Fourier-Bessel expansions [27]. However, a simple Taylor expansion recovers the same result as a Fourier-Bessel expansion when approximating Bessel functions in their small-signal limit.

We take separate series expansions of the two multiplied terms of Eq. (13). First, we expand the "flux" term. We use the flux-perturbation variable ($\delta\Phi$) of Eq. (7), which was specified to be driven at the pump frequency (ω_3). To first order, we find the following.

$$I_c \cos(\pi\Phi_{ext}(t)/\Phi_0) \approx I_c\big[\cos(F) - \sin(F)\delta f \cos(\omega_3 t + \theta_3)\big] \tag{14}$$

In some cases, such as when we consider *saturation effects* due to large flux amplitudes, we will expand this term to higher order.

Next, we expand the "phase" term of Eq. (13). We use simply $\sin[\phi(t)] \approx \phi(t)$, although we also include the cubic term in cases where we consider saturation effects due to junction phase. In the linear limit we consider the "phase" term to be the superposition of contributions from the six considered frequencies, $\phi(t) = \sum_{n=1}^{6}\phi_n(t)$, with $\phi_n(t)$ taken from Eq. (12).

The total SQUID current can now be approximated as the following.

$$i(t) \approx \underbrace{\big\{I_c\big[\cos(F) - \sin(F)\delta f \cos(\omega_3 t + \theta_3)\big]\big\}}_{\text{"flux" term}} \underbrace{\left(\sum_{n=1}^{6}\phi_n(t)\right)}_{\text{"phase" term}} \tag{15}$$

Equation (15) is central to this work. It informs us how the SQUID current mixes magnetic flux and junction phase, allowing for gain and dissipation effects at and between different frequencies. In what follows, we treat the response of the SQUID under various, specific pumping conditions. We begin by studying the *three-wave degenerate* case.

4 The three-wave degenerate case

The three-wave degenerate case was treated at length in our previous work [14]. In a three-wave parametric amplifier or converter, a pump acts as a source of power to both a signal tone and an idler tone via a nonlinear coupling (*e.g.*, a SQUID). We therefore consider tones at the signal (ω_1), pump (ω_3), and idler ($\omega_2 = \omega_3 - \omega_1$) frequencies. Energy conservation in this three-wave case gives $\omega_1 + \omega_2 = \omega_3$. As we consider this condition to be *degenerate*, the signal and idler frequencies coincide (*i.e.*, $\omega_1 = \omega_2$).

The three-wave degenerate case:

$$\omega_1 = \omega_2 = (\omega_3 - \omega_1) = \omega_3/2 \tag{16}$$

We use Eq. (15) to determine the response of the SQUID. Since the signal and idler tones are no longer distinct for degenerate conditions, the "phase" term of Eq. (15) simplifies to the following.

$$\sin[\phi(t)] \approx \phi(t) = \tilde{\phi}_1 \cos(\omega_1 t + \theta_1) \tag{17}$$

$$= \frac{1}{2}\tilde{\phi}_1 e^{j\theta_{p1}} e^{j\omega_1 t} + \frac{1}{2}\tilde{\phi}_1^* e^{-j\theta_{p1}} e^{-j\omega_1 t} \quad \text{(degenerate case)} \tag{18}$$

In this section, we depart slightly from the form of Eq. (12) in that we have assumed a cosine dependence with an explicit phase angle. The amplitude $\tilde{\phi}_1$ is therefore now real and equal to its complex conjugate $\tilde{\phi}_1^*$, although we retain the use of conjugate notation for generality.

We did not consider including $\phi_3(t)$, which is the junction phase contribution at the pump frequency (ω_3). This is because we are interested in the *signal* response. For frequency *mixing* to occur, components at different frequencies must be *multiplied*. As long as the approximation $\sin[\phi(t)] \approx \phi(t)$ is valid, $\phi_3(t)$ does not contribute to the SQUID current at the signal frequency.

We apply the degenerate condition of Eq. (16) and the "phase" term of Eq. (18) to Eq. (15). From the resulting expression, we find the terms proportionate to the frequency component at $e^{j\omega_1 t}$. We consider the signal current to be of the form of Eq. (11),

$$i_1(t) = \underbrace{\frac{1}{2}I_1 e^{j\omega_1 t}}_{i_1(t)_+} + \underbrace{\frac{1}{2}I_1^* e^{-j\omega_1 t}}_{i_1(t)_-} \tag{19}$$

such that we can match its $e^{j\omega_1 t}$ component, $i_1(t)_+$, to the following form.

$$i_1(t)_+ = \frac{1}{2}I_1 e^{j\omega_1 t} \tag{20}$$

$$= \frac{1}{2}I_c\tilde{\phi}_1 \left[e^{j\theta_1}\cos(F) - \frac{\delta f}{2}\sin(F)e^{j(\theta_3-\theta_1)} \right] e^{j\omega_1 t} \tag{21}$$

Now considering a *voltage* based on the ac Josephson relation applied to the phase response, we find the following component which is also proportionate to $e^{j\omega_1 t}$.

$$v_1(t)_+ = \frac{1}{2}V_1 e^{j\omega_1 t} = \frac{\Phi_0}{2\pi}\frac{d}{dt}\left[\frac{1}{2}\tilde{\phi}_1 e^{j\theta_1} e^{j\omega_1 t} \right]$$

$$= \frac{1}{2}\left[\frac{\Phi_0 \tilde{\phi}_1 \omega_1}{2\pi}\left(je^{j\theta_1}\right) \right] e^{j\omega_1 t} \tag{22}$$

By dividing Eq. (21) by Eq. (22), we can define a *signal admittance*, $Y(\omega_1)$.

$$Y(\omega_1) = \frac{i_1(t)_+}{v_1(t)_+} \tag{23}$$

$$= (j\omega_1 L_{3d,0})^{-1} + (j\omega_1 L_{3d,1})^{-1} \tag{24}$$

Above, we have defined the following identities.

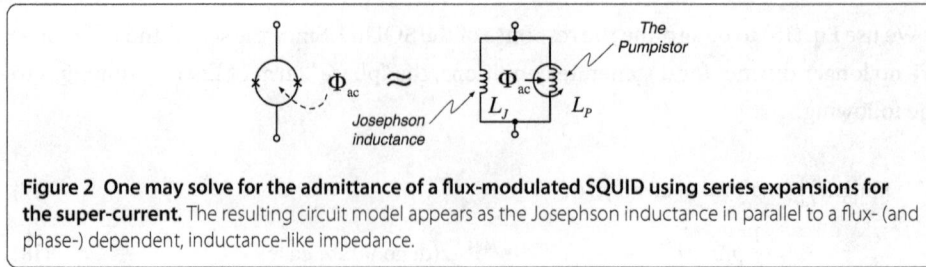

Figure 2 One may solve for the admittance of a flux-modulated SQUID using series expansions for the super-current. The resulting circuit model appears as the Josephson inductance in parallel to a flux- (and phase-) dependent, inductance-like impedance.

Three-wave degenerate amplifier:

$$
\begin{aligned}
L_{3d,0} &= L_J \\
L_{3d,1} &= -L_J \frac{2}{\tan(F)}\left(\frac{1}{\delta f}\right)e^{+j\Delta\theta_{3d}} \\
\Delta\theta_{3d} &= 2\theta_1 - \theta_3
\end{aligned}
\tag{25}
$$

The subscript "3d" denotes the *three-wave degenerate* case. We identify the Josephson inductance, L_J, from Eq. (6) for the unperturbed flux ($\Phi_{ext} = \Phi_{dc}$) and small phase ($\phi \approx 0$) conditions. We therefore consider $L_J = \Phi_0/[2\pi I_c \cos(F)]$ for the remainder of this work. From these definitions, Eq. (24) shows that the admittance appears as the parallel combination of the Josephson inductance and a perturbation inductance with an ac-flux dependence (*i.e.*, "the pumpistor" [14]).

Note that this extra inductance, $L_{3d,1}$, has a dependence on the effective pump phase, $\Delta\theta_{3d}$. Depending on the value of $\Delta\theta_{3d}$, the inductance $L_{3d,1}$ has both real and imaginary contributions, which may be either positive or negative. Our amplifier topology will be able to supply signal gain when $L_{3d,1}$ has a substantial negative and real impedance. This depicts the mechanism which allows the SQUID to inject power back into the external circuit at the signal frequency. A diagram of this equivalent circuit is demonstrated in Figure 2.

Here we have treated the degenerate case to first order both in pump flux and in signal phase. We recover the Josephson inductance in combination with a component representing the perturbation to the signal response. This extra impedance, as defined by its frequency dependence, is an inductor. However, its phase dependence allows it to take on complex amplitudes.

It is important to point out that, mathematically, this relation only holds at *precisely* the frequency $\omega_1 = \omega_3/2$. When this condition is not met, we need to resort to the general form of the *nondegenerate* case, which we shall treat in Sections 6 and 7.

Now, we consider some saturation arguments for this three-wave degenerate case.

4.1 Saturation of the pump flux for the three-wave degenerate case

As in the theory of mixers [29] and other nonlinear devices, the nonlinear properties of the driven SQUID lead also to saturation effects. These effects include the amplitude-dependent modifications of the Josephson inductance, as well as the *gain compression* of the incremental response.

If we extend the degenerate treatment as in Eq. (23), we can find higher-order parallel inductance terms by expanding the "flux" term of Eq. (14) to higher orders in ac flux. Taking the series expansion to third-order, we find the following extension to Eq. (24).

$$L_{3d,0} = L_J \tag{26}$$

$$L_{3d,1} = -L_J \frac{2}{\tan(F)} \left(\frac{1}{\delta f}\right) e^{+j\Delta\theta_{3d}} \tag{27}$$

$$L_{3d,2} = -4L_J \left(\frac{1}{\delta f}\right)^2 \tag{28}$$

$$L_{3d,3} = L_J \frac{16}{\tan(F)} \left(\frac{1}{\delta f}\right)^3 e^{+j\Delta\theta_{3d}} \tag{29}$$

$$\Delta\theta_{3d} = 2\theta_1 - \theta_3 \tag{30}$$

We find that the terms corresponding to the even powers of ac flux contribute to modifying the standard Josephson inductance. Meanwhile, the odd powers modify the phase-dependent term. Knowing that $L_{d,1}$ is responsible for gain, we can compare it to its higher-order correction, $L_{d,3}$. So by equating $|L_{d,1}|$ to $|L_{d,3}|$ we can estimate the pump ac-flux amplitude "intercept point." This is only a rough estimate of saturation, and the effects of gain compression would start to become apparent at ac-fluxes considerably smaller than this. To ensure that operation is far from this condition, we would say that the following should always be true.

$$\frac{\Phi_{ac}}{\Phi_0} \ll \frac{2\sqrt{2}}{\pi} \approx 0.90 \tag{31}$$

This is not a particularly useful constraint, as we already knew that we wish to keep the total external flux below $\Phi_0/2$. However, we could say that this constraint reinforces the notion that, for properly linearized behavior, Φ_{ac} should be maintained at some small fraction of Φ_0.

4.2 Saturation in the signal phase (or voltage) for the three-wave degenerate case

If we now substitute the *phase term* of Eq. (15) with an expansion to higher order, we can estimate nonlinear effects due to the magnitude of the *signal phase*. Here, we assume $\sin(\phi) \approx \phi - \frac{1}{6}\phi^3$, with $\phi = \phi_1(t)$ from Eq. (18). If we again combine the terms which occur at $e^{j\omega_1 t}$, we find the 3rd-order correction to the Φ_{ac}-independent term, $L_{3d,0}$, to be the following.

$$1/L_{3d,0} \rightarrow 1/L_J \left(1 - \frac{\tilde{\phi}_1^2}{8}\right) \tag{32}$$

We also find a 3rd-order correction to the $L_{3d,1}$ inductance term, which was the term inversely proportionate to δf.

$$1/L_{3d,1} \rightarrow 1/L_{3d,1} \left[1 + \tilde{\phi}_1^2 \left(\frac{1}{24} e^{j2\Delta\theta_{3d}} - \frac{1}{8}\right)\right] \tag{33}$$

To estimate an "intercept point" due to saturation of the phase amplitude, we can take the maximum of the corrected $1/L_{3d,1}$ of Eq. (33), at $\Delta\theta_{3d} = \pi/2$. It is straightforward to see

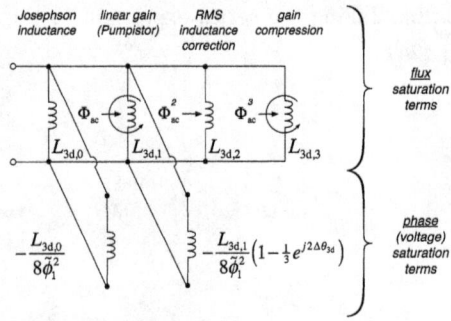

Figure 3 The admittance expansion to higher order, both in external flux and in junction phase, gives a more complete model for the three-wave degenerate case. The even harmonics in the flux expansion serve to only modify the net inductance value. The odd harmonics modify the potential gain and phase-sensitivity. This allows for estimation of the saturation effects. As it is in the theory of mixers, we see in the ac-flux expansion that the third-order term compresses the gain-providing first-order term.

that the contribution of $\tilde{\phi}_1^2$ should be negligible when the following is true.

$$|\tilde{\phi}_1| \ll \sqrt{6} \tag{34}$$

As in the previous consideration of the nonlinearity due to Φ_{ac}/Φ_0, this is not particularly a remarkable constraint. The phase amplitude $\sqrt{6}$ is obviously already a large fraction of π. It only reinforces the point that $|\tilde{\phi}_1|$ should be quite small compared to this value. Perhaps, though, it is worthwhile to point out this limit also corresponds directly to a limit on the junction *voltage*, by way of the ac Josephson effect.

$$|V_1| = \tilde{\phi}_1 \left(\frac{\Phi_0 \omega_1}{2\pi} \right) \ll \frac{\sqrt{2}}{\pi} \Phi_0 \omega_1 \tag{35}$$

Concluding discussion of saturation effects due to flux and to signal phase, we turn to Figure 3. Here, we combine the effects of gain compression into a common model. As in the theory of mixers, we see that the odd terms in the expansion account for both gain and its saturation.

5 The four-wave degenerate case

Next, we take interest in the SQUID with zero dc flux. When the dc flux is zero, the first derivative of inductance as a function of flux is also zero. We notice from Eq. (27) that $L_{3d,1}$ becomes infinite (an "open") and no longer contributes to the circuit. In fact, all of the odd powers of Φ_{ac} will disappear from the "flux" term of Eq. (15). The reason for this can be attributed to the symmetric behavior of the unbiased device. Yet it is still possible to achieve parametric amplification among the even harmonics of the admittance expansion in flux, in a degenerate case without an idler tone distinct from a signal ($\omega_1 = \omega_2$). In this case one must use *four-wave degenerate* mixing, where we can consider this as *two* pump photons interacting with a signal photon and an idler photon (*i.e.*, $\omega_1 + \omega_2 = 2\omega_3$).

The four-wave degenerate case:

$$\omega_1 = \omega_2 = (2\omega_3 - \omega_1) = \omega_3 \tag{36}$$

As in the three-wave degenerate case, both idler and signal tones occur at identically the same frequency and we consider only the disturbance of their combined response. Again, we treat this degenerate tone as the *signal* (ω_1) response.

When the external magnetic flux is comprised of solely the ac component, we mentioned that the device behaves symmetrically around zero flux. To find the relevant dynamical response, we need to expand the "flux" term of Eq. (14) to *2nd-order* for this *zero flux-bias case*.

$$I_c \cos(\pi \delta \Phi / \Phi_0) \approx I_c - I_c \frac{\pi^2}{2\Phi_0^2} (\delta \Phi)^2$$

$$= I_c - \frac{I_c}{2} (\delta f)^2 \left[\cos(\omega_3 t + \theta_3) \right]^2 \tag{37}$$

As in Eq. (15), we find the total current at the signal frequency by multiplying our "flux" approximation by the "phase" term approximation. We use the approximation for the signal phase as in Eq. (18). The resulting signal current, analogous to Eq. (21) but with $\omega_1 = \omega_3$, becomes

$$i_1(t)_+ = \frac{1}{2} I_c \tilde{\phi}_1 \left[1 - \frac{1}{4} (\delta f)^2 \right] e^{j\theta_1} e^{j\omega_1 t} - \frac{1}{16} I_c \tilde{\phi}_1^* (\delta f)^2 e^{j2\theta_3 - j\theta_1} e^{j\omega_1 t} \tag{38}$$

Considering the small-signal voltage of Eq. (22), we find the signal admittance in the four-wave degenerate case to be

$$Y_{4d}(\omega_1) = \frac{2I_c \pi}{j\omega_1 \Phi_0} - \frac{I_c \pi}{j2\omega_1 \Phi_0} (\delta f)^2 - \frac{I_c \pi}{j4\omega_1 \Phi_0} (\delta f)^2 e^{j2\theta_3 - j2\theta_1} \tag{39}$$

$$= (j\omega_1 L_{4d,0})^{-1} + (j\omega_1 L_{4d,2a})^{-1} + (j\omega_1 L_{4d,2b})^{-1} \tag{40}$$

In this case, we have defined the following.

Four-wave degenerate amplifier:

$$\boxed{\begin{aligned} L_{4d,0} &= L_J \\ L_{4d,2a} &= -4L_J \left(\frac{1}{\delta f} \right)^2 \\ L_{4d,2b} &= -8L_J \left(\frac{1}{\delta f} \right)^2 e^{j\Delta\theta_{4d}} \\ \Delta\theta_{4d} &= 2(\theta_1 - \theta_3) \end{aligned}} \tag{41}$$

So we find that the admittance which is proportionate to $(\delta f)^2$ has both phase-insensitive and phase-sensitive terms. Note also the dependence on the pump phase in $\Delta\theta_{4d}$ is different by 2 compared to the phase angle $\Delta\theta_{3d}$ of Eq. (25). Also in this four-wave degenerate case, we can produce a negative resistance, and consequently gain, from the $L_{4d,2b}$ term by adjusting $\Delta\theta_{4d}$ accordingly.

In the next sections, we turn to the more general case of *nondegenerate* operation. There, the idler response must now be considered separately from the signal response.

6 General conditions for nondegenerate parametric effects using the small-signal admittance matrix

We now consider specifically *nondegenerate* mixing conditions. Here, "nondegenerate" asserts its standard meaning that all frequency terms under consideration are unique, *i.e.*, $\omega_i \neq \omega_j$ for all $j \neq i$. Where any of our six considered mixing frequencies (Table 1) may contribute to a flux-pumped SQUID circuit, we work within our typical small-signal limit using a linearized system of equations. From this, we will develop an equivalent admittance matrix.

As before, the SQUID current is directly related to the junction phase by the dc Josephson effect as in Eq. (15). However, we include a 2nd-order expansion of the "flux" term of Eq. (14), which also includes dc flux. In this case, expanding the "flux" term to 2nd-order ensures nontrivial couplings to most frequency components. We wish to find the contributions of the current at different frequencies, given by the form $i(t) = \sum_{n=1}^{6} i_n(t)$ as in Eq. (11). For a single frequency component of the junction phase, we find the current amplitudes at all considered frequencies.

Although we could present a matrix of frequency couplings by what we have just described, we wish to find an *admittance matrix* relating current and voltage amplitudes. We therefore translate *junction phase* amplitudes into *voltage* amplitudes by way of the ac Josephson effect. Starting from Eq. (10), we find the following.

$$\phi_n(t) = \frac{2\pi}{\Phi_0} \int v_n(t)\, dt = -j\frac{\pi}{\Phi_0 \omega_n} V_n e^{j\omega_n t} + j\frac{\pi}{\Phi_0 \omega_n} V_n^* e^{-j\omega_n t} \tag{42}$$

Taking into consideration how frequency components of the voltage couple to both conjugate and non-conjugate terms of the current, we arrive at our desired small-signal admittance matrix. Rather than a basis set of physical ports as in a multi-terminal device, here the admittance matrix "ports" (indices) represent the frequencies from Table 1.

$$
\begin{pmatrix} I_1 \\ I_2^* \\ I_4 \\ I_5^* \\ I_6 \end{pmatrix} = \frac{1}{jL_J}
\begin{pmatrix}
\frac{\epsilon_0}{\omega_1} & -\frac{\epsilon_1^*}{\omega_2} & \frac{\epsilon_1}{\omega_4} & -\frac{\epsilon_2^*}{\omega_5} & \frac{\epsilon_2}{\omega_6} \\
\frac{\epsilon_1}{\omega_1} & -\frac{\epsilon_0}{\omega_2} & \frac{\epsilon_2}{\omega_4} & -\frac{\epsilon_1^*}{\omega_1} & 0 \\
\frac{\epsilon_1^*}{\omega_1} & -\frac{\epsilon_2^*}{\omega_2} & \frac{\epsilon_0}{\omega_4} & 0 & \frac{\epsilon_1}{\omega_6} \\
\frac{\epsilon_2}{\omega_1} & -\frac{\epsilon_1}{\omega_2} & 0 & -\frac{\epsilon_0}{\omega_5} & 0 \\
\frac{\epsilon_2^*}{\omega_1} & 0 & \frac{\epsilon_1^*}{\omega_4} & 0 & \frac{\epsilon_0}{\omega_6}
\end{pmatrix}
\begin{pmatrix} V_1 \\ V_2^* \\ V_4 \\ V_5^* \\ V_6 \end{pmatrix} \tag{43}
$$

We do not list in this matrix the pump current amplitude, I_3, as it couples to none of the other six frequencies but its own (ω_3).

This admittance matrix holds true as long as the pump frequency is larger than the signal frequency ($\omega_3 > \omega_1$) so that the "three-wave difference idler" frequency remains positive ($\omega_2 > 0$). In the case of $\omega_3 < \omega_1$, some matrix elements appear instead with conjugate quantities. Similar redefinitions are also necessary if frequency $\omega_5 = 2\omega_3 - \omega_1$ were also to become negative. We consider the conditions ($\omega_2 > 0$) and ($\omega_5 > 0$) to be the standard situation.

We find again the quiescent Josephson inductance, $L_J = \frac{\Phi_0}{2\pi I_c \cos(F)}$. Some new, flux-dependent terms ϵ_0, ϵ_1, and ϵ_2 also appear, which are not indexed by frequency. Rather, their indices indicate the order of the series expansion in flux for which they first become nontrivial. Their expressions are the following.

$$\epsilon_0 = 1 - \frac{1}{4}\delta f^2 \tag{44}$$

$$\epsilon_1 = \frac{\delta f}{2}\tan(F)e^{-j\theta_3} \tag{45}$$

$$\epsilon_2 = \frac{\delta f^2}{8}e^{-j2\theta_3} \tag{46}$$

To note, for vanishing $\delta f = \frac{\pi \Phi_{ac}}{\Phi_0}$, the limit of ϵ_0 is unity, while ϵ_1 and ϵ_2 tend to zero.

The importance of the matrix equation (Eq. (43)) should be emphasized. This tells us the response of a flux-pumped SQUID between all relevant frequency components, but yet it can be used in the same form as any other n-port admittance matrix from circuit theory. So for this very general degenerate case, we may now consider a large number of three-wave and four-wave mixing devices, both as negative-resistance amplifiers and as frequency converters. It further allows us to describe a number of next-order effects which also occur in these devices.

The elements of the admittance matrix (Eq. (43)), are specifically the *short-circuit* admittance parameters [30]. This is defined as the following,

$$Y_{kl} = \frac{I_k}{V_l}\bigg|_{V_m=0, m \neq l} \tag{47}$$

where $V_m = 0$ with $m \neq l$ is a condition met by shorting all ports, m other than the port of interest, l.

In the next section, we begin by considering a special case of Eq. (43) where the desired harmonics form a subset of the admittance matrix. The unwanted components are assumed to be zero (*i.e.*, shorted). We will then find necessary corrections for when unwanted harmonics are instead *open-circuited*.

7 The three-wave nondegenerate negative-resistance parametric amplifier

When the signal frequency under consideration is not degenerate relative to the pump frequency, the findings of Sections 4 and 5 break down. We now return to considerations of three-wave mixing, but for the *nondegenerate* case where $\omega_1 \neq \omega_3/2$. In this case, it is necessary to provide for the presence of an *idler* junction phase (voltage) at ω_2. The idler comes about due to the nonlinear frequency coupling between the signal and pump terms. A response at the idler frequency need not be induced at the input, nor measured as an output variable, for it to play an important role as an internal state variable.

In this section, we consider the following conditions on the signal and idler frequencies. *The three-wave nondegenerate case*:

$$\omega_2 = (\omega_3 - \omega_1) \neq \omega_1 \tag{48}$$

We consider the matrix subset of Eq. (43) corresponding to a signal at ω_1 and the idler at ω_2. The circuit at all other harmonics is assumed to be shorted.

$$\begin{pmatrix} I_1 \\ I_2^* \end{pmatrix} = \frac{1}{jL_J} \begin{pmatrix} \frac{\epsilon_0}{\omega_1} & -\frac{\epsilon_1^*}{\omega_2} \\ \frac{\epsilon_1}{\omega_1} & -\frac{\epsilon_0}{\omega_2} \end{pmatrix} \begin{pmatrix} V_1 \\ V_2^* \end{pmatrix} \tag{49}$$

This provides the current and voltage relations directly across the SQUID at the signal and idler frequencies. Next, we generalize the circuit such that we take into account the possible effects of other generator and load admittances.

7.1 Understanding this three-wave nondegenerate model as a circuit topology

To conceptualize the system we have just described, consider the flux-pumped SQUID as the primary element of a multi-frequency circuit. This is depicted in Figure 4(a). We assume this circuit to be sourced by a signal current $i_s(t)$ of the form of Eq. (11), such that

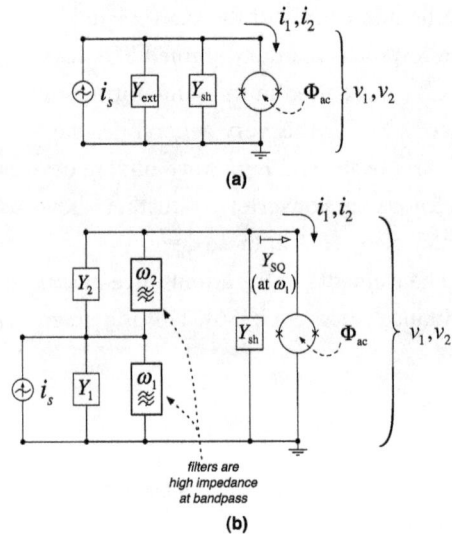

Figure 4 Considerations of the external circuit for the three-wave nondegenerate case. (a) This figure depicts the physical, general circuit considered in this section, containing a flux-pumped SQUID. This circuit accounts for an external source current, i_s, as well as external (Y_{ext}) and local shunt (Y_{sh}) admittances. **(b)** It is possible to represent the general circuit in an equivalent way that separates the external loading effects of the circuit at the "signal" (ω_1) and "idler" (ω_2) frequencies. This is done by introducing hypothetical, ideal bandpass filters at ω_1 and ω_2. These filters act open-circuited at their respective frequency, but short-circuited for all other frequencies. In this representation, the external admittance, Y_{ext}, is represented at different frequencies by Y_1 and Y_2. The input admittance (Y_{SQ}) is then based on the signal voltage response to current i_1, which depends upon the idler in a way that may allow for gain.

its amplitude is I_s and its frequency is ω_1. This external current source may be loaded by an external admittance, Y_{ext}. The currents $i_1(t)$ and $i_2(t)$, with amplitudes I_1 and I_2, continue to indicate the currents directly into the SQUID at frequencies ω_1 and ω_2, respectively. We account for either parasitic or intentional admittances directly across the SQUID by the term $Y_{sh}(\omega)$, which may be frequency dependent. We make a distinction between Y_{ext} and $Y_{sh}(\omega)$ since the definition of available power from the external source involves only $\text{Re}[Y_{ext}]$.

In Figure 4(b), we depict how we can think of the effects of the external load at different frequencies by recasting this circuit in an equivalent representation. In this case, we separate Y_{ext} into distinct impedances Y_1 at frequency ω_1 and Y_2 at frequency ω_2. We introduce hypothetical bandpass filters which isolate Y_1 and Y_2 to their respective frequencies outside of the pumped SQUID. These ideal filters work by providing a high-impedance (open) at their intended frequencies, while at all other frequencies they serve as a perfect short. This topology ensures that all unwanted frequencies short the SQUID, preventing any voltage at those frequencies to accumulate. Thus, we are able to reduce the general admittance matrix of Eq. (43) to the much simpler matrix of Eq. (49). While we do not actively source the idler current, we will find that the external admittance at the idler frequency, Y_2, effects response of the SQUID at the signal frequency in an important way.

7.2 The voltage and current ratios of the three-wave nondegenerate parametric amplifier

We are not quite ready to understand how gain appears in this system. This nondegenerate case is complicated by the appearance of an idler response distinct from the signal. For instance, the idler-to-signal voltage ratio $\frac{V_2^*}{V_1}$ will become important. To find a relation for V_2^*, we examine the second line of the matrix equation (49). While it is clear that we need to solve for V_2^*, what is I_2^*? Unlike the signal response, we are not sourcing or measuring an idler current. The idler current is the result of voltage disturbances at the idler frequency, coupled to the external circuit of the surrounding electrical system. Consequently, we must specify how the circuit is loaded at the idler frequency. This is why specifying some external (conjugate) idler admittance, Y_2^*, was necessary in the previous section. In what follows, we complete an analysis of our generalized circuit to solve for the idler voltage and current in terms of the signal.

Regarding the general circuit as depicted in Figure 4, we use Kirchhoff's node equations for both the signal and idler.

$$I_s - V_1 Y_1' - I_1 = 0 \tag{50}$$

$$-V_2^* Y_2'^* - I_2^* = 0 \tag{51}$$

Above, we defined the grouped admittances $Y_1' = Y_1 + Y_{sh}(\omega_1)$ and $Y_2' = Y_2 + Y_{sh}(\omega_2)$. To go further, the coupled subsystem of Eq. (49) allows us to eliminate I_1 and I_2^*, giving the following.

$$I_s = V_1 \left(Y_1' + \frac{\epsilon_0}{j\omega_1 L_J} \right) - \frac{V_2^* \epsilon_1^*}{j\omega_2 L_J} \tag{52}$$

$$0 = \frac{V_1 \epsilon_1}{j\omega_1 L_J} + V_2^* \left(Y_2'^* - \frac{\epsilon_0}{j\omega_2 L_J} \right) \tag{53}$$

Equations (52) and (53) now represent the current and voltage response of the generalized circuit depicted in Figure 4. Since the *signal current* is sourced in this model, what remains to be solved are the voltage disturbances V_1 and V_2^*. We define the impedances $Z_{L1} = j\omega_1 L_J/\epsilon_0$ and $Z_{L2} = j\omega_2 L_J/\epsilon_0$. The voltage amplitudes are then found to be

$$V_1 = L_J^2 Z_{L1} \omega_1 \omega_2 \left(Y_2'^* Z_{L2} - 1 \right) \left(\frac{I_s}{\Delta} \right) \tag{54}$$

$$V_2^* = j L_J Z_{L1} Z_{L2} \epsilon_1 \omega_2 \left(\frac{I_s}{\Delta} \right) \tag{55}$$

where the denominator term, Δ, is proportionate to the determinant formed by the matrix of Eqs. (52) and (53).

$$\Delta = L_J^2 \omega_1 \omega_2 (Y_1 Z_{L1} + 1) \left(Y_2'^* Z_{L2} - 1 \right) - Z_{L1} Z_{L2} |\epsilon_1|^2 \tag{56}$$

When we consider the *voltage ratio* between the idler and signal, the cumbersome denominator cancels, providing the more simple relation

$$\frac{V_2^*}{V_1} = \frac{\omega_2 \epsilon_1}{\omega_1 \epsilon_0} \frac{1}{1 + Z_{L2}^* Y_2'^*} \tag{57}$$

Here $Z_{L2}^* Y_2'^* \ll 1$, we see Eq. (57) go to the limit

$$\lim_{Z_{L2}^* Y_2'^* \to 0} \left(\frac{V_2^*}{V_1} \right) = \frac{\omega_2 \epsilon_1}{\omega_1 \epsilon_0} = \frac{\omega_2}{2\omega_1} \frac{\delta f}{1 - \frac{1}{4}\delta f^2} \tan(F) e^{-j\theta_3} \tag{58}$$

On the other hand, when this quantity becomes large such that $Z_{L2}^* Y_2'^* \gg 1$, we see

$$\lim_{Z_{L2}^* Y_2'^* \to \infty} \left(\frac{V_2^*}{V_1} \right) = 0 \tag{59}$$

So the voltage of the idler response is of course a function of how well the external circuit is being kept "open" at the idler frequency, ω_2.

We can also find the idler-to-signal *current ratio*. For this we revisit the system represented by Eq. (49), and divide its second equation by its first. We substitute the signal and idler voltage amplitudes found in Eqs. (54) and (55). This gives the following.

$$\frac{I_2^*}{I_1} = \frac{\epsilon_1}{\epsilon_0} \frac{1}{1 + \frac{1}{Z_{L2}^* Y_2'^*} + \frac{|\epsilon_1|^2}{\epsilon_0^2} \frac{\omega_1}{\omega_2 Z_{L1} Y_2'^*}} \tag{60}$$

We can look at the admittance limits of the current ratio as well. When the external admittance is small, we see

$$\lim_{Z_{L2}^* Y_2'^* \to 0} \left(\frac{I_2^*}{I_1} \right) = 0 \tag{61}$$

Conversely, when external admittance is large, we see

$$\lim_{Z_{L2}^* Y_2'^* \to \infty} \left(\frac{I_2^*}{I_1} \right) = \frac{\epsilon_1}{\epsilon_0} = \frac{1}{2} \frac{\delta f}{1 - \frac{1}{4}\delta f^2} \tan(F) e^{-j\theta_3} \tag{62}$$

These limits are intuitive. We can see the idler current will be inhibited when the external circuit is comparatively more "open," representing a small external admittance. Note the similar behavior indicated between Eqs. (58) and (61), as well as between Eqs. (59) and (62).

These quantities depict the response of the circuit at the idler frequency, ω_2, relative to the circuit behavior at the signal frequency, ω_1. We will now utilize this understanding in the next section to find how this system acts as a negative-resistance amplifier.

7.3 The input impedance of the nondegenerate three-wave parametric amplifier

To understand how this system works as an amplifier, we must find how it provides a negative resistance at the signal frequency. To this end, we seek to find the input admittance as seen at ω_1.

The input admittance as seen into the device at the signal frequency we can say is $Y_{SQ} = I_1 V_1$, giving

$$Y_{SQ} = \frac{I_1}{V_1} = \frac{1}{Z_{L1}} - \frac{V_2^*}{V_1} \frac{\epsilon_1^*}{\epsilon_0} \frac{1}{Z_{L1}} \tag{63}$$

Recall that $\epsilon_0 \approx 1$ to first order. To interpret Eq. (63) as an *impedance*, this is the Josephson inductance again in parallel to some other term. To find this other term, which is represented (as an admittance) by the second term on the right-hand side of Eq. (63), we must incorporate the ratio $\frac{V_2^*}{V_1}$ from Eq. (57). Substituting this term into Eq. (63), we arrive at

$$Y_{SQ} = \frac{1}{Z_{L1}} - \frac{|\epsilon_1|^2}{\epsilon_0^2} \frac{1}{Z_{L1}} \frac{1}{1 + Z_{L2}^* Y_2'^*} \tag{64}$$

$$= (j\omega_1 L_{n,0})^{-1} + (j\omega_1 L_{n,2})^{-1} \tag{65}$$

We have therefore represented the input admittance again as inductive terms. We determined a parallel inductance model before, in the *degenerate* case, but here the dependence

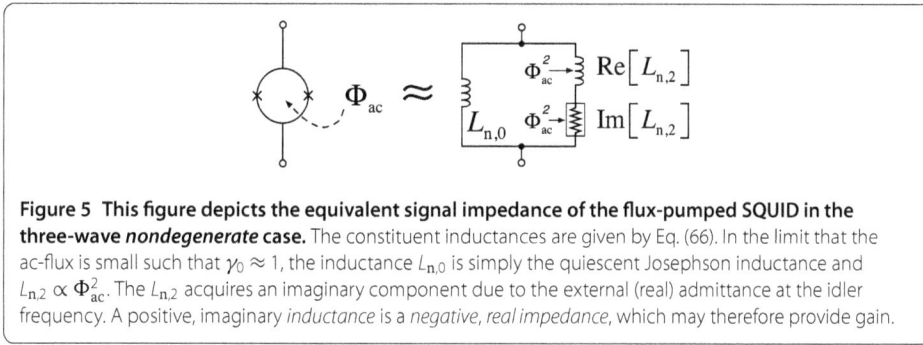

Figure 5 This figure depicts the equivalent signal impedance of the flux-pumped SQUID in the three-wave *nondegenerate* case. The constituent inductances are given by Eq. (66). In the limit that the ac-flux is small such that $\gamma_0 \approx 1$, the inductance $L_{n,0}$ is simply the quiescent Josephson inductance and $L_{n,2} \propto \Phi_{ac}^2$. The $L_{n,2}$ acquires an imaginary component due to the external (real) admittance at the idler frequency. A positive, imaginary *inductance* is a *negative, real impedance*, which may therefore provide gain.

on the *pump phase is no longer present*. This *nondegenerate* amplifier, therefore, is *phase insensitive*. The following terms for inductances are used.

Three-wave nondegenerate amplifier:

$$
\begin{aligned}
L_{n,0} &= L_J/\varepsilon_0 \\
L_{n,2} &= -\frac{L_J}{|\varepsilon_1|^2}\left(\varepsilon_0 - j\omega_2 L_J Y_2'^*\right)
\end{aligned}
\tag{66}
$$

Above, the "$L_{n,0}$" inductance is once again simply the Josephson inductance in the small-signal limit. The "$L_{n,2}$" inductance, in parallel to $L_{n,0}$, contains two terms which are both proportionate to $|\epsilon_1|^{-2}$. The first term is negative and simply modifies the net inductance by a small correction, making the net inductance appear bigger as the ac flux increases. The second term of $L_{n,2}$ depends on $Y_2'^*$ in an important way, providing the possibility for gain in this scenario. If $Y_2'^*$ has a real and positive component, this allows the impedance represented by $L_{n,2}$ to acquire a *negative* and real component. Therefore the $Y_2'^*$ term of $L_{n,2}$ acts as an active impedance converter, allowing the impedance external to the SQUID at the idler frequency to appear, transformed, at the signal frequency. We may think of the input admittance (or input impedance $Z_{SQ} = 1/Y_{SQ}$) directly into the three-wave nondegenerate pumped SQUID as depicted in Figure 5.

We comment on the *frequency dependence* of $L_{n,2}$. If we subscribe to axiomatic circuit theory [31–33], our linearized inductances should have a dependence strictly proportional to $j\omega$. The second term in $L_{n,2}$, which is the same term that may act as a negative resistance, also contains an extra factor, $j\omega_2 = j(\omega_3 - \omega_1)$. This gives a maximum of the product $\omega_1\omega_2$ at $\omega_3/2$, which for this reason is why $\omega_3/2$ is the frequency of maximum parametric amplification (or nearly so) in a three-wave nondegenerate amplifier. Between an uncommon frequency dependence and negative-resistance behavior, it may be logical to consider this second term of $L_{n,2}$ as relating to something other than an inductance. Yet we choose keep the terminology of an inductance only for consistency.

To conclude this section, we repeat that we have found the *negative resistance* that provides gain in this three-wave nondegenerate amplifier. This appears in the imaginary component of the term $L_{n,2}$ from Eq. (66). Although the frequency mixing between the idler and signal is provided for by the pump, the negative resistance occurs as an effect of mapping the idler's external (real) load admittance back onto the signal as a negative resistance.

7.4 The three-wave nondegenerate amplifier: transducer gain

The common readout implementation for a parametric amplifier (*e.g.*, a flux-pumped SQUID) is as a reflection device coupled to a circulator and a 2nd-stage amplifier (*e.g.*, a high electron mobility transistor (HEMT)) [7, 9, 14, 22–24]. It is therefore important that the first-stage gain of a parametrically flux-pumped SQUID be adequate to overcome the noise of subsequent gain stages. An insightful quantity in this context (in addition, say, to other quantities such as a *noise figure*) is the *transducer gain* of the device. It is straight-forward to specify the transducer gain, considering the simplified circuit we have so far described in this section.

The transducer gain is the ratio of the output power to the available input power. We consider the source admittance as Y_1. For the *(rms) available input power* at the signal frequency, we find

$$P_{a,1} = \frac{I_s^2}{8\,\text{Re}[Y_1]} \tag{67}$$

We consider the output signal to be reflected back onto the input admittance, such that we say the *(rms) output power* is

$$P_{o,1} = \frac{V_1^2}{2}\,\text{Re}[Y_1] \tag{68}$$

The transducer gain is then

$$G_T = \frac{P_{o,1}}{P_{a,1}} = \frac{4V_1^2(\text{Re}[Y_1])^2}{I_s^2} = 4\frac{(\text{Re}[Y_1])^2}{|Y_1' + Y_{SQ}|^2} \tag{69}$$

This can be expressed as

$$G_T = \frac{4\,\text{Re}[Y_1]^2}{\left(\frac{1}{\omega_1 L_{n,0}} + \frac{\text{Re}[L_{n,2}]}{\omega_1|L_{n,2}|^2} - \text{Im}[Y_1']\right)^2 + \left(\frac{\text{Im}[L_{n,2}]}{\omega_1|L_{n,2}|^2} - \text{Re}[Y_1']\right)^2} \tag{70}$$

where $L_{n,0}$ and $L_{n,2}$ are from Eq. (66).

7.5 Adding open-circuited terms

As an *admittance* model, as opposed to an *impedance* model, the ideal case is for all non-intentional harmonics to be subject to an infinite admittance external to the pumped SQUID (*e.g.*, to have a shorted external load at frequencies other than the signal and idler). This prevents voltages at these other frequencies from accumulating across the SQUID, thereby removing their influence from the admittance matrix and the resulting mixed currents. Conversely, when the external impedance is nontrivial at other frequencies, other harmonics will modify the description we have just presented.

Here, we treat the case opposite from before, where we now consider frequencies other than ω_1 and ω_2 to be *open-circuited*. Therefore, we consider the last three rows of the admittance matrix of Eq. (43) to represent no current flow, setting currents I_4, I_5^*, and I_6 to zero. We solve for the voltage amplitudes of these harmonics, which are now nontrivial. We substitute these voltage amplitudes into the expressions for current at the signal (ω_1)

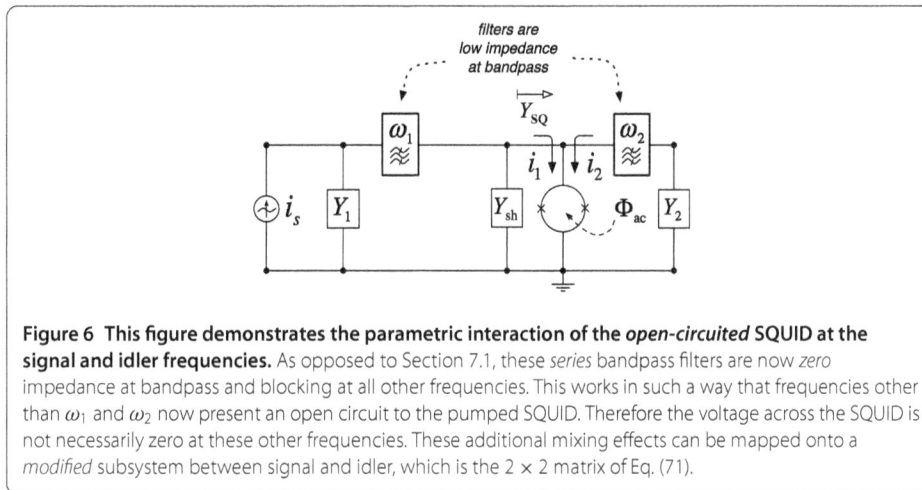

Figure 6 This figure demonstrates the parametric interaction of the *open-circuited* SQUID at the signal and idler frequencies. As opposed to Section 7.1, these *series* bandpass filters are now *zero* impedance at bandpass and blocking at all other frequencies. This works in such a way that frequencies other than ω_1 and ω_2 now present an open circuit to the pumped SQUID. Therefore the voltage across the SQUID is not necessarily zero at these other frequencies. These additional mixing effects can be mapped onto a *modified* subsystem between signal and idler, which is the 2 × 2 matrix of Eq. (71).

and idler (ω_2) frequencies. To reach a manageable solution, we assume the limiting conditions $\epsilon_0 \approx 1$ and $\epsilon_2 \approx 0$ for currents at ω_4, ω_5, and ω_6. If we keep terms up to δf^2, we find the signal-idler subset matrix has the simple form,

$$\begin{pmatrix} I_1 \\ I_2^* \end{pmatrix} = \frac{1}{jL_J} \begin{pmatrix} \frac{\varepsilon_0}{\omega_1}[1 - \frac{|\varepsilon_1|^2}{\varepsilon_0}] & -\frac{\varepsilon_1^*}{\omega_2} \\ \frac{\varepsilon_1}{\omega_1} & -\frac{\varepsilon_0}{\omega_2}[1 - \frac{|\varepsilon_1|^2}{\varepsilon_0}] \end{pmatrix} \begin{pmatrix} V_1 \\ V_2^* \end{pmatrix} \tag{71}$$

We find this system identical to that of Eq. (49), except for the multiplicative correction factor in square brackets, $[1 - \frac{|\varepsilon_1|^2}{\varepsilon_0}]$, appearing in the two matrix elements of the main diagonal. This correction factor may become significant even for reasonably small δf as the dc flux, F, approaches $\pi/2$. This is the notable difference between this open-circuited case and the short-circuited case we treated in Section 7.3.

We illustrate the open-circuited case as an equivalent circuit in Figure 6. We depict signal and idler circuits now directly in parallel to the pumped SQUID. As opposed to the short-circuited case depicted in Figure 4(b), here the ideal filters are accomplished *in series* such that only the permitted frequency is allowed to pass, while all other frequencies see an open-circuit.

In this section, we have determined the response of the three-wave *nondegenerate* amplifier as an impedance model. This is analogous to the "pumpistor" models we found for the three-wave and four-wave degenerate cases treated in Sections 4 and 5. A notable difference in this nondegenerate case is that the external admittance at the idler frequency now determines the negative resistance. As can be seen by Eq. (66), for a negative resistance to occur at the signal frequency, it is necessary that the circuit external to the SQUID at the idler frequency appear as a positive and real admittance. By treating both a "short-circuited" and an "open-circuited" model, we found that a finite external admittance at harmonics other than the signal and idler frequencies may also affect amplifier performance.

8 Conclusions

In conclusion, we have substantially extended the equivalent impedance models of a flux-pumped SQUID which we first put forth for the three-wave degenerate case [14]. For all general classes of parametric driving, a flux-pumped SQUID can be described at the signal frequency as a Josephson inductance in parallel to an effective, flux-dependent circuit element, "the pumpistor." Parametric amplification can be intuitively understood within this framework, as the pumpistor impedance manifests in whole or in part as a negative resistance.

We reviewed three-wave degenerate pumping, which explains why gain in this case should be *phase sensitive* between the signal and pump. For this case, we also extended our impedance approximation to demonstrate how the SQUID saturates both by pump flux and by junction phase (or voltage). We also depicted the four-wave degenerate case which is appropriate when the device is biased with zero-flux. Here, the pumpistor element is inversely proportionate to the *square* of the ac flux. We found this case also to be phase sensitive, but with a slightly different signal-to-pump difference than in the three-wave degenerate case.

We also depicted nondegenerate pumping in a very general sense, using a matrix equation formalism. This formalism accounts for the presence of one or up to four "idler" frequencies which occur as mixing tones between the pump and the signal response. Many three- and four-wave nondegenerate parametric phenomena can be interpreted from this matrix, including effects such as frequency up- and down-conversion. Using a subset of these matrix equations, we treated the three-wave nondegenerate amplifier, where the signal and single idler are considered. By solving for an idler distinct from the signal, we found that the pumpistor impedance was now *phase insensitive*. We found the negative resistance responsible for gain was now dependent on the external circuit admittance at the idler frequency. With regards to the other, higher harmonics, we treated the three-wave nondegenerate amplifier in both the "short-circuited" and "open-circuited" approximations. While all of these models operate under a classical, circuit-theoretic framework rather than a quantum optics framework, they should be of great benefit for future designs of experiments using superconducting circuits for quantum information purposes.

Competing interests
The authors declare that they have no competing interests.

Authors' contributions
KMS derived most of the equations. Both authors developed the concept and wrote this paper together.

Author details
[1] Electrical and Computer Engineering, Texas A&M University, College Station, TX 77843, USA. [2] Microtechnology and Nanoscience, Chalmers University of Technology, Göteborg, SE-412 96, Sweden.

Acknowledgements
We acknowledge support from the EU through the ERC and the projects SOLID, SCALEQIT, and PROMISCE, as well as from the Swedish Research Council and the Wallenberg Foundation. We are also grateful for fruitful discussions with Chris Wilson, Seckin Kintaş, Michaël Simoen, Philip Krantz, Martin Sandberg, and Jonas Bylander.

References
1. Caves CM: Quantum limits on noise in linear amplifiers. *Phys Rev D* 1982, **26**(8):1817-1839.
2. Feldman MJ, Parrish PT, Chiao RY: Parametric amplification by unbiased Josephson junctions. *J Appl Phys* 1975, **46**(9):4031-4042.
3. Taur Y, Richards PL: Parametric amplification and oscillation at 36 GHz using a point-contact Josephson junction. *J Appl Phys* 1977, **48**(3):1321-1326.
4. Feldman MJ: The thermally saturated SUPARAMP. *J Appl Phys* 1977, **48**(3):1301-1310.
5. Wahlsten S, Rudner S, Claeson T: Parametric amplification in arrays of Josephson junctions. *Appl Phys Lett* 1978, **30**:298-300.
6. Wahlsten S, Rudner S, Claeson T: Arrays of Josephson tunnel junctions as parametric amplifiers. *J Appl Phys* 1978, **49**(7):4248-4263.

7. Yurke B, Kaminsky P, Miller R, Whittaker E, Smith A, Silver A, Simon R: **Observation of 4.2-K equilibrium-noise squeezing via a Josephson-parametric amplifier**. *Phys Rev Lett* 1988, **60**(9):764.

8. Yurke B, Corruccini LR, Kaminsky PG, Rupp LW, Smith AD, Silver AH, Simon RW, Whittaker EA: **Observation of parametric amplification and deamplification in a Josephson parametric amplifier**. *Phys Rev A* 1989, **39**(5):2519-2533.

9. Castellanos-Beltran M, Lehnert KW: **Widely tunable parametric amplifier based on a superconducting quantum interference device array resonator**. *Appl Phys Lett* 2007, **91**:083509.

10. Eichler C, Bozyigit D, Lang C, Baur M, Steffen L, Fink JM, Filipp S, Wallraff A: **Observation of two-mode squeezing in the microwave frequency domain**. *Phys Rev Lett* 2011, **107**(11):113601.

11. Hatridge M, Vijay R, Slichter DH, Clarke J, Siddiqi I: **Dispersive magnetometry with a quantum-limited squid parametric amplifier**. *Phys Rev B* 2011, **83**(13):134501.

12. Yamamoto T, Inomata K, Watanabe M, Matsuba K, Miyazaki T, Oliver WD, Nakamura Y, Tsai JS: **Flux-driven Josephson parametric amplifier**. *Appl Phys Lett* 2008, **93**(4):042510.

13. Wilson CM, Duty T, Delsing P: **Parametric oscillators based on superconducting circuits**. In: *Fluctuating Nonlinear Oscillators*. Edited by Dykman M. Oxford: Oxford University Press; 2012.

14. Sundqvist KM, Kintas S, Simoen M, Krantz P, Sandberg M, Wilson CM, Delsing P: **The pumpistor: a linearized model of a flux-pumped superconducting quantum interference device for use as a negative-resistance parametric amplifier**. *Appl Phys Lett* 2013, **103**(10):102603.

15. Tholen EA, Ergul A, Doherty EM, Weber FM, Gregis F, Haviland DB: **Nonlinearities and parametric amplification in superconducting coplanar waveguide resonators**. *Appl Phys Lett* 2007, **90**:253509.

16. Ho Eom B, Day PK, LeDuc HG, Zmuidzinas J: **A wideband, low-noise superconducting amplifier with high dynamic range**. *Nat Phys* 2012, **8**(8):623-627.

17. Mallet F, Ong FR, Palacios-Laloy A, Nguyen F, Bertet P, Vion D, Esteve D: **Single-shot qubit readout in circuit quantum electrodynamics**. *Nat Phys* 2009, **5**(11):791-795.

18. Vijay R, Macklin C, Slichter DH, Weber SJ, Murch KW, Naik R, Korotkov AN, Siddiqi I: **Stabilizing rabi oscillations in a superconducting qubit using quantum feedback**. *Nature* 2012, **490**:77-80.

19. Flurin E, Roch N, Mallet F, Devoret MH, Huard B: **Generating entangled microwave radiation over two transmission lines**. *Phys Rev Lett* 2012, **109**(18):183901.

20. Steffen L, Salathe Y, Oppliger M, Kurpiers P, Baur M, Lang C, Eichler C, Puebla-Hellmann G, Fedorov A, Wallraff A: **Deterministic quantum teleportation with feed-forward in a solid state system**. *Nature* 2013, **500**:319-322.

21. Abdo B, Schackert F, Hatridge M, Rigetti C, Devoret MH: **Josephson amplifier for qubit readout**. *Appl Phys Lett* 2011, **99**(16):162506.

22. Blackwell LA, Kotzebue KL: *Semiconductor-Diode Parametric Amplifiers*. Englewood Cliffs: Prentice Hall; 1961.

23. Decroly JC: *Parametric Amplifiers*. New York: Wiley; 1973.

24. Howson DP, Smith RB: *Parametric Amplifiers*. London: McGraw-Hill; 1970.

25. Josephson BD: **Possible new effects in superconductive tunnelling**. *Phys Lett* 1962, **1**:251-253.

26. Zagoskin AM: *Quantum Engineering: Theory and Design of Quantum Coherent Structures*. Cambridge: Cambridge University Press; 2011.

27. Van Duzer T, Turner CW: *Principles of Superconductive Devices and Circuits*. 2nd edition. Upper Saddle River: Prentice Hall; 1999.

28. Tinkham M: *Introduction to Superconductivity*. 2nd edition. New York: McGraw-Hill; 1996.

29. Maas SA: *Microwave Mixers*. Dedham: Artech House; 1986. [*The Artech House Microwave Library*.]

30. Chen W-K: *Active Network and Feedback Amplifier Theory*. New York: McGraw-Hill; 1980.

31. Chua LO: *Introduction to Nonlinear Network Theory*. New York: McGraw-Hill; 1969.

32. Chua LO, Desoer CA, Kuh ES: *Linear and Non-linear Circuits*. New York: McGraw-Hill; 1987. [*McGraw-Hill Series in Electrical and Computer Engineering*.]

33. Chua LO: **Nonlinear circuit foundations for nanodevices**. *Proc IEEE* 2003, **91**(11):1830-1859.

Method for validating radiobiological samples using a linear accelerator

Muriel Brengues[1*], David Liu[2], Ronald Korn[2] and Frederic Zenhausern[1,2*]

* Correspondence:
brengues@email.arizona.edu;
fzenhaus@email.arizona.edu
[1]Center for Applied NanoBioscience
and Medicine, The University of
Arizona College of Medicine, 425 N.
5th Street, Phoenix, AZ 85004, USA
[2]Scottsdale Clinical Research
Institute, Scottsdale Healthcare,
10510 N. 92nd Street, Scottsdale,
AZ 85258

Abstract

There is an immediate need for rapid triage of the population in case of a large scale exposure to ionizing radiation. Knowing the dose absorbed by the body will allow clinicians to administer medical treatment for the best chance of recovery for the victim. In addition, today's radiotherapy treatment could benefit from additional information regarding the patient's sensitivity to radiation before starting the treatment. As of today, there is no system in place to respond to this demand. This paper will describe specific procedures to mimic the effects of human exposure to ionizing radiation creating the tools for optimization of administered radiation dosimetry for radiotherapy and/or to estimate the doses of radiation received accidentally during a radiation event that could pose a danger to the public. In order to obtain irradiated biological samples to study ionizing radiation absorbed by the body, we performed *ex-vivo* irradiation of human blood samples using the linear accelerator (LINAC). The LINAC was implemented and calibrated for irradiating human whole blood samples. To test the calibration, a 2 Gy test run was successfully performed on a tube filled with water with an accuracy of 3% in dose distribution. To validate our technique the blood samples were *ex-vivo* irradiated and the results were analyzed using a gene expression assay to follow the effect of the ionizing irradiation by characterizing dose responsive biomarkers from radiobiological assays. The response of 5 genes was monitored resulting in expression increase with the dose of radiation received. The blood samples treated with the LINAC can provide effective irradiated blood samples suitable for molecular profiling to validate radiobiological measurements via the gene-expression based biodosimetry tools.

Keywords: Linac accelerator; Blood samples; Biodosimetry; Gene expression
PACS numbers: 87.53.-j; 87.53.Bn

Background

Since September 11, 2001, the possibilities of a radiological or nuclear terrorist attack have been a central focus for both governmental agencies and communities throughout the world. There is an urgent need to have the adequate infrastructure to rapidly assess radiation injury in such a mass casualty scenario. Biodosimetry measurements after a radiation incident will be an immensely helpful tool in order to perform screenings, triages and management of clinical facilities in a mass casualty incident [1]. In case of an accidental radiation exposure, the effects of ionizing radiation can be wide-ranging and involve either the entire body or specific organs. Depending of the dose received, different medical treatments can be adapted and in most cases, the deadly effects of high radiation exposure (>2 Gy) can be mitigated by early triage and treatment decision. Unfortunately, as of today, no single time point measurement that is diagnostic of radiation exposure can reliably or rapidly discriminate the different levels of radiation received. Different platforms are currently under development along with the discovery and evaluation of radiation induced biomarkers to achieve that same goal of being able to measure the dose of radiation absorbed by the body [2-5]. The dicentric chromosome assay (DCA) is currently considered as the goal standard method because of its high sensitivity and accuracy [6]. The dose estimation of this method is based on the frequency of radiation specific aberrant chromosomes with two centromeres (dicentrics) in an irradiated individual's peripheral blood lymphocytes. But this assay is not applicable in mass casualty incident because it is very labor intensive and time consuming (several days) [7]. The γ-H2AX assay also used in radiation biodosimetry is a direct measure of the number of DNA double strand breaks (DSB) induced by ionizing radiation [8,9]. The yield of γ-H2AX foci has been shown to be linearly related to dose over a very wide dose range. The number of foci per cell in macaque lymphocytes after total body irradiation with doses of 1, 3.5, 6.5 and 8,5 Gy increase linearly with the irradiation dose (especially at doses greater than 1 Gy) [8]. This assay gives a same day result, but requires that the blood samples are available within about 36 hours of irradiation. Gene expression based assay have been extensively used by many laboratories for biodosimetry measurement of radio-responsive genes in human peripheral blood lymphocytes and is a potential method that fits the criteria to provide high-throughput data with an accurate measurement and in a timely manner [10-18].

The monitoring of the dose administered and received in radiotherapy is also needed as the possibility of radiation induced cancer exists for patients exposed intentionally to radiation. Being also able to measure the radio-sensitivity of each individual before starting radiotherapy treatment would be of great benefit not only to help control the late toxicity effect of treatment but also to integrate a personalized approach to tumor treatment based upon the chances of recovery and recidivism. One common tool needed for all of these studies is appropriate clinical samples in order to test the platforms under development. When studying the effects of ionizing radiation, it is not always possible to use *in vivo* samples. Several governmental research programs including NASA, Armed Forces, and BARDA agencies are pursuing work with non human primates. In addition to costs ranging in the millions of dollar and ethical issues, there is not really a standardized animal model in place to compare the results between studies [19]. While there is a need for animal models in radiation research, it might not always be necessary and not for every stage of the research. Very often cancer patients themselves are volunteering to participate in these studies to provide irradiated blood samples while on radiotherapy treatment.

The present study focused on the technique to provide irradiated samples to study biodosimetry. The main sources to irradiated samples are either a Gammacell-40 with a Cesium-137 source or a Cobalt-60 source, as well as X-rays or electron beams for a linear accelerator which does not contain radionuclide sources. The use of cesium-137 has been discontinued for practical reason and safety concerns in radiotherapy. Cobalt-60 is currently used in external beam radiotherapy devices found mostly in developing countries. Based on a report in 2006, there were several thousand radiotherapy devices in the United States in over 2,400 institutions and clinics [20]. Fewer than 250 cobalt-

60 teletherapy devices are licensed in the United States and most of those are thought to be in storage for decay, in use for other purposes (such as fixed radiography), or in use for teaching. This is because the linear accelerator (LINAC) is considered a better, more accurate and versatile radiotherapy tool, and has largely supplanted cobalt-60 teletherapy devices in the United States and other developed countries [21]. Another advantage of the LINAC is that it does not require replacement of the radiation source such as the cobalt or cesium sources that have an issue with radioactive decay contributing to dose inhomogeneities and errors in dose calculation. The LINAC is the device most commonly used for external beam radiation treatments for patients with cancer. It is used to treat all parts/organs of the body by delivering high-energy X-rays or electron beams to the region of the patient's tumor. These treatments can be designed in such a way that they destroy the cancer cells while sparing the surrounding normal tissue. More recently, the LINAC has been utilized to irradiate blood components [22]. Blood component irradiation is the only method of preventing a risk of transfusion associated graft versus host disease [23]. It has been demonstrated that by utilizing the LINAC, the internal irradiation procedures has been proven to be safe and feasible, and along with the significant cost/time reduction suggested that it is more advantageous than external procedures in hospitals without dedicated devices [22]. Our goal being to provide samples that will mimic real life situation as radiotherapy treatment, we irradiated our samples using the linear accelerator (LINAC). Using the LINAC for *ex-vivo* irradiation to study biodosimetry will allow a direct comparison with future data obtained from *in-vivo* irradiation. Comparing samples irradiated with the same source will eliminate false interpretations that could rise from using different irradiation methodologies.

This paper describes the process to perform irradiation on human blood samples using the LINAC in order to provide the appropriate irradiated samples as a tool for radiation biology study. To validate our method, the *ex-vivo* irradiated samples were analyzed to monitor the changes due to irradiation. We chose gene expression assay to analyze the data as it has been extensively used by many laboratories for biodosimetry measurement of radio-responsive genes in human peripheral blood lymphocytes, it has also been the method of choice in our laboratory for many years and it provides a result in less than 3 hours, however any assay could be used to analyze the data once the samples are irradiated with the LINAC. Validation of the use of LINAC can expand the opportunity to use this instrument for other biodosimetry studies and help to standardize the methodology for radiation trials in the future.

Results and discussion

LINAC calibration

Linear accelerators (LINACs) are the most popular equipment to deliver radiation therapy in hospitals. A Varian 21EX (S/N 1847) linear accelerator at the department of radiation oncology was used for the blood irradiation in this study. The Varian 21EX has two photon-beam energies (6 MV and 18 MV beams) and five electron-beam energies (6 MeV, 9 MeV, 12 MeV, 16 MeV and 20 MeV beams) that can be selected by the operator. A radiation beam was selected to provide the most homogeneous dose to blood samples. The depth dose curve shows the dose deposition along the beam direction and the radiation beam profile indicates the dose distribution perpendicular to the beam direction (Figures 1 and 2). These curves are for a field size of 20 cm × 20 cm and were measured in water. Our experimental setup was designed to have the radiation beam directed vertically down toward the floor at the blood tube holder that was lying horizontally on the table. With this geometry, we need to select a radiation beam with a uniform depth dose for about 1.5 cm (slightly more than the diameter of the tube) and a uniform profile for the central 5 cm (slightly longer than the length of the test tube). The radiation beam that best meets these requirements is 20 MeV electron beam at a depth of 2 to 3.5 cm (Table 1). The output of the radiation beam was calibrated based on the AAPM TG-51 protocol using a PTW Farmer-type ionization chamber (model 30013) (PTW, Freiburg, Germany) and a CNMC 1100 electrometer (CNMC Co, Nashville, TN) [24]. Both the ionization chamber and the electrometer were calibrated by Accredited Dosimetry Calibration Laboratory (ADCL) at the University of Wisconsin. Based on the calibration with this set up, the cone output factor for 20 cm cone with 20E is 0.953 cGy/monitor unit (MU) and the profile correction is 0.99. To deliver 2 Gy to the blood tube the required MU setting is 200/0.953/0.99 = 212 MU. To deliver 4 and 6 Gy, MU settings are 424 and 636 MU respectively. A rigorous quality assurance program has been already established to check the beam characteristics for this LINAC, including daily check with a Nuclear Associate QA phantom (Inovision Co. Carle Place, NY) and monthly calibration by a medical physicist certified by the American Board of Radiology using an ADCL-calibrated ionization chamber and electrometer. The outputs of all beams were also verified regularly by Radiological Physics Center at M D Anderson Cancer Center in Houston through mailed dosimeter program.

Dose measurement on blank sample

The sample holder was made of Plastic Water® and Superflab® material, which may have slightly different radiation attenuation property from water. The introduction of the test tube, as well as the air space in and around the tube, further perturbs the dose distribution. Thermoluminescent dosimeters (TLD) and film dosimetry were used to study these effects. For this purpose, a blank measurement using a vacutainer tube filled with water

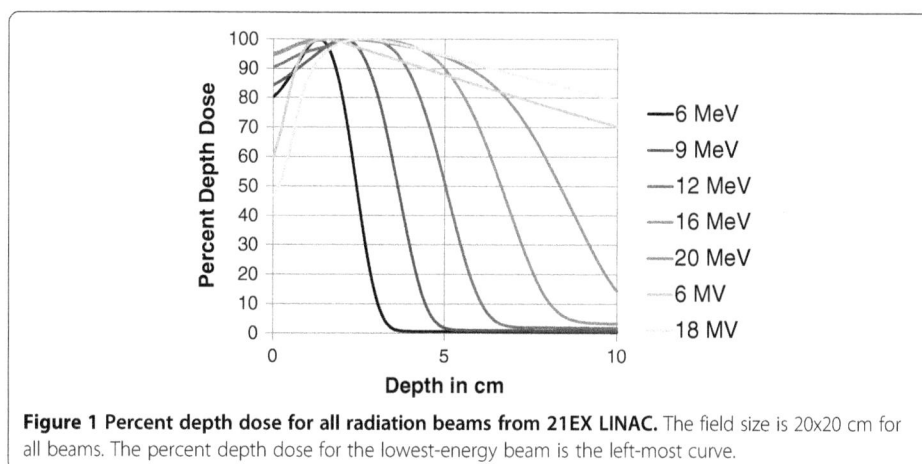

Figure 1 Percent depth dose for all radiation beams from 21EX LINAC. The field size is 20x20 cm for all beams. The percent depth dose for the lowest-energy beam is the left-most curve.

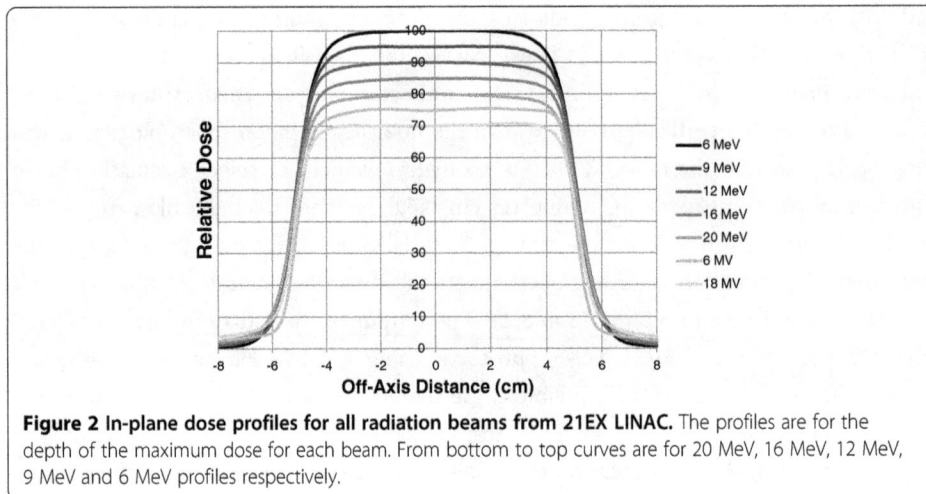

Figure 2 In-plane dose profiles for all radiation beams from 21EX LINAC. The profiles are for the depth of the maximum dose for each beam. From bottom to top curves are for 20 MeV, 16 MeV, 12 MeV, 9 MeV and 6 MeV profiles respectively.

and placed in the irradiation phantom was performed using LiF thermoluminescent dosimeters (TLDs) (Radiation Products Design, Inc. Albertville, MN) and EDR2 radiographic film (Carestream Health, Inc. Rochester, NY). TLD measures radiation dose by measuring the light output during the process of heating a crystal which has been exposed to ionization radiation. The amount of light emitted is dependent upon the radiation exposure. EDR films are often used to measure and verify the dose distributions. In addition, they are often used to perform various mechanical and dosimetric tests of the linear accelerator as part of routine quality assurance. Harshaw TLD-100 ribbons were taped around the tube (data not shown) and EDR-2 films were placed above and below the test tube (Figure 3). After irradiation with 212 MU of 20 MeV beam, TLDs were read with a Victoreen 2800 TLD reader and the EDR2 films were scanned with a Vidar 16 film scanner (Vidar Systems Corp., Herndon, VA). The radiation doses based on TLD results, which has an accuracy of 5%, were 2.03, 1.97 and 2.03 Gy for the top, bottom and the side of test tube respectively (data not shown). These readings were in agreement with the expected dose of 2 Gy. The dose profile along the vertical line is shown in the green curve; the one along the horizontal line is represented by the red curve. The dose distribution from the EDR2 film placed on the top of the test tube is quite uniform for both curves (Figure 3A). For the film placed below the test tube, there are very prominent dose inhomogeneities due to air gaps, tube cap and TLDs (Figure 3B). The shadow of the test tube is evident. The four peaks are due to air gaps surrounding the test tube and the filler Superflabs®. In

Table 1 Dose variations due to depth doses and profiles for 21EX radiation beams

	Depth of Max Dose (dmax)	Depth dose variation (dmax + 1.5 cm)	Profile variation (±4 cm from center axis)	Total variation
6 MV	1.4 cm	4.9%	2%	6.9%
18 MV	2.7 cm	3.3%	3%	6.6%
6 MeV	1.5 cm	86.5%	1%	87.5%
9 MeV	2.2 cm	49.3%	2%	51.3%
12 MeV	2.9 cm	19.7%	0.5%	20.2%
16 MeV	2.0 cm	2.5%	1%	3.5%
20 MeV	2.1 cm	0.4%	2%	2.4%

Total variation is the smallest for 20 MeV electron beam.

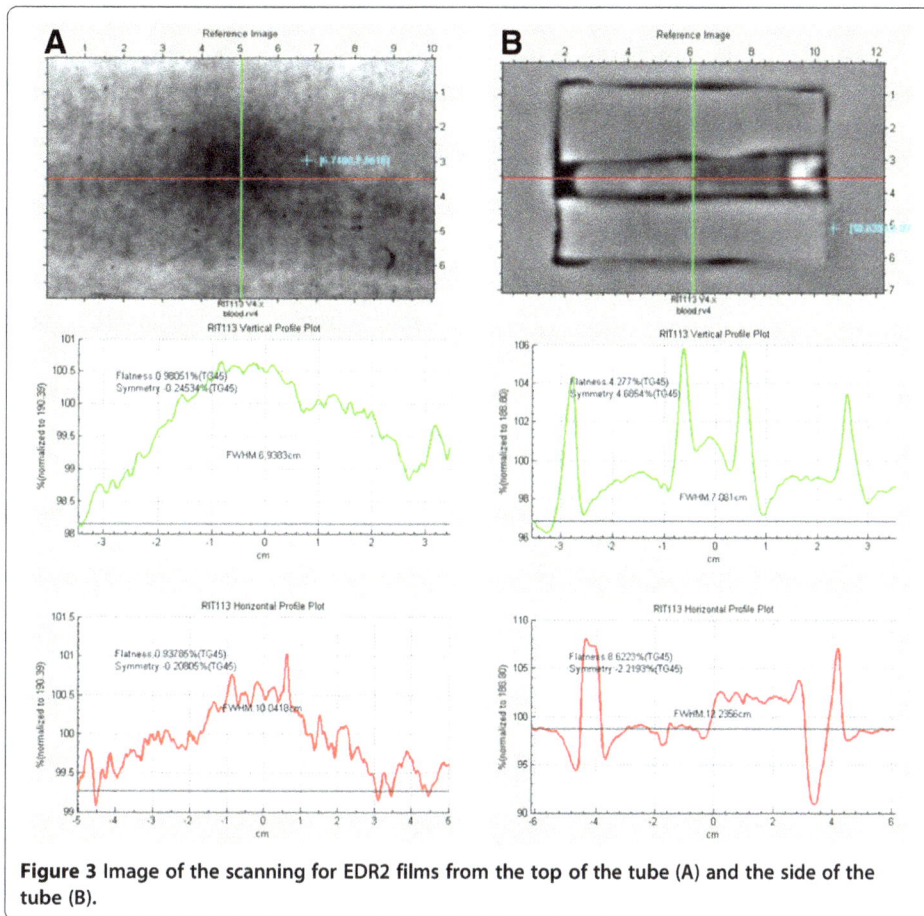

Figure 3 Image of the scanning for EDR2 films from the top of the tube (A) and the side of the tube (B).

addition to the air gaps at both ends, it also shows attenuation from the cap and the bottom of the test tube. Because the diameter of the cap is larger than the test tube, there is less water near the end of the cap, resulting in higher dose to the film at this end. Nonetheless, based on the film study, the dose pattern directly below the tube is within 3%.

Gene expression analysis

The Chemical Ligation Dependent Probe Amplification assay (CLPA) is a Non Enzymatic Assay Technology (NEAT) based on a non-enzymatic oligonucleotide probe ligation method [25]. The assay enables for rapid, simple and inexpensive analysis of gene signature compatible with capillary electrophoresis (CE) system for rapid readout (e.g. multiple capillaries ABI 3130). Another advantage of the technology is its capability to work directly with whole blood samples which eliminate the need of RNA preparation and only require 25 μl of blood. Another reason for choosing this assay is the fact that the manufacturer specially formulated a commercial kit to measure gene expression for acute radiation exposure (http://redidx.com/index.php). The assay can be performed in less than 3 hours using equipment commonly found in research laboratories. The amount of each ligation product is proportional to the concentration of its RNA target sequence, and is proportionally amplified by PCR to allow quantification relative to the transcript of a housekeeping gene. The PCR products from the CLPA were analyzed on the CE instrument and the raw data from the CE were transferred into a statistical data spreadsheet file (e.g. Microsoft Excel) to provide the normalized data using the housekeeper (Figure 4). Blood samples from 3 volunteers were irradiated and we monitored the dose response after 6, 24 and 48 hours exposure of 5 genes DDB2-3, BAX-2, TMEM30A-1, PCNA-3 and BBC3-1 known as being biodosimetry markers [14,15,17]. These graphs represent the average signal intensity of the gene expression response of 3 volunteers with the error bars reflecting the standard deviation among the 3 individuals. It is important to mention that we are measuring the gene expression response from the leukocytes which vary in numbers from one person to the other (factor 3 among healthy individuals) which explain the size of the error bars. The trend of the expression pattern increased in response to radiation ranging from greater than 0 Gy to 6 Gy within the time periods of 6 hours, 24 hours and in particular 48 hours post-exposure where the gene signature signal remains stable and strong. For all the exposure times, a good dose response was observed between 0 and 2 Gy and also between 2 and 4 Gy for most of the genes. PCNA-3 is the lowest expressed gene measured with the CLPA assay among the 5 genes studied in this paper and when looking at the

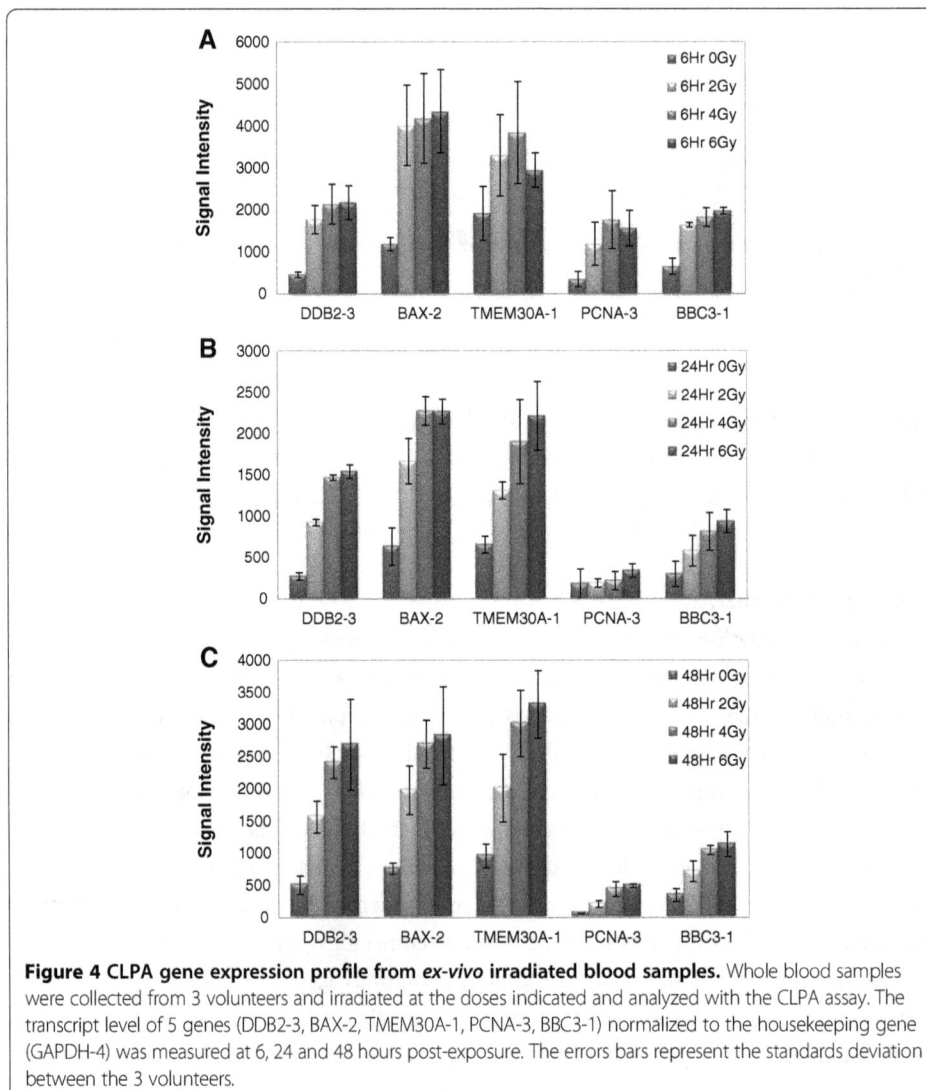

Figure 4 CLPA gene expression profile from *ex-vivo* irradiated blood samples. Whole blood samples were collected from 3 volunteers and irradiated at the doses indicated and analyzed with the CLPA assay. The transcript level of 5 genes (DDB2-3, BAX-2, TMEM30A-1, PCNA-3, BBC3-1) normalized to the housekeeping gene (GAPDH-4) was measured at 6, 24 and 48 hours post-exposure. The errors bars represent the standards deviation between the 3 volunteers.

average signal for these 3 persons, a dose response was not observed as strong as if looking at the individual response (data not shown). If we look at the average signal without considering the standard deviation, we can see the dose response for PCNA-3 as well, as expected. For the dose response between 4 and 6 Gy, the differences among the signal intensity for all exposure times are much smaller than at lower doses or even nonexistent as it is for BAX-2. The fact that the separation by exposure dose is clearest between the lowest doses, with some overlap evident between the highest doses of 5 and 8 Gy has been reported previously in the literature [17]. In addition an overlap of the standard deviation is observed when the data is averaged over the 3 volunteers between 4 and 6 Gy, which indicates that such gene expression assay might not necessarily correlate with the cellular response at higher doses. The genes DDB2-3, BAX-2, TMEM30A-1, PCNA-3 and BBC-3 were anticipated to be up-regulated in response to radiation with the LINAC and it was observed that their expression increases in response to radiation between 0 and 2 Gy by a fold change up to 3.8 and between 2 and 4 Gy by a fold change up to 2.1. This data shows that the linear accelerator (LINAC) can be used as a characterization methodology providing *ex vivo* irradiation of blood samples to validate genomic-based radiation induced responses using assay chemistries suitable for small volumes processing.

Conclusion

The authors described in detail how a LINAC at a standard Radiation Oncology Department from a hospital has been implemented in order to provide samples for validating the performance of a radiogenomic test that could be used for guiding medical countermeasures against acute exposure to ionizing radiations. A customized tube holder phantom was made to position the blood sample in a way that will mimic a body material heterogeneity and have an irradiation dose distributions and delivery process as accurate and similar as possible to an *in-vivo* irradiation. The LINAC was also calibrated in order to distribute the most homogenous dose to the sample tubes and the result showed a 3% variation in dose distribution. The phantom is actually designed to irradiate five tubes at a time but can be adapted to irradiate a larger number of tubes if needed.

By utilizing the LINAC installed in most radiotherapy departments, it is possible to provide irradiated samples that can be used to optimize and validate the radiosensitive biomarkers panel and assay chemistry platform for gene expression measurement. Not only does the use of the LINAC for that purpose allow for unlimited quantity of irradiated blood samples, it also avoids collecting extra blood from cancer patients and the use of animal models at multiple validation phases of the study. Our methodological approach using ex-vivo irradiation of blood samples could also be an advantage to evaluate relative biological effectiveness (RBE). The general procedure for evaluating RBE of most beam equipments is standardized with survival curves on cell culture. Using ex-vivo irradiation directly on blood samples instead of cell culture would allow accessing additional biomarkers representative of the biological interactions at the tissue or organ level. The use of the LINAC is a simple, fast and reliable way to provide *ex-vivo* irradiated samples with a similar dose accuracy as the one delivered during radiotherapy treatment on patients. In the future, we will assess samples from total body irradiation patients to validate various gene expression signatures and assay chemistries. Since the LINAC is the instrument used for radiotherapy treatment, these *in vivo* samples will have the same irradiation source as our preliminary study, which is important in order to compare data. The combined data will allow correlating the expression patterns of genes that are involved in biochemical and cellular responses to the irradiation dose by using the *ex-vivo* LINAC approach reported in this paper.

Methods

Design of experiment

Whole blood samples were collected from volunteers and *ex-vivo* irradiated at 0, 2, 4, and 6 Gy using the Varian Millennium Linear Accelerator (LINAC) with a customized phantom. The irradiated blood samples were placed in culture for allowing gene expression in response to radiation after 6, 24 and 48 hours. Following exposure, the samples were harvested and the samples were processed with the CLPA and the gene expression was analyzed on the CE instrument.

Customized phantom for blood sample irradiation

To facilitate and standardize the irradiation of multiple blood tube vacutainers using a linear accelerator, a custom phantom was build, made of Plastic Water® (CIRS, Norfolk, VA) and Superflab® slabs (Mick Radio-nuclear, Mount Vernon, NY) (Figure 5A and B). The Superflab® material is flexible and would not crush the blood tube. A slab of 5 cm of Plastic Water® was used as base; for the top, a 1.5 cm Superflab® slab with a hole cut at the center was created to accommodate the blood tubes (Figure 5B). Another 1 cm of Superflab® was placed to cover the blood tubes. Finally, another 1 cm of Plastic Water® was put on top of the 1 cm Superflab®. To irradiate the blood tubes vacutainers, the phantom is placed on the treatment couch with its center aligned with the central axis of the radiation beam at 100 cm SSD (Figure 5C).

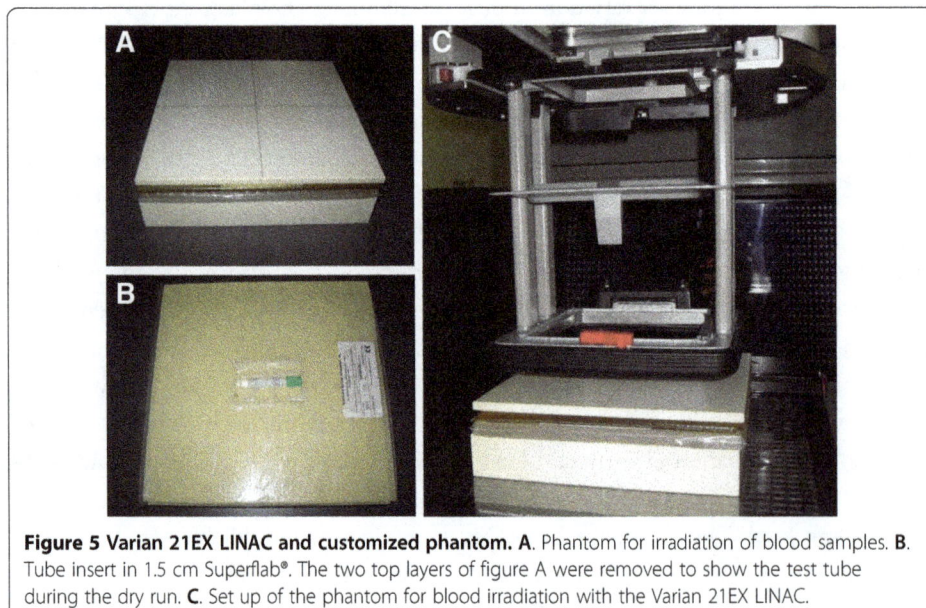

Figure 5 Varian 21EX LINAC and customized phantom. A. Phantom for irradiation of blood samples. **B**. Tube insert in 1.5 cm Superflab®. The two top layers of figure A were removed to show the test tube during the dry run. **C**. Set up of the phantom for blood irradiation with the Varian 21EX LINAC.

Blood collection, irradiation and culture

These experiments were approved by the institutional review board (IRB#2008-036) and were conducted according to the principles expressed in the Declaration of Helsinki. Written consent for participation in the study was obtained from all the subjects voluntarily. A total of 3 healthy donors, 1 male and 2 female between the age of 21 and 55 years old participated in this study. On the day of blood irradiation, informed consent was obtained from the volunteers. Prior to phlebotomy, a single certified

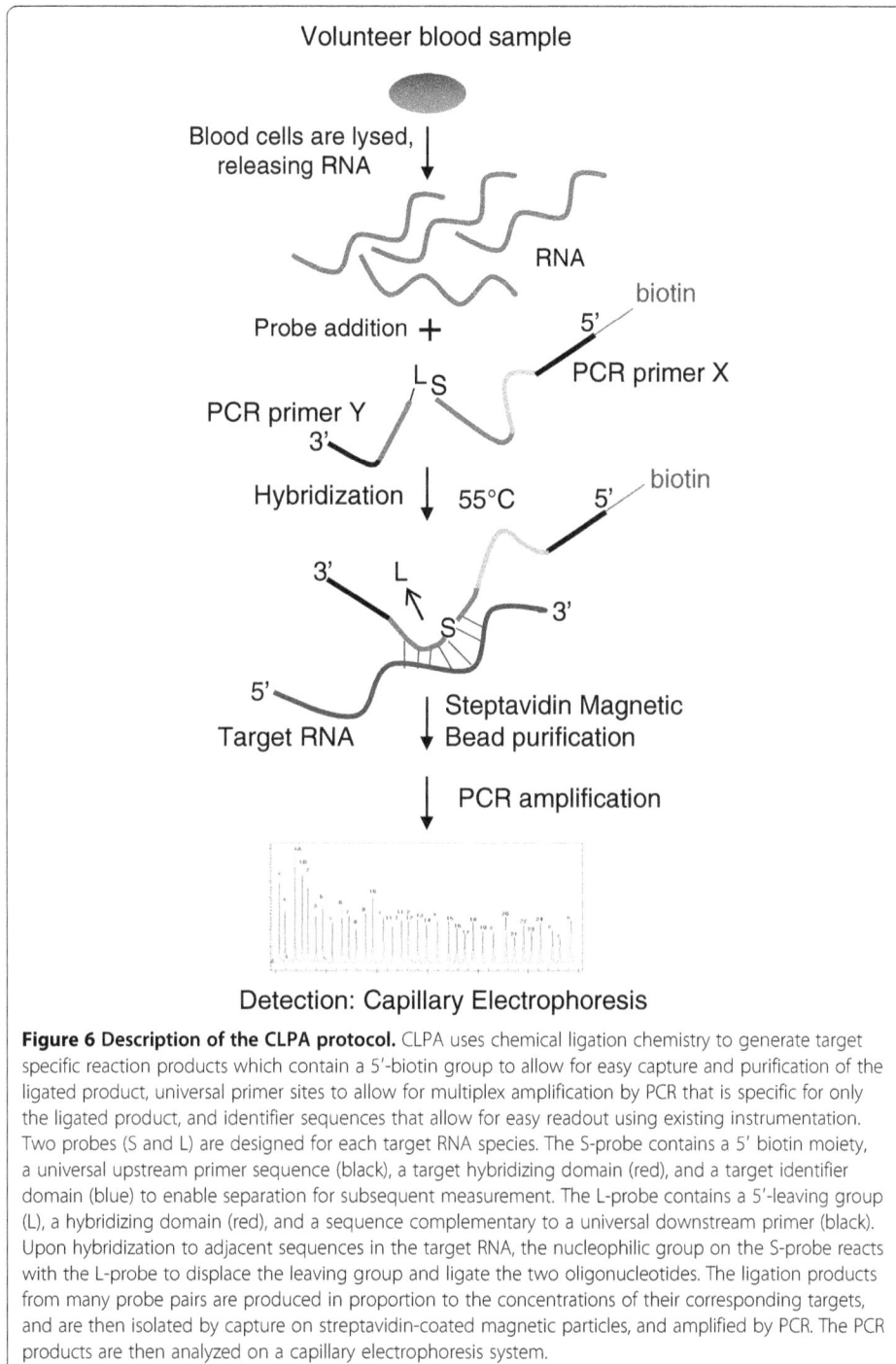

Figure 6 Description of the CLPA protocol. CLPA uses chemical ligation chemistry to generate target specific reaction products which contain a 5'-biotin group to allow for easy capture and purification of the ligated product, universal primer sites to allow for multiplex amplification by PCR that is specific for only the ligated product, and identifier sequences that allow for easy readout using existing instrumentation. Two probes (S and L) are designed for each target RNA species. The S-probe contains a 5' biotin moiety, a universal upstream primer sequence (black), a target hybridizing domain (red), and a target identifier domain (blue) to enable separation for subsequent measurement. The L-probe contains a 5'-leaving group (L), a hybridizing domain (red), and a sequence complementary to a universal downstream primer (black). Upon hybridization to adjacent sequences in the target RNA, the nucleophilic group on the S-probe reacts with the L-probe to displace the leaving group and ligate the two oligonucleotides. The ligation products from many probe pairs are produced in proportion to the concentrations of their corresponding targets, and are then isolated by capture on streptavidin-coated magnetic particles, and amplified by PCR. The PCR products are then analyzed on a capillary electrophoresis system.

radiation physicist programmed the Varian 21 EX (3100 Hansen Way, Palo Alto, CA 94306) at the following settings.: 20 MeV, 20 cone, and the following Monitor Unit (MU): 212, 414 and 636 MU for 2, 4 and 6 Gy. The blood irradiation phantom was put in the center of the table and raised to 100 cm source to surface distance (SSD). Samples containing 4 ml of peripheral blood samples were collected from healthy volunteers in glass vacutainer tubes (12.35 mg Sodium Citrate, 2.21 mg Citric Acid) (VWR International, Pittsburg, PA). The vacutainers were transported by batches of 5 for irradiation. The blood was exposed to 0, 2, 4 and 6 Gy X-rays using the Varian 21EX LINAC. After irradiation, blood samples were diluted 1:1 with RPMI 1640 medium (Invitrogen) supplemented with 10% heat inactivated fetal bovine serum (Invitrogen) and incubated for 6, 24 and 48 hours at 37°C in a humidified incubator with 5% $CO2$. After the indicated exposure times, the sample tubes were removed from the incubator and set up in a sterile field. The blood samples were aliquoted in 2 ml Eppendorf tubes and an equal volume of CLPA reaction buffer was added (1:1). The samples were stored at 4°C.

Gene expression assay – chemical ligation dependent probe amplification (CLPA)

We selected a commercially available assay chemistry test (CLPA, DxTerity Diagnostics, Rancho Vista, CA) for multiplex gene expression analysis that combines a robust chemical ligation process and a sample stabilization method (Figure 6) (http://dxterity. com/dx_direct.php). The assay requires 25 µl of whole blood which corresponds to 100 µl of the blood sample obtained previously. For each reaction, 20 µl of S probe (probe with a nucleophillic group) mix was added to the 100 µl blood sample and placed in a thermal cycler for 10 min at 80°C. During the incubation, a master mix containing 60 µl CLPA Buffer 2 and 20 µl L-probe (probe with an electrophillic leaving group) mix per sample was prepared and kept at 55°C. The master mix was immediately added to each sample and gently mixed before incubation for 3 hours at 55°C for hybridization. When the hybridization was done, 4 µl of M-270 Streptavidin Dynabeads was added to each sample and incubated for 5 min at 55°C. Three washes were then performed using a magnetic block for 10 sec and 100 µl CLPA wash buffer. A complete PCR mixed was prepared (20 µl per reaction) by combining an equal volume of PCR primer mix (Dxterity) and 2X Dynamo probe master mix (New England Biolabs, F-450 L). After the final wash, the beads were re-suspended in 20 µl CLPA complete PCR solution and the PCR was performed using the following cycling conditions: 94°C for 10 min followed by 30 cycles [94°C for 10 sec, 60°C for 20 sec, 72°C for 20 sec]. 1 µl of PCR product was used to perform the analysis on the multiple capillaries electrophoresis (CE) system (ABI 3130 instrument, Life Technologies, CA).

Competing interests
The authors declare that they have no competing interests.

Authors' contributions
MB designed and coordinated the study, collected and analyzed the data, carried out data interpretation and wrote the manuscript. DL designed and performed the sample irradiation and participated in drafting the manuscript. The study was overseen and directed by FZ. RK was a clinical consultant for the project. FZ and RK gave critical comments. All authors read and approved the final manuscript.

Authors' information
MB: Research Associate Scientist – Biologist – University of Arizona.
DL: Medical Physicist responsible for patient radiotherapy treatment delivery at Scottsdale Healthcare.
RK: Radio oncologist and Chief Medical Officer and Director of Virginia Piper Cancer Center at Scottsdale Healthcare.

FZ: Professor of Basic Medical Sciences - University of Arizona College of Medicine and Director of the Center for Applied Nanobioscience and Medicine (ANBM) at the University of Arizona.

Acknowledgements

We thank the members of Scottsdale Medical Imaging Ltd (SMIL) Research Institute and Scottsdale Healthcare Research Institute especially Brenda Culver for the samples irradiation, Deandra O'Connor and Joanne Saczynski for collecting the blood samples and all the volunteers who enrolled in the study. We also thank Carla Brooks for help in the laboratory and Dxterity Diagnostics for providing the assay reagents. This work was supported by the Center for High-Throughput Minimally-Invasive Radiation Biodosimetry (National Institute of Allergy and Infectious Diseases Grant U19 AI067773).

Author details

[1]Center for Applied NanoBioscience and Medicine, The University of Arizona College of Medicine, 425 N. 5th Street, Phoenix, AZ 85004, USA. [2]Scottsdale Clinical Research Institute, Scottsdale Healthcare, 10510 N. 92nd Street, Scottsdale, AZ 85258, USA.

References

1. Blakely WF, Salter CA, Prasanna PG: **Early-response biological dosimetry–recommended countermeasure enhancements for mass-casualty radiological incidents and terrorism.** *Health Phys* 2005, **89:**494–504.
2. Rana S, Kumar R, Sultana S, Sharma RK: **Radiation-induced biomarkers for the detection and assessment of absorbed radiation doses.** *J Pharm Bioallied Sci* 2010, **2:**189–196.
3. Blakely WF, Ossetrova NI, Whitnall MH, Sandgren DJ, Krivokrysenko VI, Shakhov A, Feinstein E: **Multiple parameter radiation injury assessment using a nonhuman primate radiation model-biodosimetry applications.** *Health Phys* 2010, **98:**153–159.
4. DiCarlo AL, Jackson IL, Shah JR, Czarniecki CW, Maidment BW, Williams JP: **Development and licensure of medical countermeasures to treat lung damage resulting from a radiological or nuclear incident.** *Radiat Res* 2012, **177:**717–721.
5. Garty G, Chen Y, Turner HC, Zhang J, Lyulko OV, Bertucci A, Xu Y, Wang H, Simaan N, Randers-Pehrson G, Lawrence Yao Y, Brenner DJ: **The RABiT: a rapid automated biodosimetry tool for radiological triage. II. Technological developments.** *Int J Radiat Biol* 2011, **87:**776–790.
6. Pinto MM, Santos NF, Amaral A: **Current status of biodosimetry based on standard cytogenetic methods.** *Radiat Environ Biophys* 2010, **49:**567–581.
7. Romm H, Wilkins RC, Coleman CN, Lillis-Hearne PK, Pellmar TC, Livingston GK, Awa AA, Jenkins MS, Yoshida MA, Oestreicher U, Prasanna PG: **Biological dosimetry by the triage dicentric chromosome assay: potential implications for treatment of acute radiation syndrome in radiological mass casualties.** *Radiat Res* 2011, **175:**397–404.
8. Redon CE, Nakamura AJ, Gouliaeva K, Rahman A, Blakely WF, Bonner WM: **The use of gamma-H2AX as a biodosimeter for total-body radiation exposure in non-human primates.** *PLoS One* 2010, **5:**e15544.
9. Redon CE, Nakamura AJ, Martin OA, Parekh PR, Weyemi US, Bonner WM: **Recent developments in the use of γ-H2AX as a quantitative DNA double-strand break biomarker.** *Aging* 2011, **3:**168–174.
10. Amundson SA, Do KT, Shahab S, Bittner M, Meltzer P, Trent J, Fornace AJ Jr: **Identification of potential mRNA biomarkers in peripheral blood lymphocytes for human exposure to ionizing radiation.** *Radiat Res* 2000, **154:**342–346.
11. Amundson SA, Bittner M, Meltzer P, Trent J, Fornace AJ Jr: **Induction of gene expression as a monitor of exposure to ionizing radiation.** *Radiat Res* 2011, **156:**657–661.
12. Amundson SA, Grace MB, McLeland CB, Epperly MW, Yeager A, Zhan Q, Greenberger JS, Fornace AJ Jr: **Human in vivo radiation-induced biomarkers: gene expression changes in radiotherapy patients.** *Cancer Res* 2004, **64:**6368–6371.
13. Dressman HK, Muramoto GG, Chao NJ, Meadows S, Marshall D, Ginsburg GS, Nevins JR, Chute JP: **Gene expression signatures that predict radiation exposure in mice and humans.** *PLoS Med* 2007, **4:**e106.
14. Brengues M, Paap B, Bittner M, Amundson S, Seligmann B, Korn R, Lenigk R, Zenhausern F: **Biodosimetry on small blood volume using gene expression assay.** *Health Phys* 2010, **98:**179–185.
15. Kabacik S, Mackay A, Tamber N, Manning G, Finnon P, Paillier F, Ashworth A, Bouffler S, Badie C: **Gene expression following ionising radiation: identification of biomarkers for dose estimation and prediction of individual response.** *Int J Radiat Biol* 2011, **87:**115–129.
16. Sunirmal P, Barker CA, Turner HC, McLane A, Wolden SL, Amundson SA: **Prediction of In Vivo radiation dose status in radiotherapy patients using Ex Vivo and In Vivo gene expression signatures.** *Radiat Res* 2011, **175:**257–265.
17. Paul S, Amundson SA: **Development of gene expression signatures for practical radiation biodosimetry.** *Int J Radiat Oncol Biol Phys* 2008, **71:**1236–1244.
18. Riecke A, Rufa CG, Cordes M, Hartmann J, Meineke V, Abend M: **Gene expression comparisons performed for biodosimetry purposes on in vitro peripheral blood cellular subsets and irradiated individuals.** *Radiat Res* 2012, **178:**234–243.
19. Williams JP, Brown SL, Georges GE, Hauer-Jensen M, Hill RP, Huser AK, Kirsch DG, Macvittie TJ, Mason KA, Medhora MM, Moulder JE, Okunieff P, Otterson MF, Robbins ME, Smathers JB, McBride WH: **Animal models for medical countermeasures to radiation exposure.** *Radiat Res* 2010, **173:**557–578.
20. Ballas LK, Elkin E, Schrag D, Minsky B: **Radiation therapy facilities in the United States.** *Int J of Radiat Oncol Biol Phys* 2006, **66:**1204–1211.

21. Committee on Radiation Source Use and Replacement, National Research Council. "7 RADIOTHERAPY": **Radiation source use and replacement.** Abbreviated Version. Washington, DC: The National Academies Press 2008, **7:**117–134.

22. Pinnarò P, Soriani A, D'Alessio D, Giordano C, Foddai ML, Pinzi V, Strigari L: **Implementation of a new cost efficacy method for blood irradiation using a non dedicated device.** *J Exp Clin Cancer Res* 2011, **30:**7.

23. Weiss B, Hoffmann M, Anders C, Hellstern P, Schmitz N, Uppenkamp M: **Gamma-irradiation of blood products following autologous stem cell transplantation: surveillance of the policy of 35 centers.** *Ann Hematol* 2004, **83:**44–49.

24. Almond PR, Biggs PJ, Coursey BM, Hanson WF, Huq MS, Nath R, Rogers DW: **AAPM's TG-51 protocol for clinical reference dosimetry of high-energy photon and electron beams.** *Med Phys* 1999, **26:**1847–1870.

25. Abe H, Kool ET: **Destabilizing universal linkers for signal amplification in self-ligating probes for RNA.** *J Am Chem Soc* 2004, **126:**13980–13986.

3

On-chip cavity optomechanical coupling

Bradley D Hauer, Paul H Kim, Callum Doolin, Allison JR MacDonald, Hugh Ramp and John P Davis[*]

*Correspondence:
jdavis@ualberta.ca
Department of Physics, University of
Alberta, T6G 2E1 Edmonton, AB,
Canada

Abstract

Background: On-chip cavity optomechanics, in which strong co-localization of light and mechanical motion is engineered, relies on efficient coupling of light both into and out of the on-chip optical resonator. Here we detail our particular style of tapered and dimpled optical fibers, pioneered by the Painter group at Caltech, which are a versatile and reliable solution to efficient on-chip coupling. A brief overview of tapered, single mode fibers is presented, in which the single mode cutoff diameter is highlighted.

Methods: The apparatus used to create a dimpled tapered fiber is described, followed by a comprehensive account of the procedure by which a dimpled tapered fiber is produced and mounted in our system. The custom-built optical access vacuum chambers in which our on-chip optomechanical measurements are performed are then discussed. Finally, the process by which our optomechanical devices are fabricated and the method by which we explore their optical and mechanical properties is explained.

Results: Using this method of on-chip optomechanical coupling, angular and displacement noise floors of 4 nrad/$\sqrt{\text{Hz}}$ and 2 fm/$\sqrt{\text{Hz}}$ have been demonstrated, corresponding to torque and force sensitivities of 4×10^{-20} N·m/$\sqrt{\text{Hz}}$ and 132 aN/$\sqrt{\text{Hz}}$, respectively.

Conclusion: The methods and results of our on-chip optomechanical coupling system are summarized. It is our expectation that this manuscript will enable the novice to develop advanced optomechanical experiments.

Keywords: Cavity optomechanics; Nanoscale transduction; Dimpled fiber; Tapered fiber; Nanomechanics

PACS codes: 07.60.-j; 07.10.Cm; 42.50.Wk

Background

State-of-the-art nanofabrication technologies have allowed for a drastic reduction in the size, and increase in quality, of nanomechanical systems, which have been the driving force behind radically increasing the sensitivity of numerous devices. Examples include accelerometers [1], mass sensors [2-6], electrometers [7], temperature sensors [8,9], force transducers [10,11] and biosensors [12-15]. The mass of a sensor and its ability to precisely measure physical quantities are intimately related, with smaller devices having superior sensitivity [5,6,11]. However, as we continue to reduce device volume, it is difficult to find transduction methods that scale appropriately. Furthermore, it is generally the case that the target quantity is measured through the nanomechanical device's motion, hence as we move to more sensitive devices, we require a detection method with comparable precision. A solution to these issues has been found in the field of cavity optomechanics, which allows for quantum-limited, sub-am/$\sqrt{\text{Hz}}$ displacement sensitivity [16,17] and device masses down to the pico/femto-gram range [18-20].

Optomechanics describes the coupling of the mechanical motion of a device to an optical field, often to manipulate or detect its motion. It is advantageous to use an optical cavity, such as a whispering gallery mode (WGM) resonator [21] (see Figure 1b), to provide this field, as the light in the optical cavity is able to sample the mechanics many times due to its long photon lifetime and leads to resonantly enhanced optomechanical coupling. In such a system, the motion of the mechanical device shifts the resonance frequency and phase of the optical cavity. By detecting this signal, it is possible to infer the motion of the device. At the same time, photons inside the cavity apply a radiation pressure to the mechanical resonator [22], which can be used to optomechanically dampen or amplify its motion, leading to a large number of interesting phenomena [23]. Cavity optomechanical systems have been realized in a number of different geometries, including photonic crystal cavities [20], Fabry-Pérot etalons [24,25], WGM resonators [21,26,27] and electronic microwave cavities [28].

It is advantageous to fabricate cavity optomechanical devices on-chip, as it is therefore possible engineer both the optical and mechanical resonators to certain desired specifications. Modern nanofabrication technologies make it possible to produce devices with extremely accurate dimensions, allowing feature sizes as small as 100 nm for foundry-based deep ultraviolet (DUV) optical lithography [29] and 2 nm with electron beam lithography [30]. This enables precise tailoring of important device parameters, such as the gap between mechanical and optical resonators, which controls the optomechanical coupling in our devices [21]. Furthermore, devices fabricated using top-down lithography can be integrated into electronic on-chip devices [31] and, in the case of optical lithography, can easily be mass produced.

However, difficulties arise when trying to couple light into these on-chip devices, as a high on-chip density and planar geometry require a precise optical probe which can couple exclusively to a particular device. This problem has been solved by using a dimpled tapered fiber, which is created by introducing a small protrusion to a straight tapered

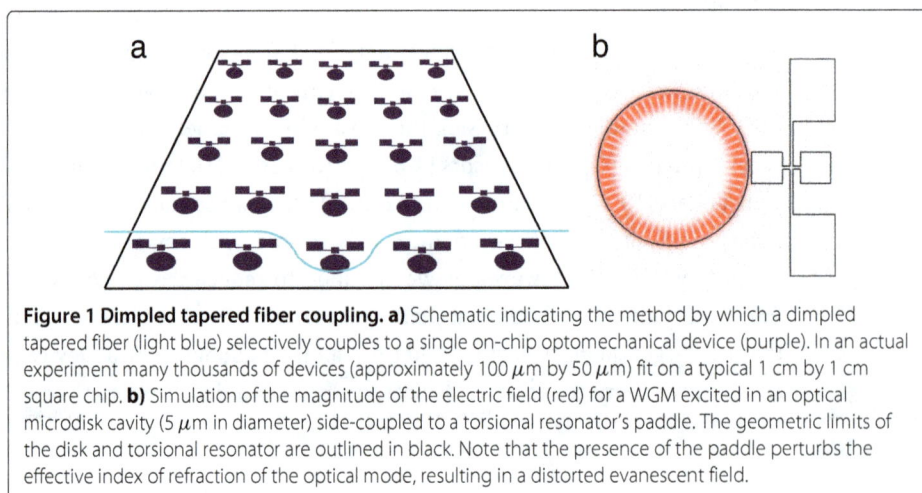

Figure 1 Dimpled tapered fiber coupling. a) Schematic indicating the method by which a dimpled tapered fiber (light blue) selectively couples to a single on-chip optomechanical device (purple). In an actual experiment many thousands of devices (approximately 100 μm by 50 μm) fit on a typical 1 cm by 1 cm square chip. **b)** Simulation of the magnitude of the electric field (red) for a WGM excited in an optical microdisk cavity (5 μm in diameter) side-coupled to a torsional resonator's paddle. The geometric limits of the disk and torsional resonator are outlined in black. Note that the presence of the paddle perturbs the effective index of refraction of the optical mode, resulting in a distorted evanescent field.

fiber [21,32]. Such a waveguide provides an efficient and maneuverable probe, allow-ing selective coupling to on-chip optomechanical devices. This procedure is illustrated schematically in Figure 1a.

In this article, we describe a process by which such a coupling system is produced, out-lining the necessary steps while assuming no special knowledge *a priori*. We begin by investigating the fundamentals of tapered optical fibers, as well as describing an appa-ratus which can be used to fabricate and dimple them. Following this is a discussion of the procedure by which dimpled tapered fibers are produced. Custom-built optical access vacuum chambers, which are used for optomechanical coupling to on-chip devices, are also detailed. Finally, we explain our method of coupling to optomechanical devices with tapered fibers. Using these systems, we have been able to demonstrate the first ever on-chip optomechanical torsional sensors [21], as well as multidimensional detection of high frequency microcantilevers [18] suitable for force sensing applications.

Single mode tapered optical fibers

A crucial element in any optomechanical device is the method by which the optical field is injected, and subsequently collected, from the optical resonator in the system. While a number of different options exist, including free space optical coupling [33], grating couplers [34] and fiber-to-waveguide coupling [35,36], we have chosen to use direct cou-pling from tapered optical fibers [20,37-40]. Tapered fibers are more efficient, and require less on-chip space, than grating couplers, while free-space coupling is inconsistent with on-chip devices. It may prove that fiber-to-waveguide coupling [36] is more efficient and stable than tapered fibers, but the versatility and maneuverability of tapered fibers remains a significant advantage.

A tapered fiber is a standard optical fiber (silica core surrounded by a higher index cladding) that has had its initial diameter adiabatically reduced over a small length known as the *tapered region*. This can be performed either through hydrofluoric acid etching of an optical fiber [41,42], or by the heat-and-pull method [43-46]. In this latter method, a small region of an optical fiber, known as the *hot-zone*, is heated to the point of melting and subsequently stretched to reduce its diameter. The final tapered fiber will then consist of three regions, the initial unstreched fiber, the taper transition, and the taper waist, all of which are detailed in Figure 2. From a conservation of mass argument, it can be shown that for a constant hot-zone of length L, which is produced in the case of a stationary flame, the taper transition is exponential [47]. Using a constant pull speed v, this results in a taper waist diameter d that decreases with pull time t according to

$$d = d_0 e^{-vt/L}, \tag{1}$$

where d_0 is the diameter of the initial untapered fiber.

Following the heat-and-pull process, a new air-clad core exists in the taper waist, com-prised of a composite material with an effective index determined by the indices and relative sizes of the initial core and cladding. This region can be modeled as a long, dielec-tric cylinder, for which Maxwell's equations can be solved analytically to determine the electromagnetic modes of the core (cladding) in terms of Bessel (modified Bessel) func-tions of the first (second) kind, as described in [48]. In general, such a structure will support many modes, lending to the description of a multimode fiber. However, once the fiber's diameter drops below a critical value, known as the *single mode cut-off diameter*,

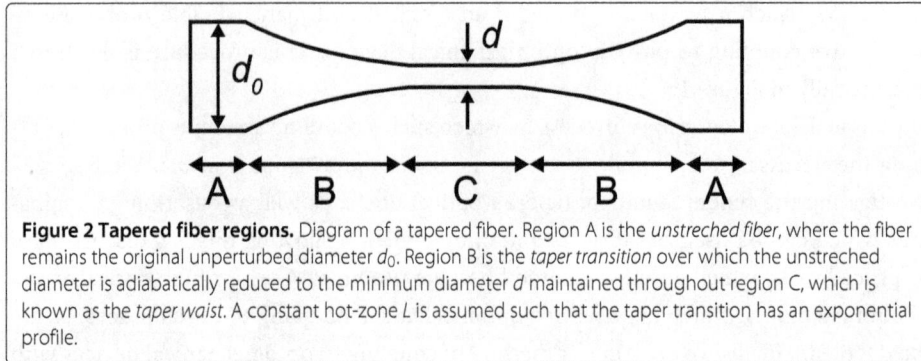

Figure 2 Tapered fiber regions. Diagram of a tapered fiber. Region A is the *unstreched fiber*, where the fiber remains the original unperturbed diameter d_0. Region B is the *taper transition* over which the unstreched diameter is adiabatically reduced to the minimum diameter d maintained throughout region C, which is known as the *taper waist*. A constant hot-zone L is assumed such that the taper transition has an exponential profile.

only a single guided mode remains in the fiber, labeled the hybrid HE_{11} mode [49], as all other spatial modes decay evanescently. We look to determine this critical diameter for light with a free space wavelength, λ, traveling in a fiber with a core of index of refraction, n_{co}, surrounded by a cladding with index, n_{cl}. This is done by matching the electromagnetic fields within the core and cladding according to the boundary conditions given by Maxwell's equations [50], resulting in the following expression for the single remaining mode

$$\left[\frac{J_1(x)}{xJ_0(x)} + \frac{K_1(y)}{yK_0(y)} \right] \left[\frac{n_{co}^2}{n_{cl}^2} \frac{J_1(x)}{xJ_0(x)} + \frac{K_1(y)}{yK_0(y)} \right] = 0. \tag{2}$$

In the above equation, $J_\nu(x)$ is the Bessel function of the first kind and $K_\nu(x)$ is the modified Bessel function of the second kind. As well, $x = \frac{d}{2}\sqrt{k_{co}^2 - \beta^2}$ and $y = \frac{d}{2}\sqrt{\beta^2 - k_{cl}^2}$, where $k_{co} = 2\pi n_{co}/\lambda$ and $k_{cl} = 2\pi n_{cl}/\lambda$ are the magnitudes of the wavevector in the core and cladding, respectively, and β is the fiber's propagation constant. From these definitions, we can immediately derive the expression

$$x^2 + y^2 = \frac{\pi^2 d^2 (n_{co}^2 - n_{cl}^2)}{\lambda^2}. \tag{3}$$

Using this relationship between x and y, we are able to determine the fiber diameter d_c such that Eq. 2 has only one solution for $d < d_c$, indicating the point at which the penultimate mode ceases to exist. This value is the single mode cut-off diameter and is determined by numerically calculating the solutions to Eq. 2 while iteratively increasing d until a second solution emerges.

It is also possible to determine an analytic expression for d_c in the weakly-guiding approximation (WGA) [51]. In this case, we take $n_{co} \approx n_{cl}$, so that Eq. 2 becomes

$$\frac{xJ_0(x)}{J_1(x)} = -\frac{yK_0(y)}{K_1(y)}. \tag{4}$$

For the single mode cut-off, $y = 0$ (*i.e.* $\beta = \pm k_{cl}$), which is physically interpreted as the mode evanescently decaying into the cladding. Using $\lim_{y\to 0} \frac{yK_0(y)}{K_1(y)} = 0$, we see that Eq. 4 has solutions when $xJ_0(x) = 0$. One solution will always exist for $x = 0$, corresponding to the single remaining mode below cut-off. The penultimate mode comes into existence when $J_0(x) = 0$ for the first time, which occurs at $x = 2.4048$. Therefore, we can find d_c by rearranging Eq. 3 to get [49]

$$d_c = \frac{2.4048\lambda}{\pi\sqrt{n_{co}^2 - n_{cl}^2}}. \tag{5}$$

The validity of the WGA is confirmed by comparing the results of Eq. 5 to numerically calculated cutoff diameters, which are summarized for a number of situations in Table 1.

At these diameters there exists a significant evanescent field surrounding the waist region of the tapered fiber. This allows for substantial overlap between an optical resonator's modes and the fiber's guided light when it is approached to an optical cavity. Likewise, light trapped inside the cavity will couple back into the fiber, which will be carried away as optomechanical signal.

While this type of straight tapered fiber is useful for coupling to a single off-chip device, such as a microsphere [37], it is difficult to use as probe of on-chip devices, although it can be done if the device is cleaved to hang over the edge of the chip [26] or isolated using a mesa [52]. Instead, it is useful to introduce a small dimpled region to the fiber, which when oriented towards the sample chip produces a portion of the taper waist that can be used as a probe of an individual on-chip optomechanical device [32]. By combining this probe with a precise positioning system, numerous devices can be sampled using the localized coupling region at the tip of the dimple of the tapered fiber.

Methods

Tapered fiber puller

To produce tapered fibers, we use a heat-and-pull method in which a flame from a hydrogen torch is used to soften or melt an optical fiber while simultaneously stretching it at a constant speed. In our system, we produce this flame using a custom-built mountable hydrogen torch, as seen in Figure 3a, which is threaded using a 7/16″-24 die (McMaster-Carr, Part No. 26005A128) producing standard threads that allow for interchangeability of torch tips. The tips we use are the HT and OX series purchased from National Torch (see Figure 3d), which provide a wide variety of flame sizes useful for producing different sizes of tapered fibers. The hydrogen torch is fed by a needle valve-controlled line,

Table 1 Single mode cutoff diameters for tapered fibers in a number of situations

λ (nm)	n_{co}	n_{cl}	d_c (nm)	
			WGA	Numerical
637	1.47	1.00	452.6	452.6
	1.47	1.33	778.8	778.9
780	1.47	1.00	554.1	554.2
	1.47	1.33	953.6	953.7
1310	1.47	1.00	930.7	930.7
	1.47	1.33	1601.6	1601.7
1550	1.47	1.00	1101.2	1101.2
	1.47	1.33	1895.0	1895.1

Single mode cutoff diameter calculated using both the WGA approximation and numerical calculations for a green light (637 nm) observed in nitrogen vacancy photoluminescence [53] and near infrared light (780 nm) used in aqueous biosensing [54], as well as the dispersionless and low attenuation telecom wavelengths of 1310 nm and 1550 nm. All calculations are performed for both air-clad ($n_{cl} = 1.0$) and water-clad ($n_{cl} = 1.33$) environments. The appropriate number of digits are retained to show the difference in WGA and numerical calculations.

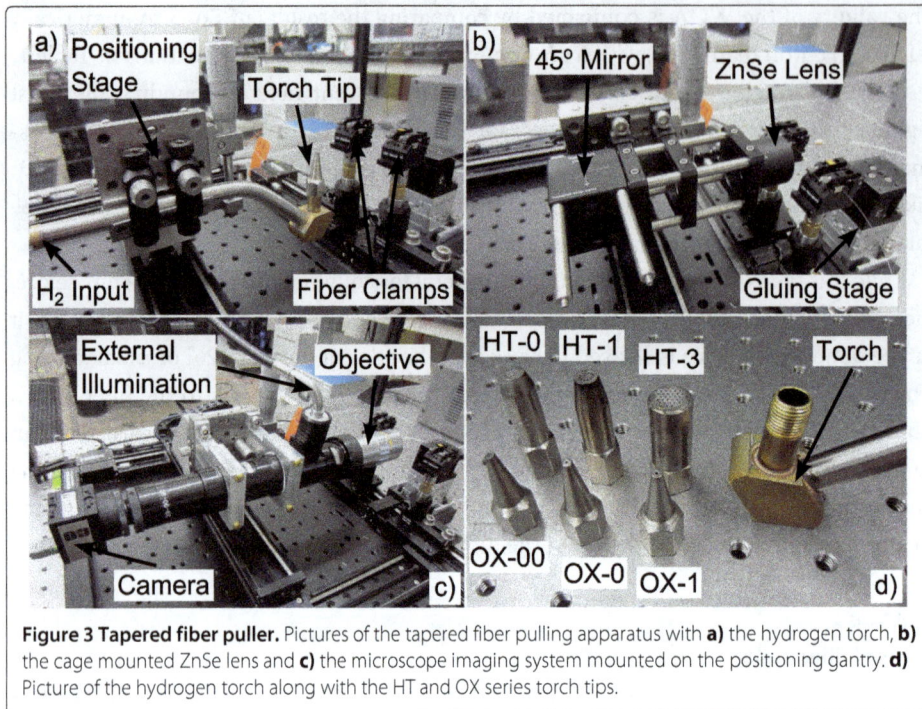

Figure 3 Tapered fiber puller. Pictures of the tapered fiber pulling apparatus with **a)** the hydrogen torch, **b)** the cage mounted ZnSe lens and **c)** the microscope imaging system mounted on the positioning gantry. **d)** Picture of the hydrogen torch along with the HT and OX series torch tips.

allowing for a very small and stable flame using the OX-00 torch tip, with a single 0.51 mm diameter hole. This tip is chosen because it produces compact tapers (less than 1 cm in total length) which are ideal for our fiber holders, while maintaining a relatively high transmission efficiency (up to ~80%).

The hydrogen torch is mounted on a three-axis positioning system, consisting of automated xy-translation in the plane of the optical table on which the apparatus is mounted, along with perpendicularly oriented manual z-adjustment. The xy-translation system is based on a Zaber T-G-LSM200A200A two-axis gantry system. Each orthogonal axis is driven by a Zaber T-LSM200A linear motorized stage, allowing for a total travel range of 200 mm in either dimension with a minimum step size of 50 nm. Manual z-adjustment is provided by a New Focus 9063-COM gothic-arch translation stage mounted using a New Focus 9063-A angle bracket. The stage is manipulated by a Mitutoyo No. 906912 micrometer, providing a 25 mm travel range with 10 μm resolution. This system is used for precise and reproducible placement of the hydrogen torch flame as it heats the fiber, which is an important element required to consistently produce high quality tapered fibers. The fiber itself is held using two Newport 466A-710 dual arm V-groove fiber holders, each of which is connected to an adjustable optical post mounted on a Zaber T-LSM100A linear motorized stage. Each stage has a travel range of 100 mm with a resolution of 50 nm and can pull the melted fiber at speeds up to 7 mm/s. All of the Zaber stages are automated in software, allowing for precise, reproducible xy-positioning of the torch gantry, as well as the ability to set a consistent pull speed. The adjustable optical posts help to ensure that the fiber is level, as proper alignment is crucial for producing a low-loss taper. This entire setup is surrounded by a protective box, built from optical rails and acrylic sheets, which helps reduce flame instability due to air currents, as well as preventing contaminants from entering the system.

Another method by which a fiber can be tapered is using a CO_2 laser, which produces radiation with a wavelength ranging from 10.2–10.8 μm [44]. Absorption of these photons by an optical fiber causes it to heat in proportion to the intensity of the beam and the cross section of the fiber being irradiated. Therefore, the power of a CO_2 laser must be carefully controlled while pulling a fiber in order to ensure even heating. CO_2 lasers have also been used as a heat source for other processes, namely in the production of high-Q silica WGM resonators, such as microspheres [55] and bottles [56].

We integrate a CO_2 laser into our fiber pulling system by replacing our hydrogen torch with a cage mount system (see Figure 3b) containing a 45 degree cube-mounted silvered mirror (Thorlabs - Product No. CM1-P01) and a plano-convex ZnSe lens (Thorlabs - Product No. LA7542-F) with a focal length of 25.4 mm, which can be used to focus the intense, infrared radiation from the laser onto the fiber. Attaching our lens to the torch positioning system, we gain full control of its position. This allows for defocusing of the CO_2 laser beam, effectively controlling both the size and temperature of the hotspot on the fiber. Furthermore, the focused beam can be scanned along the fiber, allowing for a movable hotspot, which is required for bottle fabrication [54]. Finally, manual height adjustment of the cage mount focusing system allows us to ensure that the CO_2 beam will hit the center of the lens, reducing aberration.

It is also possible to attach a microscope imaging system directly to our torch positioning gantry, as shown in Figure 3c. The microscope is comprised of a 10X M Plan Apo long working distance infinity-corrected objective (Edmund Optics - Stock No. #59-877) attached to an Optem Zoom 70XL lens system, allowing for 70× magnification of the setup. This image is recorded using an Edmund Optics EO-5012C color USB webcam, providing a video feed to a nearby computer. To ensure proper lighting and image quality, light from an external Metaphase MP-LED-150 microscope LED illuminator is coupled into the lens system's coaxial illumination port using a fiber optic waveguide (Edmund Optics - Stock No. #39-368). This system is very useful, as it allows for real time imaging of our completed tapered fibers (and other fabricated optical components) when dimpling or attaching it to its holder, with full three-axis control.

As tapered fibers are quite fragile, it is difficult to move them without breaking. For this reason, we first attach the tapered fiber to a holder, creating a more robust system which can easily be relocated. To this end, we have also included a manually adjusted Newport Compact Dovetail DS40-XYZ three-dimensional linear positioning stage in our system, which allows for 1 μm sensitivity over a travel range of 14 mm in each of x and y and 5 mm in z. This stage allows us to properly position and align the fiber holder, as well as gradually approach it to the fiber for gluing. In addition, it is used to position the fiber mold used in the dimpling process, which must be approached and raised precisely at the thinnest point of the tapered fiber.

Another important aspect of the tapered fiber puller is the fiber transmission monitoring system, which allows us to determine the point at which the taper becomes single mode, as well as assess fiber losses due to tapering. To do this, we measure the transmitted power of light from a New Focus Velocity 6330 tunable diode laser through the fiber during the pulling process. To control the amount of injected power, laser light is first passed through a Thorlabs VOA50-APC variable optical attenuator (VOA) before it is coupled into the fiber using a mechanical splicer (Fiber Instrument Sales elastomeric lab splice - Part No. FIS114012), in which two straight cleaved fiber ends are butt coupled to each other with the aid of index matching gel (Fiber Instrument Sales matching gel - Part No. F10001V). Likewise, the fiber is mechanically spliced on its opposite end to a patch cable connected to a New Focus Model 1811 IR DC-125 MHz low noise photoreceiver. The DC signal from this photodiode is split off and recorded using an NI USB-6259 BNC DAQ card for the duration of a fiber pull, providing a record of transmission vs pull time, as seen in Figure 4.

Fiber tapering procedure

To create tapered fibers, we begin with a Corning SMF-28e optical fiber that has a silica core and cladding diameter of $8.2\,\mu$m and $125\,\mu$m, respectively, all of which is protected by an acrylate coating which extends out to a diameter of $245\,\mu$m. The indices of refraction and dimensions of the core and cladding are chosen such that this original fiber is single mode for wavelengths exceeding 1260 nm, which includes both the dispersionless and minimum loss wavelengths in silica of 1310 nm and 1550 nm, respectively.

To begin the tapering process, the acrylate coating is removed using a Micro-Strip® stripping tool over a region approximately 3 cm long in the center of an SMF-28e fiber around one meter in total length. This section of stripped fiber is subsequently cleaned using a solvent to remove any remaining acrylate. The tapering occurs in this stripped region, where the flammable acrylate has been removed. In addition, the two ends of the fiber are stripped of acrylate and cleaved flat using an Ericsson EFC11 fiber cleaver. Utilizing the mechanical splicers and index matching gel described above, these cleaved ends are spliced to two ends of a severed FC/APC patch cable, one of which leads to the photodiode, the other to the diode laser. This method of fiber splicing is ideal for this application, as it is quick and easy, allowing for a convenient input and removal of the tapering fiber to and from the optical circuit. Losses vary depending on fiber alignment for this splicing method, but we are only concerned with providing enough power to observe variations in fiber transmission. Once we have ensured that the splices provide sufficient power to the photodiode, the fiber is

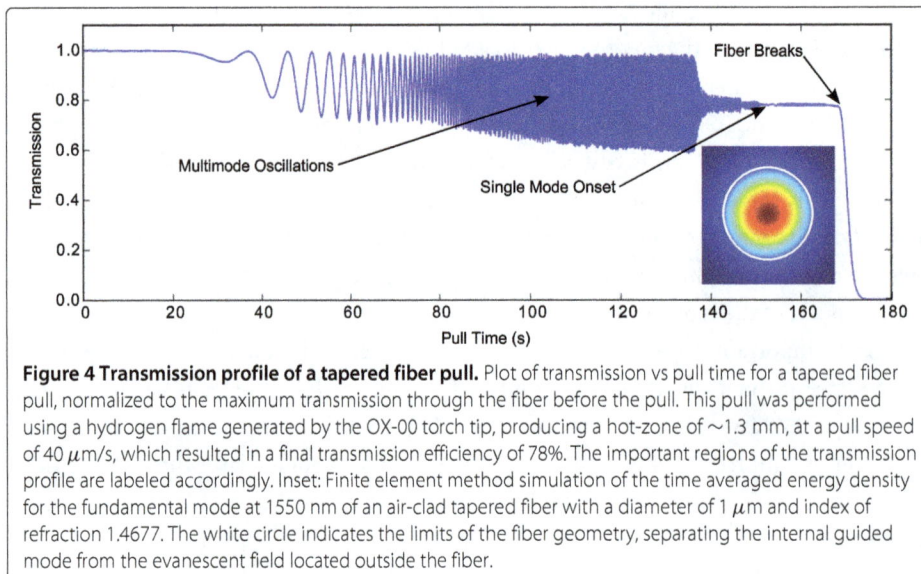

Figure 4 Transmission profile of a tapered fiber pull. Plot of transmission vs pull time for a tapered fiber pull, normalized to the maximum transmission through the fiber before the pull. This pull was performed using a hydrogen flame generated by the OX-00 torch tip, producing a hot-zone of ~1.3 mm, at a pull speed of 40 μm/s, which resulted in a final transmission efficiency of 78%. The important regions of the transmission profile are labeled accordingly. Inset: Finite element method simulation of the time averaged energy density for the fundamental mode at 1550 nm of an air-clad tapered fiber with a diameter of 1 μm and index of refraction 1.4677. The white circle indicates the limits of the fiber geometry, separating the internal guided mode from the evanescent field located outside the fiber.

placed in the V-groove fiber holders, with the region prepared for tapering centered between them.

At this point, the hydrogen torch is lit using a butane lighter and gas flow is adjusted to ensure a steady flame about 1 cm high. This flame is then approached towards the fiber until a small (a few mm) section begins to glow, indicating that the fiber is in a molten state. Once this point has been reached, the two pulling stages move in opposite directions, each at a constant speed generally chosen to be 40 μm/s.

During each pull, the transmission through the fiber vs pull time is monitored, an example of which is presented in Figure 4 for the OX-00 torch tip. By monitoring fiber transmission, it is possible to determine the point at which the fiber waist has become single mode. This will be indicated as a stabilization of the fiber transmission (which is evident in Figure 4) due to the fact that the lossy, higher order modes of the fiber have died out, leaving behind the single fundamental mode of the fiber. Using images from a scanning electron microscope (SEM - inset of Figure 5), we experimentally measured the diameters of our fibers at the single mode transition to be ~1.1 μm, consistent with the theoretically predicted diameter for an air-clad fiber with an index of 1.47 (we expect our fibers to have an index of 1.4677) at 1550 nm (see Table 1). By measuring the time required to reach this transition from a single pull, it is possible to determine a value for the hot-zone length L by inverting Eq. 1, provided that the pull speed and initial fiber diameter are known *a priori*. Using this parameter, we are able to predict the fiber waist diameter for a given pull time. Note that in order for this prediction to be accurate, care must be taken to ensure that all subsequent pulls have conditions matching the original one in order to ensure a consistent hot-zone length. This is readily accomplished using our system. A plot of fiber waist diameter vs pull time using the apparatus described here is presented in Figure 5, indicating excellent agreement between the hot-zone length of

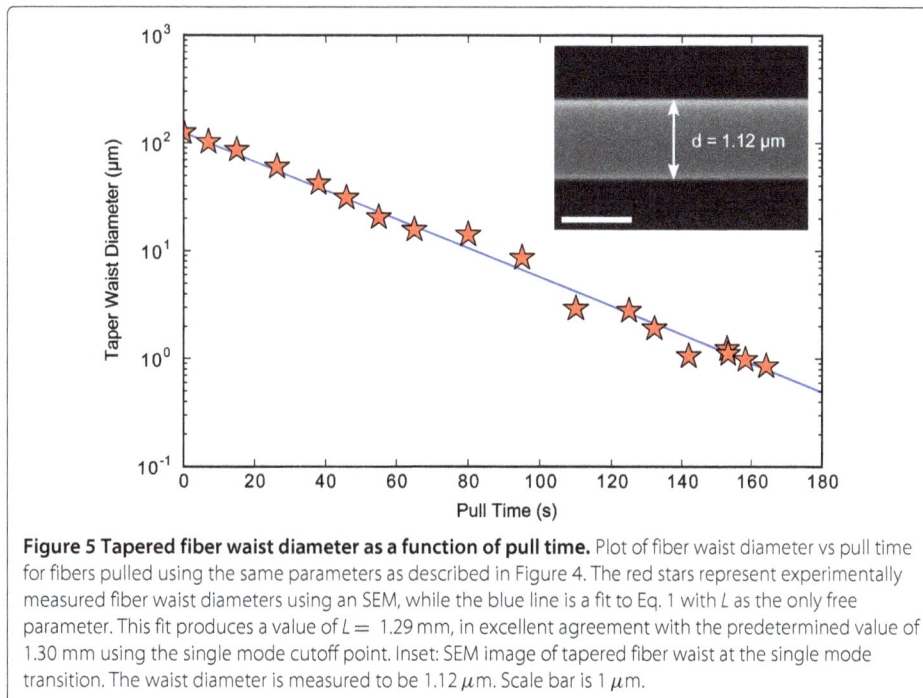

Figure 5 Tapered fiber waist diameter as a function of pull time. Plot of fiber waist diameter vs pull time for fibers pulled using the same parameters as described in Figure 4. The red stars represent experimentally measured fiber waist diameters using an SEM, while the blue line is a fit to Eq. 1 with L as the only free parameter. This fit produces a value of $L =$ 1.29 mm, in excellent agreement with the predetermined value of 1.30 mm using the single mode cutoff point. Inset: SEM image of tapered fiber waist at the single mode transition. The waist diameter is measured to be 1.12 μm. Scale bar is 1 μm.

1.30 mm determined using the single mode cutoff point and the fit value of 1.29 mm. This ability to predict the fiber waist diameter is useful, as it allows for fabrication of fibers whose diameters support a propagating mode that is phase matched with the resonance we are interested in, enhancing coupling of light from the tapered fiber to the optical resonator [37].

At the point of single mode transition, the fiber waist diameter is small enough to produce the desired evanescent field required for coupling to an optical cavity, which can be seen in the inset of Figure 4. However, it is often advantageous to continue pulling fibers to smaller diameters, further increasing the extent of the evanescent field outside the fiber geometry, allowing for a larger range of coupling before the fiber contacts the optical resonator. It is possible to create these sub-μm diameter fibers by continuing to pull for a small amount of time ($\sim$10 s) after the single mode transition has been reached. Using the OX-00 torch tip, diameters as small as 850 nm can be achieved before the fiber breaks due to the pressure of the flowing hydrogen gas from the torch. By using the HT-3, our largest torch tip, the flame size increases, nearly doubling the hot-zone to 2.4 mm, allowing for the fabrication of tapered fibers with diameters down to 500 nm and 98% transmission. This provides fibers with diameters small enough that they can be used as a probe of nitrogen vacancy center photoluminescence [55], as well as allow single mode guiding of 780 nm light (Table 1), which is used in aqueous biosensing applications [56].

By monitoring transmission before and after the pull, it is also possible to determine the losses induced in the fiber due to the tapering process. This is important for determining the amount of power injected into the optical resonator, allowing for calculation of the number of photons confined in the optical resonator. For the OX-00 tip, a tapered fiber transmission efficiency of up to $\sim$80% is achieved. By using the HT-3 tip, with its larger hot-zone, a more adiabatic taper transition region is created allowing us to produce fiber tapers with transmission efficiencies exceeding 99%, which is on par with state-of-the-art, ultralow loss fiber pullers [45].

Fiber dimpling procedure

Once a tapered fiber has been pulled, it is possible to proceed with the dimpling procedure. We begin by taping a stripped Corning SMF-28e optical fiber to the *xyz*-positioning stage located opposite the hydrogen torch, mounting it perpendicular to the tapered fiber so that it can be used as a mold in the dimpling process (see Figure 6a). The fiber mold is prepared by stripping off its acrylate coating and cleaning it with a solvent, producing a mold of 125 μm in diameter. In addition, graphite powder (SLIP Plate® Tube-O-Lube®) is applied to the fiber mold to prevent it from sticking to the tapered fiber. This graphite generally burns away when introduced to the hydrogen flame during the dimple annealing process, however, using too much graphite should be avoided as it can contaminate the dimple, inducing losses. To prevent this from happening, a fiber wipe or compressed air can be used to gently remove excess graphite.

To continue, the torch is replaced by the microscope imaging system on the torch positioning gantry so that dimpling can be observed in real time. While watching with the microscope, the tapered fiber is detensioned by approximately 10 μm to reveal its thinnest point, which appears as a small bend upwards in the fiber (see Figure 6a). The stripped fiber mold is centered on this point and manually raised to touch the tapered fiber using

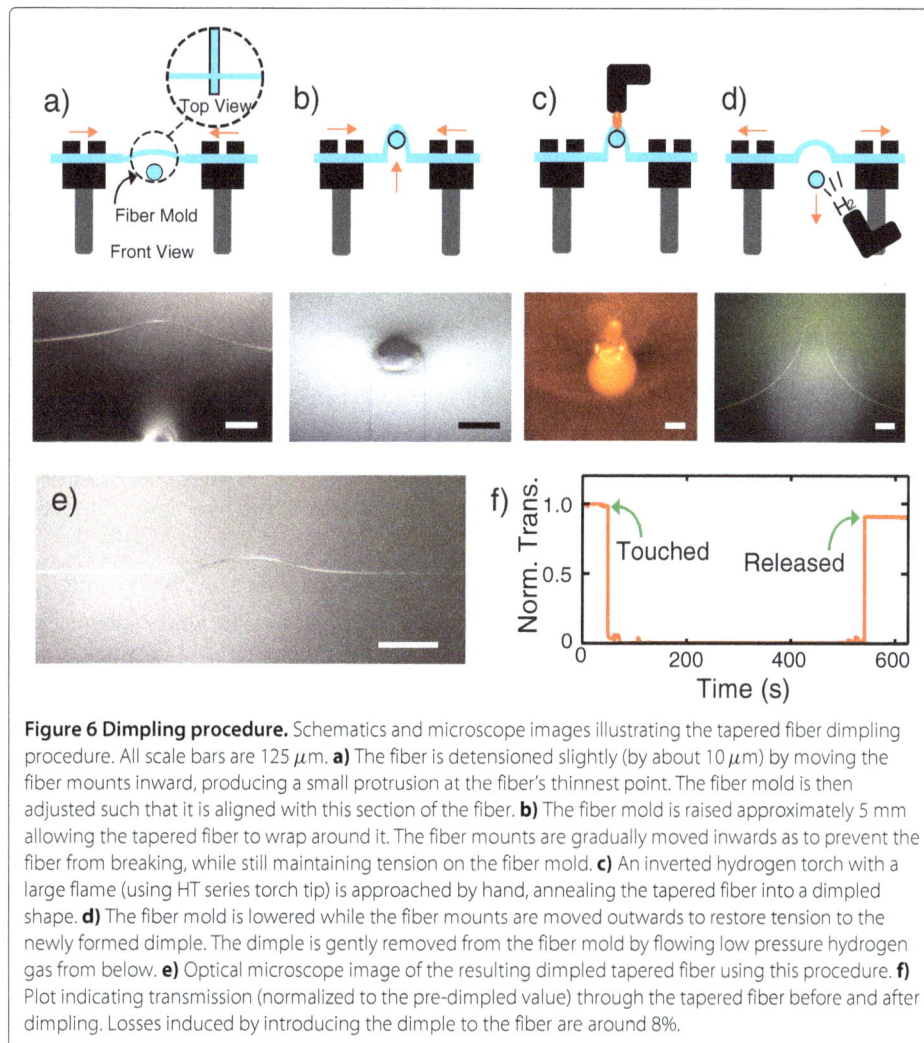

Figure 6 Dimpling procedure. Schematics and microscope images illustrating the tapered fiber dimpling procedure. All scale bars are 125 μm. **a)** The fiber is detensioned slightly (by about 10 μm) by moving the fiber mounts inward, producing a small protrusion at the fiber's thinnest point. The fiber mold is then adjusted such that it is aligned with this section of the fiber. **b)** The fiber mold is raised approximately 5 mm allowing the tapered fiber to wrap around it. The fiber mounts are gradually moved inwards as to prevent the fiber from breaking, while still maintaining tension on the fiber mold. **c)** An inverted hydrogen torch with a large flame (using HT series torch tip) is approached by hand, annealing the tapered fiber into a dimpled shape. **d)** The fiber mold is lowered while the fiber mounts are moved outwards to restore tension to the newly formed dimple. The dimple is gently removed from the fiber mold by flowing low pressure hydrogen gas from below. **e)** Optical microscope image of the resulting dimpled tapered fiber using this procedure. **f)** Plot indicating transmission (normalized to the pre-dimpled value) through the tapered fiber before and after dimpling. Losses induced by introducing the dimple to the fiber are around 8%.

the z-positioning stage. The mold fiber is then raised approximately 5 mm, while simultaneously detensioning the tapered fiber, allowing the fiber to wrap itself around the mold producing the desired dimpled shape, as shown in Figure 6b. During this process, the tapered fiber should remain tensioned tightly around the mold at all times to prevent it from twisting.

At this point, a hydrogen flame produced by the tapering torch is introduced to anneal the fiber into a dimpled shape. For this process, one of the HT series torch tips is used, producing a wide flame allowing for the increase in heat distribution required for annealing. This flame is approached to the dimple by hand, touching the mold and tapered fiber lightly (for about one second) until it glows red (see Figure 6c). The mold fiber is then slowly lowered in the same manner it was raised, this time tensioning the tapered fiber, until the mold is returned to its initial position. The dimple is then removed from the mold by using the unlit torch to flow hydrogen from below, applying a gentle pressure which releases the dimpled fiber. Typically, this process returns a dimple with minimal losses (∼8%, see Figure 6f). A microscope image of a dimpled fiber produced using this procedure is shown in Figure 6e.

Gluing procedure

Once a dimpled tapered fiber (or other optical component created by the fiber heating system) is produced, it must be carefully attached to its holder using the gluing apparatus. To begin this process, Devcon 5 Minute® epoxy gel (No. 14240) is applied to both sides of the fiber holder, which can be seen in Figure 7a. Care is taken to ensure that both droplets of epoxy are approximately the same height, ensuring that they will contact both sides of the tapered region at the same time. Once the epoxy is applied to the fiber holder, it is placed on its holding plate located on the gluing apparatus. The fiber holder is then carefully aligned beneath the fiber, ensuring that the fiber will be glued in the appropriate place. Next, the fiber holder is slowly raised using the z-axis of the positioning stages until the fiber has been enveloped in epoxy on both sides of the taper. This initial epoxy is then left to dry (for about 30 minutes) allowing the fiber to be rigidly held on the fiber holder, drastically increasing its durability. Once the initial epoxy dries, a second round of gluing is typically applied to the fiber, which increases the strength of the fiber's attachment to the holder.

This entire gluing process is monitored in real time using the microscope imaging system mounted on the positioning gantry, which is very helpful as we are able to definitively determine the point at which the fiber has been glued. As well, by imaging the tapered region, along with monitoring transmission down the fiber, we can determine whether or not the tapered fiber has survived the gluing process. Once the fiber has been properly glued in place, it can be transferred directly to the coupling chamber where it is fusion spliced to an existing optical circuit, allowing for injection of light into optomechanical devices.

Optomechanical coupling chambers

Our coupling chambers, which can be seen in Figure 8b-d, use two separate positioning stage arrangements, each of which have similar principles but different translation

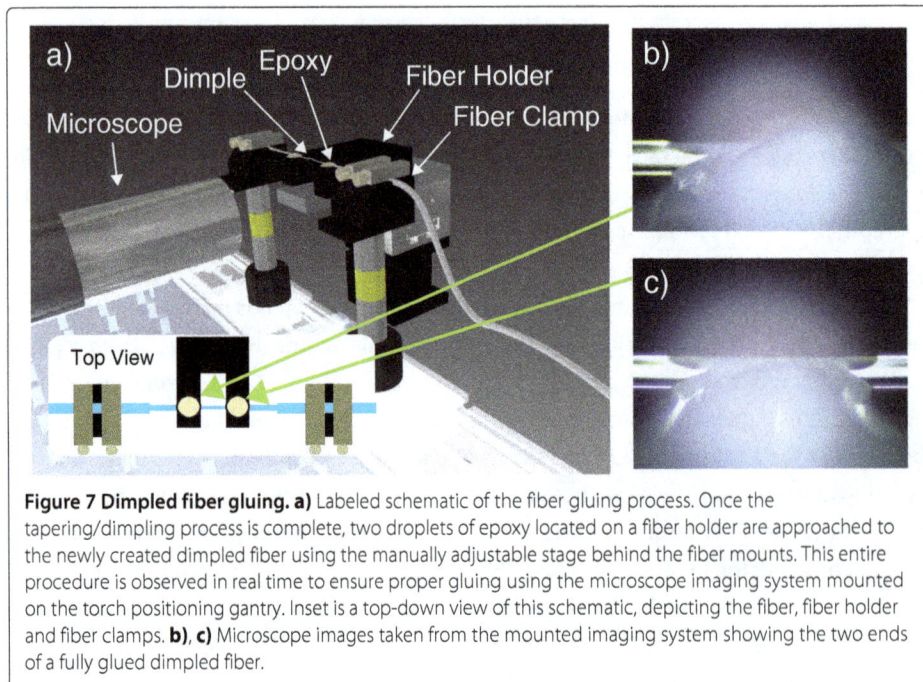

Figure 7 Dimpled fiber gluing. a) Labeled schematic of the fiber gluing process. Once the tapering/dimpling process is complete, two droplets of epoxy located on a fiber holder are approached to the newly created dimpled fiber using the manually adjustable stage behind the fiber mounts. This entire procedure is observed in real time to ensure proper gluing using the microscope imaging system mounted on the torch positioning gantry. Inset is a top-down view of this schematic, depicting the fiber, fiber holder and fiber clamps. **b), c)** Microscope images taken from the mounted imaging system showing the two ends of a fully glued dimpled fiber.

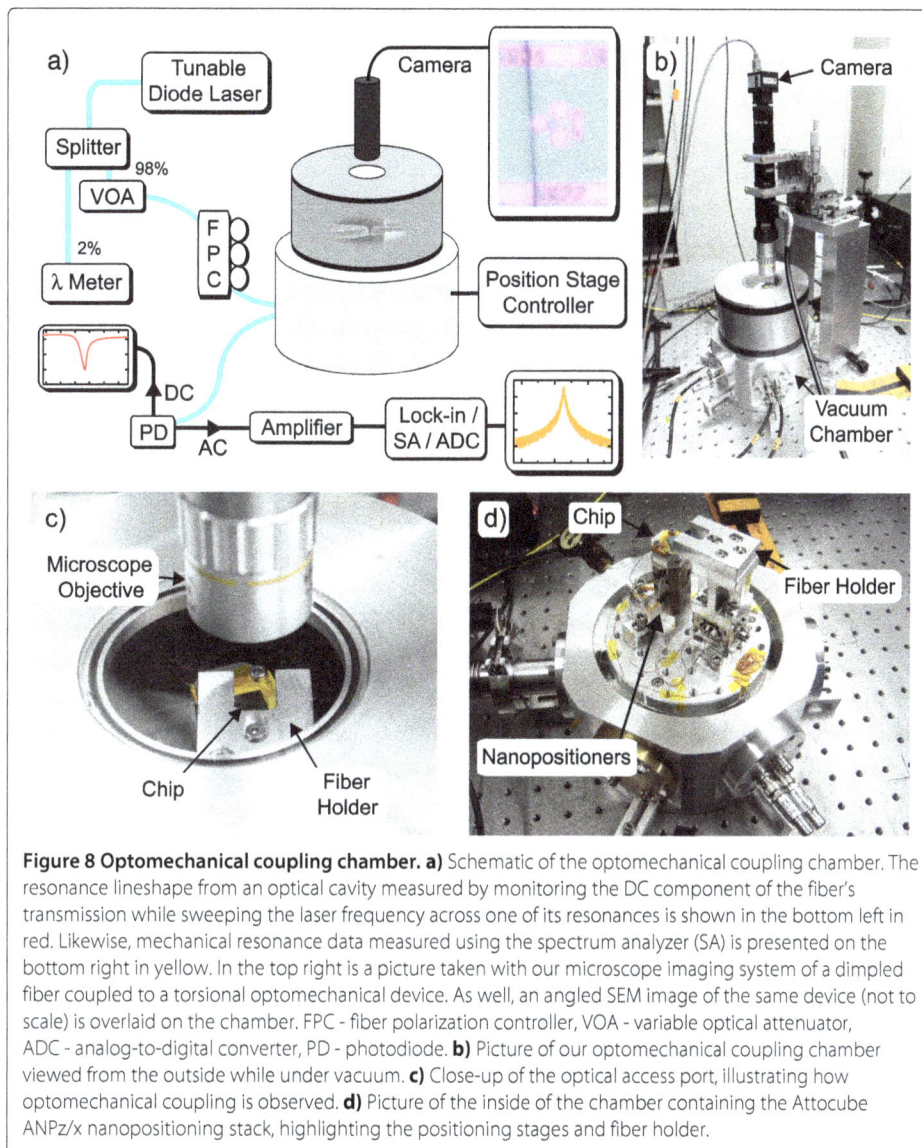

Figure 8 Optomechanical coupling chamber. a) Schematic of the optomechanical coupling chamber. The resonance lineshape from an optical cavity measured by monitoring the DC component of the fiber's transmission while sweeping the laser frequency across one of its resonances is shown in the bottom left in red. Likewise, mechanical resonance data measured using the spectrum analyzer (SA) is presented on the bottom right in yellow. In the top right is a picture taken with our microscope imaging system of a dimpled fiber coupled to a torsional optomechanical device. As well, an angled SEM image of the same device (not to scale) is overlaid on the chamber. FPC - fiber polarization controller, VOA - variable optical attenuator, ADC - analog-to-digital converter, PD - photodiode. **b)** Picture of our optomechanical coupling chamber viewed from the outside while under vacuum. **c)** Close-up of the optical access port, illustrating how optomechanical coupling is observed. **d)** Picture of the inside of the chamber containing the Attocube ANPz/x nanopositioning stack, highlighting the positioning stages and fiber holder.

stages. In each case, the positioning systems are used to approach the optomechanical devices found on the sample chip to a dimpled tapered fiber glued to a stationary custom-machined fiber holder. The fiber was chosen to remain fixed as it is far less stable than the devices on the chip, so its mechanical noise is reduced by anchoring it to an immobile fiber holder.

In the first setup, the sample chip is placed on top of a stack of Attocube linear nanopositioning stages consisting of one ANPz101 stage mounted on top of two perpendicularly oriented ANPx101 stages. The chip is attached rigidly to a custom-machined adapter, which is fastened to the top z-positioning stage. This arrangement provides positioning with sub-nm precision over a total range of 5 mm. A picture of this setup can be seen in Figure 8d. The other positioning system is built using Newport Agilis^{TM} AG-LS25V6 vacuum compatible, piezo driven linear stages. Two of these stages are stacked on top of each other, resulting in orthogonal xy-positioning, with a third mounted at a 90 degree angle for z-translation using an EQ3 Series angle bracket purchased from Newport. As above, the chip containing our optomechanical devices is mounted on a custom-built platform, which is attached to the vertical translation stage providing full three-axis control. These stages provide 50 nm stepping resolution over their entire travel range of 12 mm. Each of these systems have different strengths, with the Attocube stack providing extremely precise positioning over a relatively large range, while the Agilis stages provide a more durable, inexpensive alternative with a larger range of motion.

To allow interchangeability between our two chambers, this positioning system is fastened to a custom-machined circular plate, containing 1/4"-20 tapped holes in a square pattern with a spacing of 3/4". This plate is then screwed onto a homemade aluminum base with 6 ports, each of which is sealed with an O-ring and provide electrical and optical input/output for the setup, as well as allowing for pressure control inside the chamber.

The optical input/output port consists of fiber feedthroughs, each of which allows for both an input and output fiber, channelling light to and from the tapered fiber. Each fiber is glued in place using Varian Torr Seal high vacuum epoxy, which provides the appropriate seal required for vacuum. For the Attocube setup, the electrical port houses three hermetically sealed BNC feedthroughs. The other type of electrical port, which provides input/output for the Agilis stages, is comprised of a vacuum compatible 15-pin D-type connector housed in a KF50 feedthrough flange (Accu-Glass Products - Model No. 15D-K50).

The vacuum environment provided by our coupling chambers removes airborne contaminants which can reduce the quality of the tapered fiber and optomechanical devices over time [57]. It is also possible to remove such contaminants using a nitrogen purged environment [49], however, performing optomechanics in vacuum has the added advantage of increasing the mechanical quality factors of devices by drastically reducing viscous damping [58]. The vacuum pump port is comprised of a KF25 adapter connected to a turbo pump backed by a dry scroll pump. By using a completely dry pumping system, we ensure that no oil is ever backstreamed into our system. This connection is made using vibration isolating bellows, which are passed through a cement block to further prevent vibrations from the pump reaching the optical table where the chamber is located. Using this pumping system, we can achieve chamber pressures as low as 10^{-6} torr. There also exists a release port, consisting of a Nupro B-4HK brass bellows-sealed valve, which allows for surrounding air to enter the system, re-establishing ambient pressure inside the chamber. All unused ports on the chamber base are covered with a blank port. This entire system is leak-checked using an Adixen ASM380 dry leak detector, ensuring it is properly sealed.

On top of the chamber base is an aluminum cylinder approximately 10 cm long and 17 cm in diameter which provides the housing for the positioning stages and tapered fiber mount. An L-shaped boot gasket (Duniway Stockroom - Part No. VBJG7) is placed on each side of the cylinder providing a leak tight seal between it and both the base and its custom-machined lid. Optical access through the lid is provided by a 75 mm diameter optical flat glass window (Edmund Optics 1/4-Wave N-BK7 - Stock No. #62-606), which lays flush against an O-ring located in a recessed portion of the lid when the chamber is under vacuum. This window is located directly above the tapered fiber holder and positioning stages, which allows for real time monitoring of the optomechanical system (see Figure 8c). It is therefore possible to view the tapered fiber and on-chip devices while attempting to couple between them, which is important for this process. The fiber and chip are imaged using the exact same imaging system described above for the tapered fiber puller. This is made possible by the fact that this microscope is oriented using a three dimensional arrangement of manually positioned New Focus 9063-COM gothic-arch translation stages mounted on a two legged custom-built stand with identical mounting plate to that used for manual z-positioning of the hydrogen torch in the tapered fiber puller, allowing interchangeability of the microscope between the two systems. The Mitutoyo No. 906912 micrometers used to manipulate this positioning system, provide a 25 mm travel range with $10\,\mu$m resolution. This resolution is more than enough to view our devices in the xy-plane of the chip, as well as provide excellent focusing for our imaging setup.

On-chip optomechanical devices

Due to the small feature sizes required for our optomechanical devices, we have chosen to use foundry-based nanofabrication, which provides high throughput of devices with a minimum feature size of 100 nm using top-down DUV photolithography [29]. Each optomechanical device consists of an optical microdisk side-coupled to a mechanical nano/micro-resonator, such as a torsion paddle or a cantilever, as can be seen in Figure 9. Each device is centered in a large etched area (approximately $100\,\mu$m $\times$ $50\,\mu$m), which provides ample room for coupling using our dimpled tapered fiber method. The mask for these devices is designed using custom-programmed Python scripts utilizing the gdspy module, which generates a GDSII file containing our chip layout. This allows us to iterate through a number of device specifications, such as coupling gap, disk radius and mechanical resonator dimensions, providing a large parameter space in which we can explore different optomechanical regimes. These design files are then submitted through Canadian Microelectronic Corporation (CMC) Microsystems to the Interuniversity Microelectronics Center (IMEC) located in Leuven, Belgium. It is here our devices are fabricated on an 8 inch silicon-on-insulator (SOI) wafer, which consists of a 220 nm

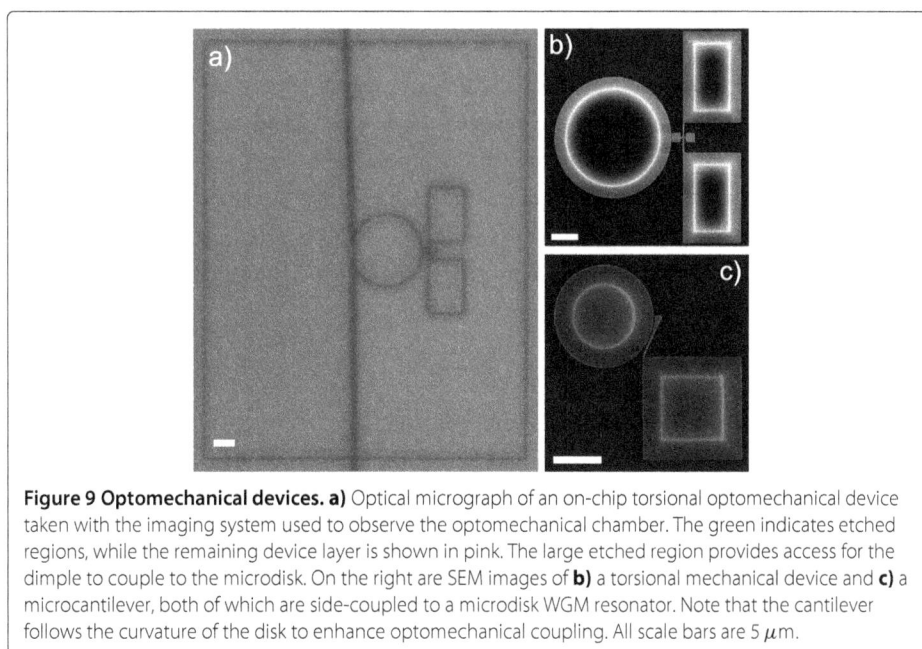

Figure 9 Optomechanical devices. a) Optical micrograph of an on-chip torsional optomechanical device taken with the imaging system used to observe the optomechanical chamber. The green indicates etched regions, while the remaining device layer is shown in pink. The large etched region provides access for the dimple to couple to the microdisk. On the right are SEM images of **b)** a torsional mechanical device and **c)** a microcantilever, both of which are side-coupled to a microdisk WGM resonator. Note that the cantilever follows the curvature of the disk to enhance optomechanical coupling. All scale bars are $5\,\mu$m.

thick layer of single crystal silicon supported by a 2 μm layer of silicon dioxide. The single crystal silicon device layer is ideal for optomechanical devices, as it has negligible absorption in the telecom band around 1550 nm and a high index of refraction ($n \approx 3.42$), enhancing the mechanical resonator's perturbation of the optical cavity's evanescent field. Our devices are patterned onto this wafer using an excimer laser (193 nm - 248 nm) and high-definition photomasks derived from our design files. These patterned wafers are then etched using either a standard or high dose recipe, producing optomechanical devices in the silicon layer which are held rigidly in place by the oxide buffer. This, along with a protective resist coating the entire wafer, help to prevent damage to the devices during transit.

After we receive these wafers, a number of post-processing procedures must be performed in order to prepare our devices for measuring. We begin by dicing the 8 inch wafer into 1 cm $\times$ 1 cm chips using a diamond saw. Each chip is then ultrasonically cleaned with acetone and rinsed with isopropyl alcohol to remove the protective coating. Once the wafer has been diced and cleaned, a buffered oxide etch (BOE) is used to selectively remove the sacrificial oxide layer beneath our mechanical devices, which releases them, allowing them to oscillate freely. It is important to note that since BOE is a wet etch, we must ensure that our devices are dried using either a critical point drier or ultralight solvents, such as n-Pentane (C_5H_{12}), to prevent stiction. Once a chip has been etched and dried, it is ready to be placed in the chamber for measuring.

Coupling procedure

Coupling to our optomechanical devices begins by locating the dimple of the tapered fiber using the imaging apparatus. This is done by searching for the portion of the fiber that is in focus at the lowest point (due to the fact that the dimple protrudes away from the rest of the fiber). Once the dimple is found, the nanopositioning stages are used to align the desired optomechanical device such that the lowest point of the dimple is able to couple light into the optical resonator.

The precision of our nanopositioning stack allows for two methods by which we can couple light into the modes of our optical resonators. We can either bring the fiber close enough to the device such that the cavity's optical modes are excited by the fiber's evanescent field or we can simply touch the fiber to the optical resonator. Hovering has the advantage that the excited optical modes are less perturbed by the fiber's presence, resulting in reduced losses. However, by touching the fiber to the resonator, the mechanical instability of the fiber is further reduced. As well, by using this coupling method, it is possible to excite a larger number of optical modes, some of which have higher Qs and larger optomechanical coupling to the mechanical resonator.

Results and discussion
Data acquisition: side-of-fringe and homodyne detection

Once we have coupled light into our optical resonators, we begin measuring our devices' mechanical motion. In cavity optomechanics, the motion of a mechanical device shifts the resonance wavelength of an optical cavity by changing its effective length. In our optomechanical systems, this is manifested by the mechanical device oscillating in the optical cavity's evanescent field, modulating its effective index of refraction. We can therefore transduce the mechanical device's motion using amplitude sensitive measurements in the "tuned-to-the-slope" regime [59], which is illustrated in Figure 10a. In this detection scheme, the cavity resonance shift due to the mechanical resonator's motion is transduced by the slope of the optical lineshape into AC transmission fluctuations in the fiber, occurring at the mechanical resonance frequency. Therefore, by tuning our probe laser wavelength to the maximal slope of our optical cavity resonance, we provide optimal optomechanical transduction efficiency, as can be seen in Figure 10b. In general, this resonant enhancement scales with the system's optical quality factor which increases the slope of the resonance, however, this is convolved with other effects such as an optical mode's volume and overlap with mechanical motion [21]. Using the experimental setup shown in Figure 8a to perform this type of measurement, we have probed devices with an angular resolution of 4 nrad/$\sqrt{\text{Hz}}$ corresponding to a torque transduction on the level of 4×10^{-20} N·m/$\sqrt{\text{Hz}}$ [21], as well as displacement noise floors of 2 fm/$\sqrt{\text{Hz}}$ and force sensitivity of 132 aN/$\sqrt{\text{Hz}}$ [18].

It is also possible to perform phase-sensitive measurements on our devices in the "tuned-to-the-peak" regime using a balanced optical homodyne detection system [60], which can be used for quadrature and entanglement measurements [61]. This method of detection also has a number of advantages, including cancellation of laser noise [62] and the ability to lock the laser to the bottom of the optical resonance [17]. Furthermore, since the laser's detuning from the cavity resonance is zero, a maximum number of photons are coupled into the cavity, which enhances the system's optomechanical coupling.

After setting up one of these detection schemes, the AC transmission signal through the fiber is sent to a spectrum analyzer (SA), which outputs its frequency power spectrum. For an optomechanical signal, this includes the power spectral density (PSD) of the mechanical motion [63].

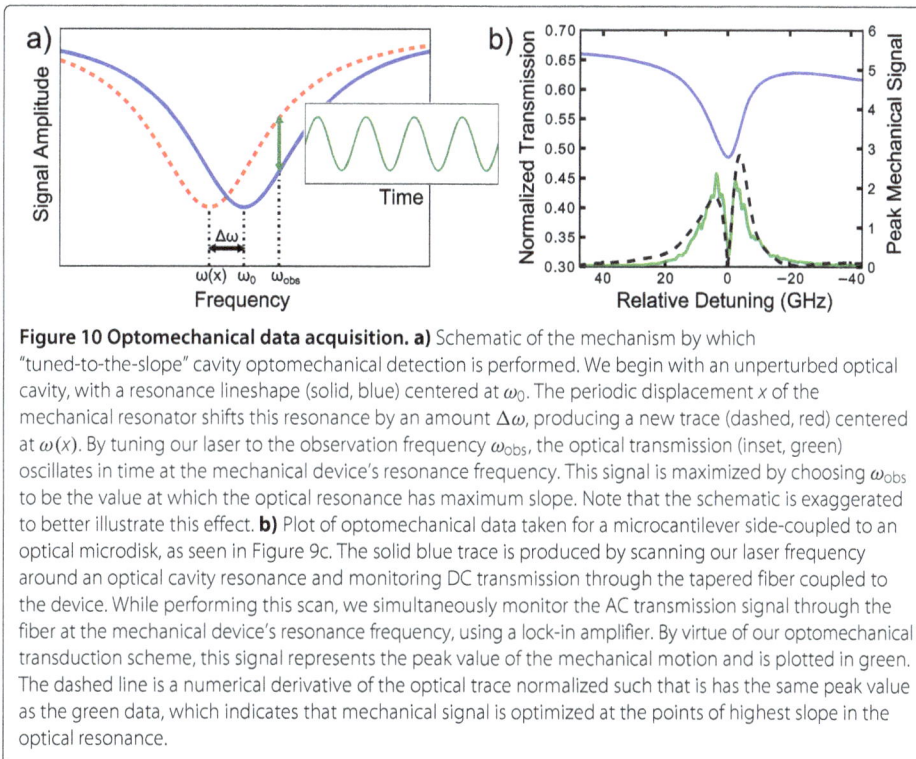

Figure 10 Optomechanical data acquisition. a) Schematic of the mechanism by which "tuned-to-the-slope" cavity optomechanical detection is performed. We begin with an unperturbed optical cavity, with a resonance lineshape (solid, blue) centered at ω_0. The periodic displacement x of the mechanical resonator shifts this resonance by an amount $\Delta\omega$, producing a new trace (dashed, red) centered at $\omega(x)$. By tuning our laser to the observation frequency ω_{obs}, the optical transmission (inset, green) oscillates in time at the mechanical device's resonance frequency. This signal is maximized by choosing ω_{obs} to be the value at which the optical resonance has maximum slope. Note that the schematic is exaggerated to better illustrate this effect. **b)** Plot of optomechanical data taken for a microcantilever side-coupled to an optical microdisk, as seen in Figure 9c. The solid blue trace is produced by scanning our laser frequency around an optical cavity resonance and monitoring DC transmission through the tapered fiber coupled to the device. While performing this scan, we simultaneously monitor the AC transmission signal through the fiber at the mechanical device's resonance frequency, using a lock-in amplifier. By virtue of our optomechanical transduction scheme, this signal represents the peak value of the mechanical motion and is plotted in green. The dashed line is a numerical derivative of the optical trace normalized such that is has the same peak value as the green data, which indicates that mechanical signal is optimized at the points of highest slope in the optical resonance.

Alternatively, a time series measurement of the voltage signal taken with an analog-to-digital converter (ADC) can be digitally analyzed to determine its spectral components. In our system, this is performed using a digital lock-in amplifier (Zurich Instruments HF2LI), which is a specialized ADC that first mixes an input voltage signal with a reference frequency, ω_{ref}, shifting the frequencies of the input signal by $\pm\omega_{\text{ref}}$. Therefore the frequency information around the reference frequency of the input signal is now the low frequency component of the mixed signal. This has the advantage of requiring a data collection rate proportional to the bandwidth of the signal measurement, as opposed to a data collection rate proportional to the maximum frequency component. For example, when bandwidth of only 100 kHz centered at 10 MHz contains important spectral information, a data collection rate of $\sim 10^5$ samples per second (SPS) can be used as opposed to a rate of $\sim 10^7$ SPS, a reduction of about 100 times the data needed to acquire the signal of interest.

After mixing, the lock-in applies a low-pass filter to reduce noise contributions from unwanted frequencies outside the measurement bandwidth, thus the time series output of a lock-in amplifier is the result of a convolution of the lock-in amplifier's filter response with the demodulated input signal, *i.e.*

$$Z(t) = X(t) + iY(t) = \{H(t) * e^{i\omega_{\text{ref}}t}V(t)\}(t), \tag{6}$$

where $X(t)$ and $Y(t)$ are the two outputs of a dual-phase lock-in amplifier, $H(t)$ is the impulse response of the lock-in amplifier's filter, and $V(t)$ is the input signal to the lock-in. Fourier transforming the output elucidates the convolution, giving

$$Z(\omega) = H(\omega)V(\omega - \omega_{\text{ref}}). \tag{7}$$

Thus the spectrum of the lock-in amplifier's output is the spectrum of the input voltage translated in frequency by the reference frequency and enveloped by the lock-in amplifier's filter. The power spectrum can then be estimated by taking $S_{ZZ}(\omega) = |Z(\omega)|^2$, or done in practice by using a PSD estimation algorithm such as Bartlett's method [64], giving

$$S_{ZZ}(\omega) = |H(\omega)|^2 S_{VV}(\omega - \omega_{\text{ref}}). \tag{8}$$

This method of data acquisition allows for real time optimization of optomechanical transduction in our devices, which facilitates sensitive probing of their mechanical motion, allowing for precise measurements of physical quantities, such as forces [18] and torques [21].

Conclusion

This article presents a method by which high efficiency optical coupling is achieved between a dimpled tapered fiber and nanofabricated on-chip optomechanical devices. By using a custom-built automated heat-and-pull fiber puller, it is possible to consistently produce tapered fibers of a predetermined diameter, which is often chosen to be less than the single mode cutoff diameter, providing ample evanescent field for coupling. Dimpling this tapered fiber, using a well-defined procedure, allows for production of an excellent localized probe of planar on-chip devices. Attaching this fiber to a robust holder permits it to be transferred to special coupling chambers. In these chambers, optomechanical coupling is performed in an optical access vacuum environment and is mediated by high precision, nanopositioning stages. By using the amplitude sensitive "tuned-to-the-slope" detection scheme, angular resolution of 4 nrad/$\sqrt{\text{Hz}}$ [21] and displacement transduction of 2 fm/$\sqrt{\text{Hz}}$ [18] have already been demonstrated. It is anticipated that using this technique for optomechanical coupling, we will be able to continue to measure increasingly sensitive devices, approaching the measurement limits imposed by quantum mechanics [59].

Competing interests

The authors declare that they have no competing interests.

Authors' contributions

BDH, PHK and JPD conceived and designed the experiment. BDH, PHK and AJRM constructed the experiment. PHK performed nanofabrication post-processing for the on-chip devices. BDH, PHK, CD, AJRM and HR collected and analyzed data. BDH and AJRM performed theoretical calculations regarding the single mode cutoff diameter for tapered fibers. CD and PHK developed and optimized the procedure for producing dimpled tapered fibers. CD and HR created software for data taking and system manipulation. BDH, PHK, CD and JPD drafted the manuscript. All authors have read, approved and provided critical revisions for the final manuscript.

Acknowledgements

The authors would like to thank Prof. Paul Barclay for numerous helpful suggestions and insight into both the theoretical and practical applications of optomechanics. We would also like to thank Don Mullin, Devon Bizuk and Greg Popowich for technical assistance. This work was supported by the University of Alberta, Faculty of Science; the Natural Sciences and Engineering Research Council of Canada; Alberta Innovates Technology Futures; the Canada Foundation for Innovation; and the Alfred P. Sloan Foundation.

References

1. Krause AG, Winger M, Blasius TD, Lin Q, Painter O: **A high-resolution microchip optomechanical accelerometer.** *Nat Photon* 2012, **6:**768–772.
2. Schmid S, Dohn S, Boisen A: **Real-time particle mass spectrometry based on resonant micro strings.** *Sensors* 2010, **10:**8092–8100.
3. Yang YT, Callegari C, Feng XL, Ekinci KL, Roukes ML: **Zeptogram-scale nanomechanical mass sensing.** *Nano Lett* 2006, **6:**583–586.
4. Jensen K, Kim K, Zettl A: **An atomic-resolution nanomechanical mass sensor.** *Nat Nanotech* 2008, **3:**533–537.
5. Chaste J, Eichler A, Moser J, Ceballos G, Rurali R, Bachtold A: **A nanomechanical mass sensor with yoctogram resolution.** *Nat Nanotech* 2012, **7:**301–304.
6. Li M, Tang HX, Roukes ML: **Ultra-sensitive NEMS-based cantilevers for sensing, scanned probe and very high-frequency applications.** *Nat Nanotech* 2007, **2:**114–120.
7. Cleland AN, Roukes ML: **A nanometre-scale mechanical electrometer.** *Nature* 1998, **392:**160–162.
8. Larsen T, Schmid S, Grönberg L, Niskanen AO, Hassel J, Dohn S, Boisen A: **Ultrasensitive string-based temperature sensors.** *Appl Phys Lett* 2011, **98:**121901.
9. Zhang XC, Myers EB, Sader JE, Roukes ML: **Nanomechanical torsional resonators for frequency-shift infrared thermal sensing.** *Nano Lett* 2013, **13:**1528–1534.
10. Gavartin E, Verlot P, Kippenberg TJ: **A hybrid on-chip optomechanical transducer for ultrasensitive force measurements.** *Nat Nanotech* 2012, **7:**509–514.
11. Moser J, Güttinger J, Eichler A, Esplandiu MJ, Liu DE, Dykman MI, Bachtold A: **Ultrasensitive force detection with a nanotube mechanical resonator.** *Nat Nanotech* 2013, **8:**493–496.
12. Fritz J, Baller MK, Lang HP, Rothuizen H, Vettiger P, Meyer E, Güntherodt H-J, Gerber Ch, Gimzewski JK: **Translating biomolecular recognition into nanomechanics.** *Science* 2000, **288:**316–318.
13. Mertens J, Rogero C, Calleja M, Ramos D, Martín-Gago JA, Briones C, Tamayo J: **Label-free detection of DNA hybridization based on hydrationinduced tension in nucleic acid films.** *Nat Nanotech* 2008, **3:**301–307.
14. Ndieyira JW, Watari M, Barrera AD, Zhou D, Vögtli M, Batchelor M, Cooper MA, Strunz T, Horton MA, Abell C, Rayment T, Aeppli G, McKendry RA: **Nanomechanical detection of antibiotic-mucopeptide binding in a model for superbug drug resistance.** *Nat Nanotech* 2008, **3:**691–696.
15. Gupta A, Akin D, Bashir R: **Single virus particle mass detection using microresonators with nanoscale thickness.** *Appl Phys Lett* 2004, **84:**1976–1978.
16. Arcizet O, Cohadon P-F, Briant T, Pinard M, Heidmann A, Mackowski J-M, Michel C, Pinard L, Français O, Rousseau L: **High-sensitivity optical monitoring of a micromechanical resonator with a quantum-limited optomechanical sensor.** *Phys Rev Lett* 2006, **97:**133601.
17. Schliesser A, Anetsberger G, Rivière R, Arcizet O, Kippenberg TJ: **High-sensitivity monitoring of micromechanical vibration using optical whispering gallery mode resonators.** *New J Phys* 2008, **10:**095015.
18. Doolin C, Kim PH, Hauer BD, MacDonald AJR, Davis JP: **Multidimensional optomechanical cantilevers for high frequency atomic force microscopy.** *New J Phys* 2014, **16:**035001.
19. Stapfner S, Ost L, Hunger D, Reichel J, Favero I, Weig EM: **Cavity-enhanced optical detection of carbon nanotube Brownian motion.** *Appl Phys Lett* 2013, **102:**151910.

20. Eichenfield M, Camacho R, Chan J, Vahala KJ, Painter O: **A picogram- and nanometre-scale photonic-crystal optomechanical cavity.** *Nature* 2009, **459:**550–555.

21. Kim PH, Doolin C, Hauer BD, MacDonald AJR, Freeman MR, Barclay PE, Davis JP: **Nanoscale torsional optomechanics.** *Appl Phys Lett* 2013, **102:**053102.

22. Hossein-Zadeh M, Hossein R, Hajimiri A, Vahala KJ: **Characterization of a radiation-pressure-driven micromechanical oscillator.** *Phys Rev A* 2006, **74:**023813.

23. Kippenberg TJ, Vahala KJ: **Cavity optomechanics: Back-action at the mesoscale.** *Science* 2008, **321:**1172–1176.

24. Jayich AM, Sankey JC, Zwickl BM, Yang C, Thompson JD, Girvin SM, Clerk AA, Marquardt F, Harris JGE: **Dispersive optomechanics: a membrane inside a cavity.** *New J Phys* 2008, **10:**095008.

25. Favero I, Stapfner S, Hunger D, Paulitschke P, Reichel J, Lorenz H, Weig EM, Karrai K: **Fluctuating nanomechanical system in a high finesse optical microcavity.** *Opt Express* 2009, **17:**12813–12820.

26. Anetsberger G, Arcizet O, Unterreithmeier QP, Rivière R, Schliesser A, Weig EM, Kotthaus JP, Kippenberg TJ: **Near-field cavity optomechanics with nanomechanical oscillators.** *Nat Phys* 2009, **5:**909–914.

27. Park Y-S, Wang H: **Resolved-sideband and cryogenic cooling of an optomechanical resonator.** *Nat Phys* 2009, **5:**489–493.

28. Teufel JD, Li D, Allman MS, Cicak K, Sirois AJ, Whittaker JD, Simmonds RW: **Circuit cavity electromechanics in the strong-coupling regime.** *Nature* 2011, **471:**204–208.

29. Selvaraja SK: **Wafer-scale fabrication technology for silicon photonic integrated circuits.** *PhD thesis*, Ghent University, Faculty of Engineering, Department of Information; 2011.

30. Manfrinato VR, Zhang L, Su D, Duan H, Hobbs RG, Stach EA, Berggren KK: **Resolution limits of electron-beam lithography toward the atomic scale.** *Nano Lett* 2013, **13:**1555–1558.

31. Winger M, Blasius TD, Alegre TPM, Safavi-Naeini AH, Meenehan S, Cohen J, Strobbe S, Painter O: **A chip-scale integrated cavity-electro-optomechanics platform.** *Opt Express* 2011, **19:**24905–24921.

32. Michael CP, Borselli M, Johnson TJ, Chrystal C, Painter O: **An optical fiber-taper probe for wafer-scale microphotonic device characterization.** *Opt. Express* 2007, **15:**4745–4752.

33. Fiore V, Yang Y, Kuzyk MC, Barbour R, Tian L, Wang H: **Storing optical information as a mechanical excitation in a silica optomechanical resonator.** *Phys Rev Lett* 2011, **107:**133601.

34. Li M, Pernice WHP, Xiong C, Baehr-Jones T, Hochberg M, Tang HX: **Harnessing optical forces in integrated photonic circuits.** *Nature* 2008, **456:**480–484.

35. Gröblacher S, Hill JT, Safavi-Naeini AH, Chan J, Painter O: **Highly efficient coupling from an optical fiber to a nanoscale silicon optomechanical cavity.** *Appl Phys Lett* 2013, **103:**181104.

36. Cohen JD, Meenehan SM, Painter O: **Optical coupling to nanoscale optomechanical cavities for near quantum-limited motion transduction.** *Opt Express* 2013, **21:**11227–11236.

37. Knight JC, Cheung G, Jacques F, Birks TA: **Phase-matched excitation of whispering-gallery-mode resonances by a fiber taper.** *Opt Lett* 1997, **22:**1129–1131.

38. Cai M, Painter O, Vahala KJ: **Observation of critical coupling in a fiber taper to a silica-microsphere whispering-gallery mode system.** *Phys Rev Lett* 2000, **85:**74–77.

39. Barclay PE, Srinivasan K, Borselli M, Painter O: **Efficient input and output fiber coupling to a photonic crystal waveguide.** *Opt Lett* 2004, **29:**697–699.

40. Srinivasan K, Miao H, Rakher MT, Davanço M, Aksyuk V: **Optomechanical transduction of an integrated silicon cantilever probe using a microdisk resonator.** *Nano Lett* 2011, **11:**791–797.

41. Zhang EJ, Sacher WD, Poon JKS: **Hydrofluoric acid flow etching of low-loss subwavelength-diameter biconical fiber tapers.** *Opt Express* 2010, **18:**22593–22598.

42. Laine J-P, Little BE, Haus HA: **Etch-eroded fiber coupler for whispering-gallery-mode excitation in high-Q silica microspheres.** *IEEE Photon Technol Lett* 1999, **11:**1429–1430.

43. Brambilla G, Finazzi V, Richardson DJ: **Ultra-low-loss optical fiber nanotapers.** *Opt Express* 2004, **12:**2258–2263.

44. Ward JM, O'Shea DG, Shortt BJ, Morrissey MJ, Deasy K, Chormaic SGN: **Heat-and-pull rig for fiber taper fabrication.** *Rev Sci Instrum* 2006, **77:**083105.

45. Ding L, Belacel C, Ducci S, Leo G, Favero I: **Ultralow loss single-mode silica tapers manufactured by a microheater.** *Appl Opt* 2010, **49:**2441–2445.

46. Tong L, Gattass RR, Ashcom JB, He S, Lou J, Shen M, Maxwell I, Mazur E: **Subwavelength-diameter silica wires for low-loss optical wave guiding.** *Nature* 2003, **426:**816–819.

47. Birks TA, Li YW: **The shape of fiber tapers.** *J Lightwave Technol* 1992, **10:**432–438.

48. Arnaud JA: *Beam and Fiber Optics.* New York: Academic Press; 1976.

49. Borselli M: **High-Q microresonators as lasing elements for silicon photonics.** *PhD thesis*, California Institute of Technology, Department of Applied Physics; 2006.

50. Jackson JD: *Classical Electrodynamics.* New York: Wiley; 1999.

51. Gloge D: **Weakly guiding fibers.** *Appl Opt* 1971, **10:**2252–2258.

52. Srinivasan K, Barclay PE, Borselli M, Painter OJ: **An optical-fiber-based probe for photonic crystal microcavities.** *EE J Sel Area Comm* 2005, **23:**1321–1329.

53. Collot L, Lefèvre-Seguin V, Brune M, Raimond JM, Haroche S: **Very high-Q whispering-gallery mode resonances observed on fused silica microspheres.** *Europhys Lett* 1993, **23:**327–334.

54. Pöllinger M, O'Shea D, Warken F, Rauschenbeutel A: **Ultrahigh-Q tunable whispering-gallery-mode microresonator.** *Phys Rev Lett* 2009, **103:**053901.

55. Fu K-MC, Barclay PE, Santori C, Faraon A, Beausoleil RG: **Low-temperature tapered-fiber probing of diamond nitrogen-vacancy ensembles coupled to GaP microcavities.** *New J Phys* 2011, **13:**055023.

56. Dantham VR, Holler S, Kolchenko V, Wan Z, Arnold S: **Taking whispering gallery-mode single virus detection and sizing to the limit.** *Appl Phys Lett* 2012, **101:**043704.

57. Fujiwara M, Toubaru K, Shigeki T: **Optical transmittance degradation in tapered fibers.** *Opt Express* 2011, **19:**8596–8601.

58. Verbridge SS, Ilic R, Craighead HG, Parpia JM: **Size and frequency dependent gas damping of nanomechanical resonators.** *Appl Phys Lett* 2008, **93:**013101.
59. Braginsky VB, Khalili FYa: *Quantum Measurement.* Cambridge: Cambridge University Press; 1995.
60. Yuen HP, Chan VWS: **Noise in homodyne and heterodyne detection.** *Opt Lett* 1983, **8:**177–179.
61. Laurat J, Keller G, Oliveira-Huguenin JA, Fabre C, Coudrea T, Serafini A, Adesso G, Illuminati F: **Entanglement of two-mode Gaussian states: characterization and experimental production and manipulation.** *J Opt B: Quantum Semiclass Opt* 2005, **7:**577–587.
62. Safavi-Naeini AH, Gröblacher S, Hill JT, Chan J, Aspelmeyer M, Painter O: **Squeezed light from a silicon micromechanical resonator.** *Nature* 2013, **500:**185–189.
63. Hauer BD, Doolin C, Beach KSD, Davis JP: **A general procedure for thermomechanical calibration of nano/micro-mechanical resonators.** *Ann Phys* 2013, **339:**181–207.
64. Bartlett MS: **Smoothing periodograms from time-series with continuous spectra.** *Nature* 1948, **161:**686–687.

4

Influence of squeeze-film damping on higher-mode microcantilever vibrations in liquid

Benjamin A Bircher[*], Roger Krenger and Thomas Braun[*]

* Correspondence:
benjamin.bircher@unibas.ch;
thomas.braun@unibas.ch
Center for Cellular Imaging and
NanoAnalytics, Biozentrum,
University of Basel, Mattenstrasse 26,
CH-4058 Basel, Switzerland

Abstract

Background: The functionality of atomic force microscopy (AFM) and nanomechanical sensing can be enhanced using higher-mode microcantilever vibrations. Both methods require a resonating microcantilever to be placed close to a surface, either a sample or the boundary of a microfluidic channel. Below a certain cantilever-surface separation, the confined fluid induces squeeze-film damping. Since damping changes the dynamic properties of the cantilever and decreases its sensitivity, it should be considered and minimized. Although squeeze-film damping in gases is comprehensively described, little experimental data is available in liquids, especially for higher-mode vibrations.

Methods: We have measured the flexural higher-mode response of photothermally driven microcantilevers vibrating in water, close to a parallel surface with gaps ranging from ~200 μm to ~1 μm. A modified model based on harmonic oscillator theory was used to determine the modal eigenfrequencies and quality factors, which can be converted into co-moving fluid mass and dissipation coefficients.

Results: The range of squeeze-film damping between the cantilever and surface decreased for eigenfrequencies (inertial forces) and increased for quality factors (dissipative forces) with higher mode number.

Conclusions: The results can be employed to improve the quantitative analysis of AFM measurements, design miniaturized sensor fluid cells, or benchmark theoretical models.

Keywords: Microcantilever; Dissipation; Squeeze-film damping; Higher eigenmode; Photothermal excitation; Eigenfrequency; Quality factor; Fluid–structure interaction

PACS: 07.10.Cm (Micromechanical devices and systems); 46.40.Ff (Resonance and damping of mechanical waves); 07.79.-v (Scanning probe microscopes and components); 07.07.Df (Sensors (chemical; optical; electrical; movement; gas; etc.); remote sensing)

Introduction

Damping is an important design criterion for micro- and nanometer sized resonators, because surface forces dominate body forces at small dimensions [1]. Immersing a resonator, e.g., a microcantilever, in fluid drastically changes its dynamic properties. The eigenfrequencies and quality factors decrease due to hydrodynamic forces, which can be decomposed into an inertial (added mass) and dissipative (viscous damping) term [2]. Additionally, placing the resonator close to a solid surface leads to squeeze-film damping, where displacement of the fluid between the resonator and the surface during each vibration period introduces additional added mass and viscous damping [3]. The damping occurring by both mechanisms has direct impact on atomic force microscopy (AFM) and dynamically operated nanomechanical sensors. With progressing miniaturization, squeeze-film damping starts to dominate other dissipative effects and, thus, needs to be considered and characterized [4].

Furthermore, higher modes of vibration are increasingly used. In multifrequency AFM imaging, higher modes allow the material characteristics, e.g., mechanical, magnetic or electrical properties, of the substrate to be measured [5]. To reduce squeeze-film damping, AFM samples have been placed on pillars [6], or cantilever geometries have been optimized by focused-ion beam milling [7]. In cantilever-based sensor applications, the use of higher vibrational modes provides increased mass sensitivity [8] and allows the elastic properties [9] and the position of adsorbates [10] to be disentangled. Squeeze-film damping needs to be considered below a certain critical dimension of the AFM cantilever tip or container in which the cantilever sensor is mounted.

To our knowledge, squeeze-film damping of micrometer-sized cantilevers vibrating in higher modes in liquid has not been measured to date. In contrast, it has been thoroughly investigated for resonators immersed in gases, because of its importance for micro-electromechanical systems (MEMS), e.g., torsional mirrors [11] or cantilevers [12]. Even though, less attention has been paid to the problem in liquids, both analytical and numerical methods have been employed to model the behavior of cantilevers immersed in liquid and vibrating in close proximity to a surface. Analytical approaches [13] account for dissipative and inertial effects in the liquid, but due to the assumption of a two-dimensional flow field higher modes of vibration were not considered. Numerical approaches can effectively describe different cantilever-surface inclination angles, vibrational modes, and varying external driving forces [2,4,14]. However, semi-analytical equations describing the hydrodynamic load acting on cantilevers under squeeze-film damping only consider the fundamental mode of vibration [13,14]. Squeeze-film damping in liquid is governed by two dimensionless quantities, the Reynolds number, Re, and the normalized gap, H [13]:

$$Re = \frac{\pi \rho_f f_n b^2}{2\eta_f}, \; H = \frac{g}{b},$$

(1)

where b is the width of the cantilever, ρ_f the fluid density, η_f the fluid viscosity, f_n the cantilever eigenfrequency in liquid, and g the gap between the cantilever and the surface (see Figure 1). The cantilever width, b, is the dominant length scale of the flow

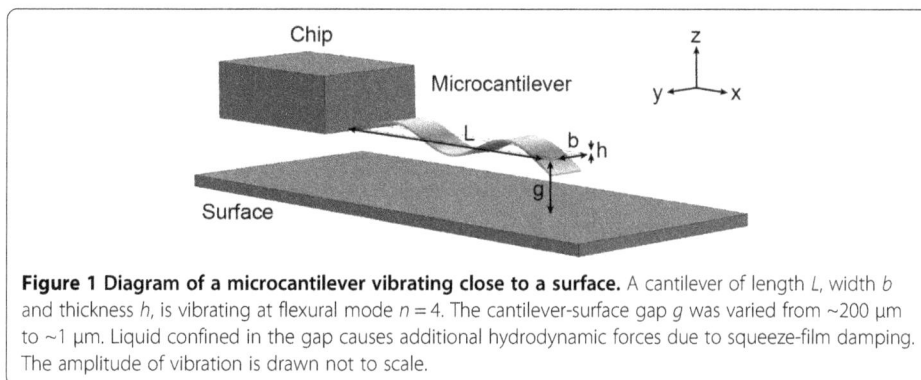

Figure 1 Diagram of a microcantilever vibrating close to a surface. A cantilever of length L, width b and thickness h, is vibrating at flexural mode $n = 4$. The cantilever-surface gap g was varied from ~200 µm to ~1 µm. Liquid confined in the gap causes additional hydrodynamic forces due to squeeze-film damping. The amplitude of vibration is drawn not to scale.

[13]. If the vibration amplitudes are orders of magnitude smaller than b, i.e., Keulegan-Carpenter numbers « 1 [15], the effect becomes independent of the amplitude [16]. Furthermore, the continuum hypothesis is valid because the mean-free-path of the molecules in liquid is very small compared to the dominant length b and the gap size g, i.e., Knudsen numbers « 1 [15].

Experimental investigations of cantilevers with dimensions ranging from centimeters to micrometers, immersed in water, buffer, organic solvents and oils are reported in the literature [3,15-19]. However, all experimental studies on microcantilevers in liquid and close to a surface, were limited to the fundamental mode ($n = 1$) [17-19]. Here we present the full spectral response of microcantilevers vibrating in water at different distances from a polydimethylsiloxane (PDMS) surface. PDMS was selected because of its abundant use for the fabrication of microfluidic devices. Spurious-free resonance spectra were obtained by driving the microcantilevers photothermally [20], and several higher flexural modes of vibration were characterized.

Results and discussion

To measure the effects of squeeze-film damping, a tipless microcantilever (250 μm × 35 μm × 2 μm) was placed close to a surface and the gap, g, was varied from ~200 μm to ~1 μm using a motorized stage (Figure 1). Experiments using longer cantilevers are reported in the Additional file 1. Photothermal excitation was employed to drive the microcantilever to resonance. Amplitude and phase spectra were acquired by sweeping an excitation frequency range from 0.5 kHz to 800 kHz and recording the corresponding cantilever response. As shown in Figure 2 for different cantilever-PDMS surface gaps, the spectra span four flexural modes of vibration. The influence of the cantilever-surface gap became substantial for $H = g/b < 1$, causing the resonance peaks to shift towards lower frequencies and broaden significantly (decreasing quality factors). A model, consisting of a sum of damped harmonic oscillators and terms considering the measurement setup (see Methods, Equation 4), described the experimental phase data with good accuracy (Figure 3). Applying this model to the data, allowed the eigenfrequency, f_n, and quality factor, Q_n, of each mode n, to be extracted at different cantilever-surface gaps, g (Figure 4).

To compare cantilevers from different chips, f_n and Q_n were normalized to the values indicated by experimental data recorded far from the surface, where its presence had no influence (see Methods). As shown in Figure 5a, due to squeeze-film damping the eigenfrequencies of all modes decrease as the cantilever-surface gaps become smaller. Further, even though some of the differences are slight, it is clear that higher-mode eigenfrequencies are less influenced by the proximity of the surface. In contrast, the higher-mode quality factors are affected when the cantilever-surface gap is still comparatively large (Figure 5b) and the fundamental mode is influenced least. To quantitatively compare the effects, a characteristic cantilever-surface gap $g_n{}^*$ was defined for the fundamental mode as $g_1{}^* = b/2 = 17.5$ μm ($H_1{}^* = 0.5$). At $g_1{}^*$ the frequencies (mean ± SD) of the fundamental vibration dropped to $(93.4 \pm 0.7)\%$ and the quality factors to $(77.8 \pm 7.5)\%$ of the initial value. Corresponding characteristic gaps ($g_n{}^*$), where the frequencies and quality factors dropped by the amounts measured for $g_1{}^*$, were then determined for the higher modes of vibration. As shown in Figure 6, the

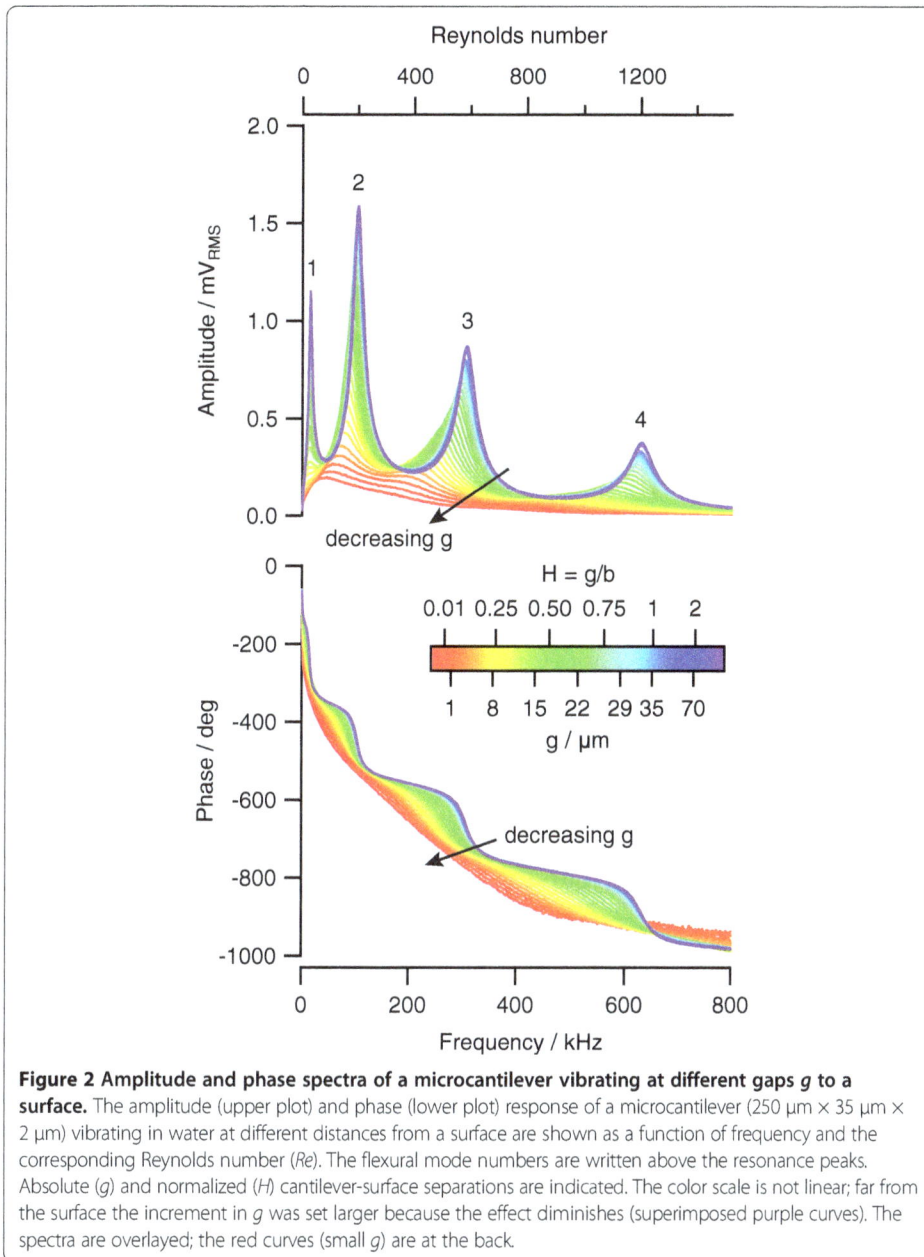

Figure 2 Amplitude and phase spectra of a microcantilever vibrating at different gaps *g* to a surface. The amplitude (upper plot) and phase (lower plot) response of a microcantilever (250 µm × 35 µm × 2 µm) vibrating in water at different distances from a surface are shown as a function of frequency and the corresponding Reynolds number (*Re*). The flexural mode numbers are written above the resonance peaks. Absolute (*g*) and normalized (*H*) cantilever-surface separations are indicated. The color scale is not linear; far from the surface the increment in *g* was set larger because the effect diminishes (superimposed purple curves). The spectra are overlayed; the red curves (small *g*) are at the back.

characteristic gap decreases for the eigenfrequencies and increases for the quality factors with increasing mode number. Similar behavior was observed for longer cantilevers (see Additional file 1), however, the effect seems to diminish with increasing cantilever length. To estimate the range of squeeze-film damping, i.e., the gap where the onset of the effect occurs, the characteristic gap was multiplied by a factor of two. Because of the definition, the range of squeeze-film damping for the fundamental mode is $H = 1$. The ranges for modes 2 to 4 (mean ± SD) were 0.93 ± 0.18, 0.84 ± 0.21, and 0.74 ± 0.21 for the eigenfrequencies and 1.22 ± 0.35, 1.55 ± 0.37, and 1.67 ± 0.20 for the quality factors. The largest critical gap for the frequencies, i.e., where the surface has no influence on the dynamics of the microcantilever, can be estimated from the fundamental mode,

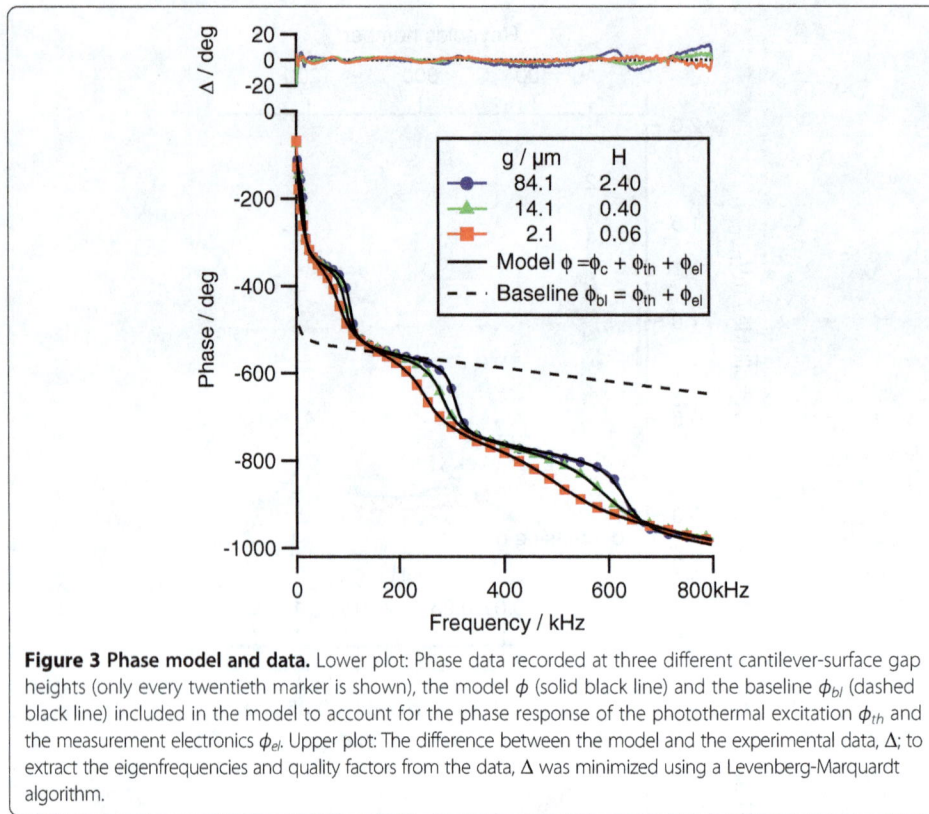

Figure 3 Phase model and data. Lower plot: Phase data recorded at three different cantilever-surface gap heights (only every twentieth marker is shown), the model ϕ (solid black line) and the baseline ϕ_{bl} (dashed black line) included in the model to account for the phase response of the photothermal excitation ϕ_{th} and the measurement electronics ϕ_{el}. Upper plot: The difference between the model and the experimental data, Δ; to extract the eigenfrequencies and quality factors from the data, Δ was minimized using a Levenberg-Marquardt algorithm.

which is affected first. In contrast, the largest critical gap for the quality factors depends on the highest mode measured.

To obtain a more general description of the results, the added mass coefficient, a_m, and the damping coefficient, c, were calculated for each mode. While the added mass coefficient a_m quantifies the co-moving fluid mass relative to the cantilever mass and is

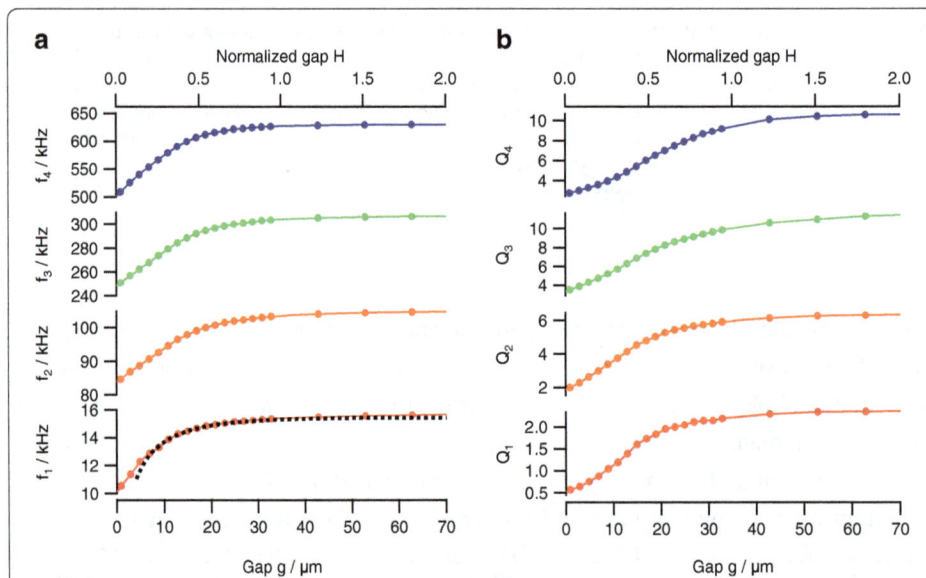

Figure 4 Absolute eigenfrequencies and quality factors. Representative measurement of the absolute **(a)** eigenfrequencies f_n and **(b)** quality factors Q_n of modes 1 to 4. The dashed line in **(a)** was calculated according to the theory describing the fundamental mode of vibration [14].

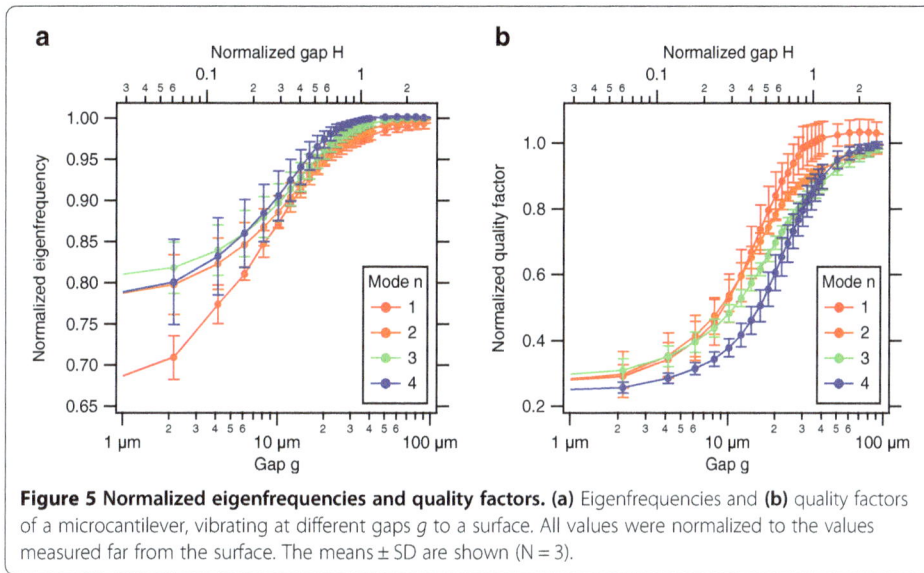

Figure 5 Normalized eigenfrequencies and quality factors. (a) Eigenfrequencies and **(b)** quality factors of a microcantilever, vibrating at different gaps g to a surface. All values were normalized to the values measured far from the surface. The means ± SD are shown (N = 3).

a measure of the inertial loading, the damping coefficient c equals the energy dissipation per unit length acting on the cantilever. The required vacuum frequencies were calculated using Equations 6 and 7 (see Methods) and the eigenfrequencies recorded in the unbounded fluid, i.e., far from the surface (Table 1). The observed vacuum frequency variations mainly originate from manufacturing-related uncertainties in the

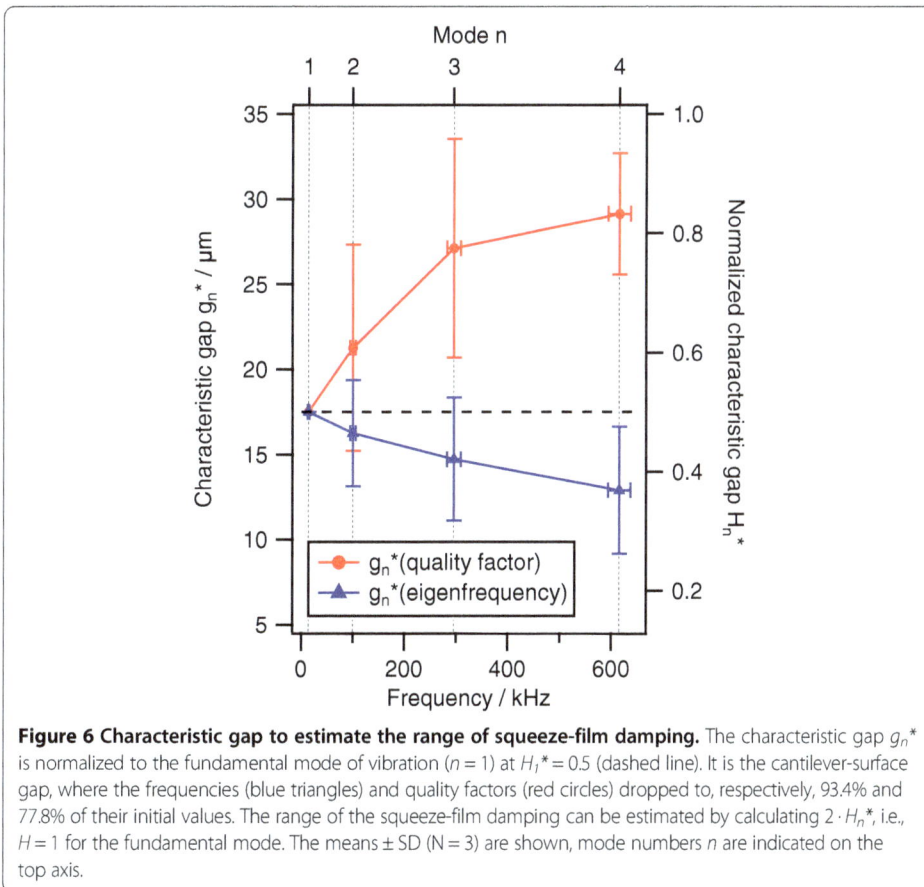

Figure 6 Characteristic gap to estimate the range of squeeze-film damping. The characteristic gap $g_n{}^*$ is normalized to the fundamental mode of vibration ($n = 1$) at $H_1{}^* = 0.5$ (dashed line). It is the cantilever-surface gap, where the frequencies (blue triangles) and quality factors (red circles) dropped to, respectively, 93.4% and 77.8% of their initial values. The range of the squeeze-film damping can be estimated by calculating $2 \cdot H_n{}^*$, i.e., $H = 1$ for the fundamental mode. The means ± SD (N = 3) are shown, mode numbers n are indicated on the top axis.

Table 1 Vacuum frequencies, added mass coefficients and damping coefficients measured far from the surface (mean ± SD)

Mode n	$f_{n,vac}$/kHz	$a_{m,H\gg1}$	$c_{H\gg1}$/mPa·s
1	44.8 ± 5.4	9.17 ± 0.20	67.8 ± 1.6
2	277 ± 15	6.78 ± 0.03	116.2 ± 11.1
3	768 ± 34	5.75 ± 0.01	186.7 ± 4.43
4	1512 ± 52	5.05 ± 0.01	372.0 ± 26.9

dimensions of the microcantilevers. Subsequently, the added mass coefficients could be determined (see Methods, Equation 6). The values without the influence of squeeze-film damping ($H \gg 1$) are provided in Table 1. Note that some authors defined the added mass coefficient as the co-moving mass relative to the fluid mass displaced by the static cantilever [3,16]. For direct comparison with their values, a_m has to be multiplied by ρ_c/ρ_f, i.e., ~2.3 in the present case. The damping coefficients c are the sum of structural, c_s, and viscous, c_v, damping. For microcantilevers immersed in liquid, structural damping is orders of magnitude smaller than viscous damping ($c_s \ll c_v$), and can thus be neglected [3]. The damping coefficients were calculated using the measured quality factors and eigenfrequencies (Equation 8, see Methods). Table 1 shows the damping coefficients without the influence of squeeze-film damping ($H \gg 1$). Even though higher-modes dissipate less energy per oscillation cycle (higher quality factors), they have larger damping coefficients due to their higher eigenfrequencies (cycles/second). Figure 7 shows how the added mass and damping coefficients increase due to squeeze-film damping. The magnitude of the observed shift in added mass coefficients decreased with mode number, whereas the shift in damping coefficients increased.

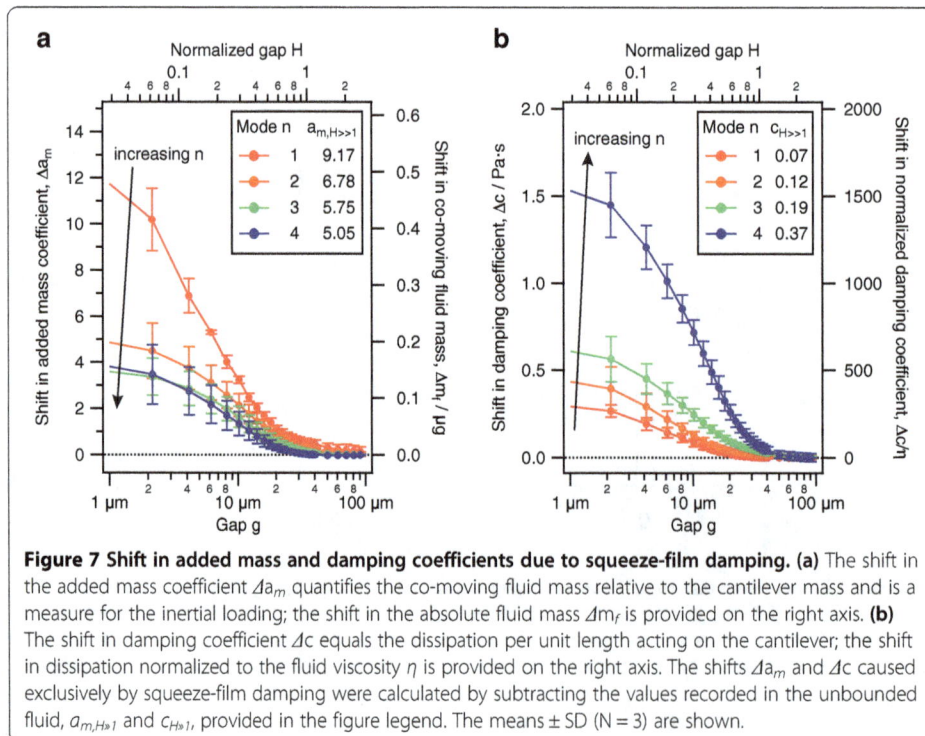

Figure 7 Shift in added mass and damping coefficients due to squeeze-film damping. (a) The shift in the added mass coefficient Δa_m quantifies the co-moving fluid mass relative to the cantilever mass and is a measure for the inertial loading; the shift in the absolute fluid mass Δm_f is provided on the right axis. **(b)** The shift in damping coefficient Δc equals the dissipation per unit length acting on the cantilever; the shift in dissipation normalized to the fluid viscosity η is provided on the right axis. The shifts Δa_m and Δc caused exclusively by squeeze-film damping were calculated by subtracting the values recorded in the unbounded fluid, $a_{m,H\gg1}$ and $c_{H\gg1}$, provided in the figure legend. The means ± SD (N = 3) are shown.

Conclusions

We have measured the squeeze-film damping on higher flexural mode vibrations of microcantilevers placed in proximity to a parallel surface in liquid. Due to the strong damping only a direct excitation method, such as the employed photothermal excitation [20], obtains spurious-free resonance spectra. A model consisting of a sum of harmonic oscillators was employed to extract the modal eigenfrequencies and quality factors from the phase spectra, and described the measured data well. Correct alignment of the data, i.e., calibration of the gap g, was crucial and limited the precision of the measurements. As predicted [13,14], strong squeeze-film damping of the fundamental mode was observed for normalized gaps $H < 1$. With increasing mode number the range of squeeze-film damping decreased for the eigenfrequencies (inertial forces) and increased for the quality factors (dissipative forces). Furthermore, the effect seems to depend on the length of the cantilever that determines the spatial wavelength of each mode. These findings should be considered for the design of sensor containers and cantilever tip geometries, because the quality factor is directly related to the sensitivity of the sensor [5]. The observed behavior is likely due to the three-dimensional nature of the flow field generated by higher modes, where gradients along the length of the cantilever must not be neglected [14]. For theoretical models, this entails the introduction of another parameter, besides the normalized gap H and the Reynolds number Re, related to the spatial wavelength of the cantilever, i.e., depending on the mode number as well as the cantilever length (similar to the normalized mode number in [21]). Finally, added mass and damping coefficients were calculated to support the comparability of the data. The shift in added mass decreased with mode number as predicted by numerical models [2]. The opposite was observed for the damping coefficients, which increased. More work is required to identify the underlying mechanisms governing squeeze-film damping acting on higher modes. Nevertheless, our data from microcantilevers with common dimensions, allows the magnitude of the squeeze-film damping effect to be assessed.

Methods

Experimental setup

Measurements were made as the upper surface of a small cavity containing water was moved closer to the immersed microcantilever. A diagram of the experimental setup is shown in Figure 8. Cantilever vibration was driven by photothermal excitation induced by an intensity-modulated laser beam (405 nm), and detected by monitoring the deflection of a second laser beam (780 nm) using the optical setup described previously [20,22]. A mirror galvanometer (GSV011, Thorlabs) was added to the setup to automatically control the low-pass filtered position ($f_{LP} = 1$ kHz) of the laser spot on the position-sensitive detector (PSD) used to monitor cantilever vibration (measurement bandwidth $\sim$ 850 kHz). A Zurich Instruments HF2 lock-in amplifier was employed to record cantilever resonance spectra by sweeping a given range of excitation frequencies and demodulating the corresponding phase and amplitude (lock-in bandwidth = 4.38 Hz, filter order = 24 dB/octave, 1000 data points). The setup was controlled using LabVIEW (National Instruments) and measurements were automated using the openBEB macro language [23]. The automation involved acquisition of spectra, adjusting the cantilever-

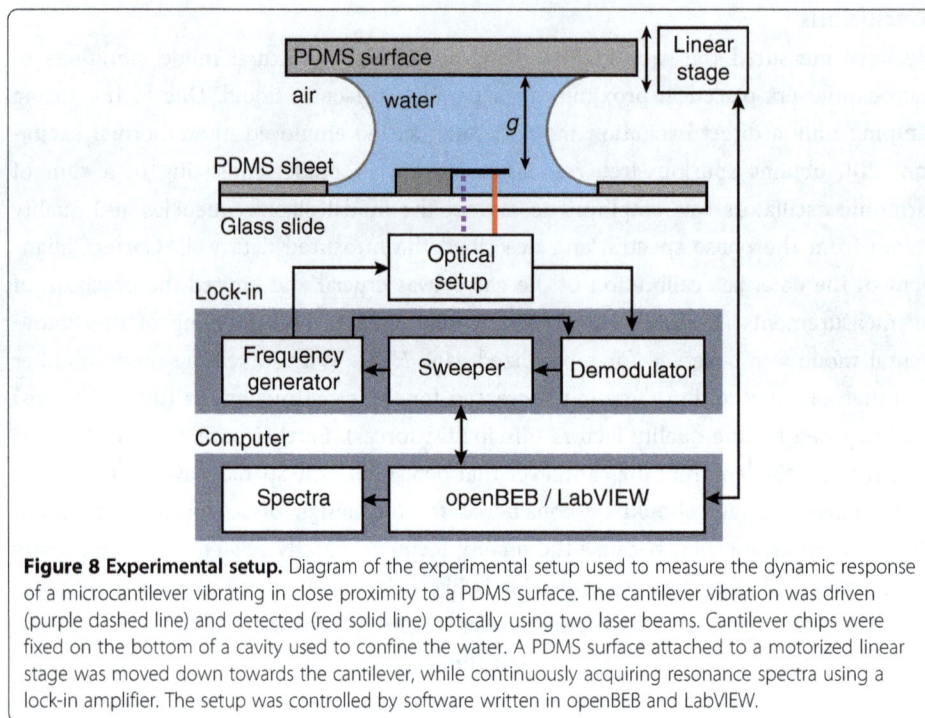

Figure 8 Experimental setup. Diagram of the experimental setup used to measure the dynamic response of a microcantilever vibrating in close proximity to a PDMS surface. The cantilever vibration was driven (purple dashed line) and detected (red solid line) optically using two laser beams. Cantilever chips were fixed on the bottom of a cavity used to confine the water. A PDMS surface attached to a motorized linear stage was moved down towards the cantilever, while continuously acquiring resonance spectra using a lock-in amplifier. The setup was controlled by software written in openBEB and LabVIEW.

surface gap, and adjusting the laser power and position of the laser spot on the position-sensitive detector using a proportional-integral-derivative (PID) controller. The whole setup was temperature controlled to 293 K within ± 0.2 K.

Tipless silicon microcantilevers (NSC12/tipless/noAl, MikroMasch) with nominal dimensions of 250 μm × 35 μm × 2 μm and calculated spring constants of 0.76 N/m were employed. The data reported in the Additional file 1 was obtained using longer microcantilevers (300 μm × 35 μm × 2 μm and 350 μm × 35 μm × 2 μm) following the same protocol. A comparison of the different cantilevers is provided in the Additional file 1: Table S1. To improve reflectivity and avoid unspecific adsorption, 20 nm gold was coated at the bottom side of the cantilevers and they were passivated with short polyethylene glycol chains, as described previously [22].

The cavity containing the water was formed using PDMS (SYLGARD 184, Dow Corning) and a glass microscope slide (AA00000112E, Menzel-Gläser), exploiting surface tension forces (see Figure 8). The base was fabricated by reversibly bonding a 150 μm-thick PDMS sheet with a 10 mm wide circular hole at its center to the glass slide. The 300 μm-thick cantilever chip was attached to the glass slide at the center of the hole using UV curable glue (F-UVE-61, Newport). The thickness of the chip was sufficient ($H = 8.6$) to exclude any influence of the glass surface on cantilever dynamics. Furthermore, as the thickness of the PDMS sheet (150 μm) was less than the thickness of the chip, access from above was retained. A flat upper cavity surface was fabricated by pouring degassed PDMS onto a silicon wafer to a thickness of about 5 mm and baking for 4 hours at 60°C. The PDMS was subsequently removed from the wafer and cut to give a circular disk with a diameter of 15 mm. The diameter exceeded all dimensions of the microcantilevers by at least an order of magnitude to avoid edge effects. The rougher surface of the disc was fixed to a kinematic mirror mount (KM05/M, Thorlabs), which was in turn mounted on an encoded piezo motor linear stage (CONEX-AG-LS25-27P, Newport) with a nominal precision of 0.2 µm. The cavity allowed the cantilever to be immersed in ~200 µL of water.

The flat upper PDMS surface was manually aligned parallel to the cantilever. To do this, a piece of silicon wafer was attached to the surface by adhesion forces to render it reflective. The read-out laser was then focused on the silicon surface and detected by the PSD otherwise used to measure the cantilever deflection. The residual angular misalignment was estimated to be less than 1 mrad (0.06°). The same procedure was repeated after rotating the PSD by 90° to align the angle perpendicular to the longitudinal axis of the cantilever.

To determine the coarse contact point, the surface was approached to the cantilever until a large deviation in the deflection signal was observed. Next, the surface was withdrawn to a distance where it had no influence on the cantilever vibration ($g \approx$ 200 µm, $H \approx 6$). To adjust the gap, the motorized linear stage was operated in a closed-loop configuration. After recording a spectrum the position was stored and the surface was moved closer to the cantilever. The step size was reduced as the gap decreased, to account for the non-linearity of squeeze-film damping. Next, for a more precise gap determination, the model of Tung et al. [14] was fitted to the frequency data of the fundamental mode (see Figure 4a) with parameters $f_{1,vac}$ and a gap offset:

$$\frac{f_1}{f_{1,vac}} = \left(1 + \frac{\pi \rho_f b}{4 \rho_c h} \Re\left(\Gamma_{\text{Tung}}(Re, 2H)\right)\right)^{-\frac{1}{2}} \tag{2}$$

The offset was then subtracted from the z-position of the measurement to align the data. We emphasize that the definition of H by Tung et al. [14] differs by a factor of two from Equation (1).

Data analysis

All data analysis was performed using custom scripts in IGOR Pro (Wavemetrics, see Additional file 1). Both amplitude and phase spectra contain the eigenfrequencies and quality factors of the vibrational modes. However, at small cantilever-surface gaps the resonance peaks in the amplitude spectrum become indistinguishable due to the strong peak broadening, i.e., low quality factors (see Figure 2). Furthermore, large differences in peak amplitude among higher modes of vibration complicate fitting and introduce dependencies on the initial parameters. In contrast, the phase shifts of each mode remain well resolved even at low quality factors. Thus, phase spectra were used to extract the modal eigenfrequencies and quality factors (see Figure 3). To weight each mode by the same amount on least squares fitting, the frequency spacing was transformed from linear ($p = 1$), i.e., equally spaced, to a power law according to

$$f^*(m) = \left(\frac{m}{M-1}\left(f^p_{m=M} - f^p_{m=0}\right) + f^p_{m=0}\right)^{\frac{1}{p}} \tag{3}$$

where m is a data point in the spectrum ranging from 0 to M-1, M the total number of points, $f_{m=0}$ the lowest and $f_{m=M}$ the highest frequency in the measured data and p the power of the transformation required for each mode of vibration to be assigned an equal number of data points. The value of p was estimated to be 0.514 from the calculated widths of the resonance peaks of all employed cantilevers in an unbounded fluid

[21]. The phase values corresponding to the transformed frequencies f^* were linearly interpolated from the measured data.

The following expression was used to extract the modal eigenfrequencies, f_n, and quality factors, Q_n, from the phase spectrum using a Levenberg-Marquardt algorithm (see Figure 3):

$$\phi(f) = \phi_c(f, f_1, Q_1, ..., f_N, Q_N) + \phi_{th}(f, \tau_{th}) + \phi_{el}(f, f_{el}, c_{off})$$
$$= \sum_{n=1}^{N} \arctan\left(Q_n \frac{f_n^2 - f^2}{f_n f}\right) - 2\pi f \tau_{th} + \arctan\left(\frac{f_{el}}{f}\right) + c_{off} \qquad (4)$$

where the cantilever response ϕ_c is the sum of damped harmonic oscillators with f_n and Q_n over all recorded modes N, ϕ_{th} is the linear thermal lag due to photothermal excitation with time constant τ_{th} [24] and ϕ_{el} (center frequency f_{el} and offset c_{off}) is an empirical first-order filter that considers the phase responses of the measurement electronics. The filter center frequency f_{el} and the time constant τ_{th} were determined on the first spectrum recorded far from the surface ($H \gg 1$) and then held constant.

The linearized equation of motion for a cantilever of length L, width b, thickness h and mass density ρ_c is [16]:

$$EI \frac{\partial^4 Z(x,t)}{\partial x^4} + \mu_c(1 + a_m) \frac{\partial^2 Z(x,t)}{\partial t^2} + c \frac{\partial Z(x,t)}{\partial t} = F_{drive}(x,t) \qquad (5)$$

where $Z(x,t)$ is the z-direction flexural displacement at position x along the cantilever beam at time point t, E and $I = bh^3/12$ the Young's modulus and area moment of inertia of the cantilever, $\mu_c = \rho_c bh$ the mass per unit length of the cantilever, a_m the added mass coefficient quantifying the co-moving fluid mass relative to the cantilever mass, c the sum of structural and viscous damping per unit length, F_{drive} an external driving

Table 2 Parameters for the employed silicon cantilevers immersed in water

Cantilever properties		
L	Length	250 μm
b	Width	35 μm
h	Thickness	2 μm
ρ_c	Mass density	2330 kg·m^{-3}
μ_c	Mass per unit length	0.163 mg·m^{-1}
E	Young's modulus	169 GPa
I	Area moment of inertia	23.3 μm^4
Q_n	Quality factor of mode n	
f_n	Eigenfrequency of mode n	Hz
$f_{n,vac}$	Vacuum frequency of mode n	Hz
a_m	Added mass coefficient	
c	Damping per unit length	Pa·s
Fluid properties		
ρ_f	Mass density	998.25 kg·m^{-3}
η_f	Viscosity	1.005 mPa·s
Gap properties		
g	Gap	m
$H = g/b$	Normalized gap	

force per unit length. The parameters used for the following calculations are provided in Table 2. The added mass coefficients a_m were calculated from the measured eigenfrequencies f_n [2]:

$$a_m = \left(\frac{f_{n,vac}}{f_n}\right)^2 - 1 \tag{6}$$

The vacuum frequencies $f_{n,vac}$ for each mode n were determined far from the surface ($H \gg 1$), where the added mass coefficient can be calculated for higher modes, with normalized mode number κ, according to the theory by Van Eysden and Sader [21]:

$$a_m = \frac{\pi\rho_f b}{4\rho_c h}\Re\left(\Gamma_{\text{VanEysden}}(Re, \kappa)\right) \text{ for } H \gg 1 \tag{7}$$

The damping coefficients per unit length were calculated as [7,16]

$$c = \mu_c(1 + a_m)\frac{2\pi f_n}{Q_n} \tag{8}$$

Additional file

Additional file 1: Information on the data analysis routine and additional data on longer microcantilevers is provided (see Additional file 1).

Abbreviations
AFM: Atomic force microscopy; PDMS: Polydimethylsiloxane; PID: Proportional-integral-derivative (controller); PSD: Position-sensitive detector; SD: Standard deviation.

Competing interests
The authors declare that they have no competing interests.

Authors' contributions
BAB and TB conceived and designed the study. RK and BAB carried out the experiments. RK, TB and BAB programmed the control software. BAB analyzed the data and drafted the manuscript. All authors revised the manuscript and approved the final version.

Acknowledgements
The authors gratefully acknowledge Henning Stahlberg (C-CINA, Biozentrum, University of Basel) for financial support and providing facilities, Shirley Müller (C-CINA, Biozentrum, University of Basel) for critically reading and discussing the manuscript, Francois Huber and Hans Peter Lang (SNI, Institute of Physics, University of Basel) for their support on cantilever preparation, Stefan Arnold and Andrej Bieri (C-CINA, Biozentrum, University of Basel) for fruitful discussions. This work was supported by ARGOVIA grant NoViDeMo and Swiss National Science Foundation grant SNF 200020_146619.

References
1. Joshi S, Hung S, Vengallatore S: **Design strategies for controlling damping in micromechanical and nanomechanical resonators.** *EPJ Tech Instrum* 2014, 1:5.
2. Basak S, Raman A, Garimella SV: **Hydrodynamic loading of microcantilevers vibrating in viscous fluids.** *J Appl Phys* 2006, **99**:114906.
3. Harrison C, Tavernier E, Vancauwenberghe O, Donzier E, Hsu K, Goodwin ARH, Marty F, Mercier B: **On the response of a resonating plate in a liquid near a solid wall.** *Sens Actuators A Phys* 2007, **134**:414–426.
4. Decuzzi P, Granaldi A, Pascazio G: **Dynamic response of microcantilever-based sensors in a fluidic chamber.** *J Appl Phys* 2007, **101**:024303.
5. Garcia R, Herruzo ET: **The emergence of multifrequency force microscopy.** *Nat Nanotechnol* 2012, **7**:217–226.
6. Kawakami M, Taniguchi Y, Hiratsuka Y, Shimoike M, Smith DA: **Reduction of the damping on an AFM cantilever in fluid by the use of micropillars.** *Langmuir* 2010, **26**:1002–1007.
7. Maali A, Cohen-Bouhacina T, Jai C, Hurth C, Boisgard R, Aimé J-P, Mariolle D, Bertin F: **Reduction of the cantilever hydrodynamic damping near a surface by ion-beam milling.** *J Appl Phys* 2006, **99**:024908.
8. Ghatkesar MK, Barwich V, Braun T, Ramseyer J-P, Gerber C, Hegner M, Lang HP, Drechsler U, Despont M: **Higher modes of vibration increase mass sensitivity in nanomechanical microcantilevers.** *Nanotechnology* 2007, **18**:445502.

9. Spletzer M, Raman A, Reifenberger R: **Elastometric sensing using higher flexural eigenmodes of microcantilevers.** *Appl Phys Lett* 2007, **91**:184103.

10. Dohn S, Sandberg R, Svendsen W, Boisen A: **Enhanced functionality of cantilever based mass sensors using higher modes.** *Appl Phys Lett* 2005, **86**:233501.

11. Bao M, Yang H: **Squeeze film air damping in MEMS.** *Sens Actuators A Phys* 2007, **136**:3–27.

12. Dareing DW, Yi D, Thundat T: **Vibration response of microcantilevers bounded by a confined fluid.** *Ultramicroscopy* 2007, **107**:1105–1110.

13. Green CP, Sader JE: **Frequency response of cantilever beams immersed in viscous fluids near a solid surface with applications to the atomic force microscope.** *J Appl Phys* 2005, **98**:114913.

14. Tung RC, Jana A, Raman A: **Hydrodynamic loading of microcantilevers oscillating near rigid walls.** *J Appl Phys* 2008, **104**:114905.

15. Grimaldi E, Porfiri M, Soria L: **Finite amplitude vibrations of a sharp-edged beam immersed in a viscous fluid near a solid surface.** *J Appl Phys* 2012, **112**:104907.

16. Naik T, Longmire EK, Mantell SC: **Dynamic response of a cantilever in liquid near a solid wall.** *Sens Actuators A Phys* 2003, **102**:240–254.

17. Rankl C, Pastushenko V, Kienberger F, Stroh CM, Hinterdorfer P: **Hydrodynamic damping of a magnetically oscillated cantilever close to a surface.** *Ultramicroscopy* 2004, **100**:301–308.

18. Kim S, Kihm KD: **Temperature dependence of the near-wall oscillation of microcantilevers submerged in liquid environment.** *Appl Phys Lett* 2007, **90**:081908.

19. Fornari A, Sullivan M, Chen H, Harrison C, Hsu K, Marty F, Mercier B: **Experimental observation of inertia-dominated squeeze film damping in liquid.** *J Fluids Eng* 2010, **132**:121201.

20. Bircher BA, Duempelmann L, Lang HP, Gerber C, Braun T: **Photothermal excitation of microcantilevers in liquid: effect of the excitation laser position on temperature and vibrational amplitude.** *Micro Nano Lett* 2013, **8**:770–774.

21. Van Eysden CA, Sader JE: **Frequency response of cantilever beams immersed in viscous fluids with applications to the atomic force microscope: arbitrary mode order.** *J Appl Phys* 2007, **101**:044908.

22. Bircher BA, Duempelmann L, Renggli K, Lang HP, Gerber C, Bruns N, Braun T: **Real-time viscosity and mass density sensors requiring microliter sample volume based on nanomechanical resonators.** *Anal Chem* 2013, **85**:8676–8683.

23. Ramakrishnan C, Bieri A, Sauter N, Roizard S, Ringler P, Müller SA, Goldie KN, Enimanev K, Stahlberg H, Rinn B, Braun T: **openBEB: open biological experiment browser for correlative measurements.** *BMC Bioinformatics* 2014, **15**:84.

24. Pini V, Tiribilli B, Gambi CMC, Vassalli M: **Dynamical characterization of vibrating AFM cantilevers forced by photothermal excitation.** *Phys Rev B* 2010, **81**:054302.

5

Photothermal cantilever deflection spectroscopy

Seonghwan Kim[1], Dongkyu Lee[2] and Thomas Thundat[2*]

* Correspondence:
thundat@ualberta.ca
[2]Department of Chemical and
Materials Engineering, University of
Alberta, T6G 2V4 Edmonton, AB,
Canada
Full list of author information is
available at the end of the article

Abstract

Microcantilever sensors offer high sensitivity in the detection of adsorbed molecules based either on resonance frequency shift or changes in cantilever deflection, as both of these signals can be detected with very high resolution. Despite the high sensitivity offered by this platform, cantilevers suffer from poor selectivity due to the lack of sufficiently selective interfacial layers which can be immobilized on cantilever surfaces. This problem can be overcome by using photothermal cantilever deflection spectroscopy (PCDS), which exploits the high thermomechanical sensitivity of bi-material microcantilevers. A bi-material cantilever responds to heat generated by the nonradiative decay process when the adsorbed molecules are resonantly excited with infrared (IR) light. The variation in the cantilever deflection as a function of illuminating IR wavelength corresponds to the conventional IR absorption spectrum of the adsorbed molecules. In addition, the mass of the adsorbed molecules can be determined by measuring the resonance frequency shift of the cantilever as an orthogonal signal for the quantitative analysis. This multi-modal PCDS offers unprecedented opportunities for obtaining very high selectivity in chemical and biological sensing without using selective interfacial layers or extrinsic labels.

Keywords: Bi-material microcantilever; Photothermal cantilever deflection spectroscopy; Nanomechanical IR spectrum; Explosives; Bitumen; Naphtha; DNA

Introduction

Recently, microcantilever sensors have attracted much attention due to their extremely high sensitivity [1,2]. These cantilevers, which can be microfabricated into arrays using conventional micromachining techniques, offer a miniature sensing platform for the real-time, simultaneous detection of multiple target analytes using a single device [3,4]. Another attractive feature of cantilever sensors is they can be operated in multiple modes. For example in the static mode, when confined to a single side of the cantilever, molecular adsorption results in cantilever deflection due to adsorption-induced forces [5]. In the dynamic mode, the resonance frequency of the cantilever varies sensitively as a function of adsorbed mass [5,6].

Molecular adsorption-induced resonance frequency variation approach

The resonance frequency, f, of a vibrating cantilever can be expressed as

$$f = \frac{1}{2\pi}\sqrt{\frac{k}{m^*}} \tag{1}$$

where k is the spring constant and m^* is the effective mass of the microcantilever. The effective mass can be related to the mass of the beam, m_b, through the relation, $m^* = n$

m_b where n is a geometric parameter. For a rectangular cantilever, n is 0.24. It is also possible that chemisorption or chemical reaction of the adsorbed ions/molecules may alter the spring constant of the cantilever [7]. Therefore, from equation 1 it is clear the resonance frequency can vary due to changes in mass as well as changes in spring constant. The variation in the resonance frequency, therefore, can be generalized as:

$$df = \frac{f}{2}\left[\frac{dk}{k} - \frac{dm^*}{m^*}\right] \qquad (2)$$

The change in spring constant can be a result of changes in Young's modulus, E, or to changes in dimensions caused by the molecular absorption-induced swelling of the chemoselective interfacial layer such as a polymer film on the cantilever [8]. If the molecular adsorption can be confined to the terminal end of the cantilever (end loading), the contribution from the spring constant variation can be neglected. In that case, equation 1 is valid. Assuming the target molecules are evenly adsorbed on the cantilever surface and do not affect the stiffness of the cantilever, the adsorbed mass of the target molecules in air, δm, can be calculated from the simple equation:

$$\delta m = \frac{k}{(2\pi)^2 n}\left(\frac{1}{f_1^2} - \frac{1}{f_0^2}\right) \qquad (3)$$

where, f_1 is the measured resonance frequency with adsorbed molecules, and f_0 is the initial resonance frequency of the cantilever [5].

Molecular adsorption-induced cantilever deflection approach

Thin microcantilevers undergo bending as a result of the mechanical forces involved in molecular adsorption. Adsorption-induced surface stress and the radius of the curvature of a cantilever can be related through Stoney's formula and the differential surface stress created by molecular adsorption can be related with the cantilever deflection as:

$$\Delta z = \frac{3L^2(1-v)}{Et^2}\delta\sigma \qquad (4)$$

where L and t are the length and the thickness of the cantilever, respectively, v is the Poisson's ratio of the cantilever, and $\delta\sigma$ is the differential surface stress between the functionalized and the passivated surfaces. Therefore, the deflection of the cantilever is directly proportional to the adsorption-induced surface stress [5]. This surface stress is expressed in units of N/m or J/m^2.

For either mass or surface stress based detection, the chemical selectivity has been obtained by immobilizing chemically selective interfaces on the cantilever surface. However, this method offers only limited selectivity since most chemical interfaces based on reversible chemical interaction are not very selective [9–13]. In spite of recent advances in microfabricated cantilever sensors with extremely high sensitivity, most sensing applications are hampered by poor selectivity. This challenge can be traced back to fundamental limitations imposed by the chemistry of the molecular interactions which forms the basis for signal generation in currently used chemical sensors. Simple chemical interactions such as hydrogen bonding are far too general for providing selectivity. In addition, immobilized chemoselective interfaces have a limited shelf-life as they degrade over time resulting even poorer selectivity [9].

We identify the lack of chemical selectivity in small molecule detection as the main challenge in accepting microcantilever sensors in practical applications. The limited selectivity and sensitivity of chemical sensors stems from the fundamental concept involved in the current sensing process which operates by the "molecule detects molecule" principle of selectivity. Adsorption of analyte molecules on the chemoselective interface immobilized on the cantilever surface results in variations in its mass and surface stress. Many current nano and micro sensors suffer from similar selectivity challenge as they are based on monitoring the changes in one of its physical properties, such as adsorbed mass, temperature, surface stress, resistance, capacitance, or refractive index, due to molecular adsorption. The transducer response is then amplified for readout and display. Despite all the advances in micro/nanotechnology, the fundamental mechanism of selectivity in sensing has not yet changed. In fact, current advances in chemical sensors are focused on developing transducers, which can detect extremely small changes in physical properties as a result of molecular adsorption. As long as the fundamental working principle for signal generation remains the same, the advantages of micro/nanotechnology will remain untouched in overcoming the challenges of sensor selectivity. Without selectivity, all other advantages (high sensitivity, fast response time, decrease in size and power consumption, and the potential for simultaneous detection of multiple analytes) are of no practical use.

Designing highly selective molecular recognition layers that can be immobilized on a sensor surface for small molecule detection is a challenging task due to the limited number of chemical interactions that can serve as the basis for synthesizing chemoselective layers while satisfying the highly desirable sensor attribute of room temperature reversibility. One way of achieving selectivity in detection is by using arrays of cantilevers where each element is modified with partially selective interfaces [10–12]. The requirement of room temperature reversibility requires the use of weak chemical interactions, for example hydrogen bonding, between the target molecules and the chemoselective interfaces. If the response from each cantilever is unique for a given chemical, it is possible to analyze the response using pattern recognition algorithms. The responses can include rise and decay times, as well as amplitudes. However, molecular recognition based on chemical interfaces relying on weak interactions is not specific enough to produce unique response pattern even in an array format. Increasing the number of sensor elements in the array for pattern recognition analysis does not improve the selectivity [13].

Imparting chemical selectivity to cantilever sensors is a challenge and will require utilizing unique properties of cantilever sensors in order to develop an entirely novel approach. One of the most under exploited properties of a cantilever sensor is its extremely high sensitivity to temperature variation when it is fabricated as a bi-material beam. This extreme high thermomechanical sensitivity of a bi-material cantilever can be exploited for measuring the extremely small thermal changes due to nonradiative decay of molecular vibrations of the adsorbed molecules [14–25]. The molecular vibrations of the adsorbed molecules can be resonantly excited by illuminating using infrared (IR) light. This change in heat energy due to molecular excitations on the cantilever sensor results in cantilever deflection. The amplitude of the bi-material cantilever deflection as a function of an illuminating wavelength shows the amount of heat energy generated by the adsorbed molecules at that wavelength and matches very well with the IR absorption peaks of the adsorbed molecules. This photothermal cantilever deflection spectroscopy (PCDS) is a unique technique that combines the extremely high thermomechanical sensitivity of a bi-material cantilever with the selectivity of mid-IR spectroscopy to achieve selectivity and sensitivity in molecular recognition of adsorbed molecules. By observing the energy location of multiple peaks in the cantilever deflection, it is possible to identify different molecules.

In this review article, we describe the working principle of PCDS, its instrumentation and methods. Quantitative results from the label-free, receptor-free detection of many analytes are presented. We also discuss potential strategies to enhance the thermomechanical sensitivity of bi-material cantilevers.

Review

The PCDS setup and its working principle are shown in Figure 1. Vapor phase analyte molecules are first allowed to adsorb on the bi-material cantilever surface. The mass of the adsorbed molecules is then calculated from the initial and final resonance frequencies of the cantilever. The cantilever is then illuminated with IR light from either a monochromator with a glow bar and a filter wheel or a broadly tunable quantum cascade laser (QCL). When IR photons of specific wavelength are resonantly absorbed by the adsorbates on the bi-material cantilever, it results in the generation of heat causing instantaneous cantilever deflection. The cantilever bends in response to this variation in temperature. By using a differential technique (cantilever with and without adsorbed molecules), it is possible to obtain a nanomechanical IR absorption spectrum of the adsorbates by plotting the differential cantilever deflection as a function of illumination wavelength. This signal depends on the thermodynamic and energy transfer properties of the adsorbates and the cantilever beam. Temperature changes resulting from the

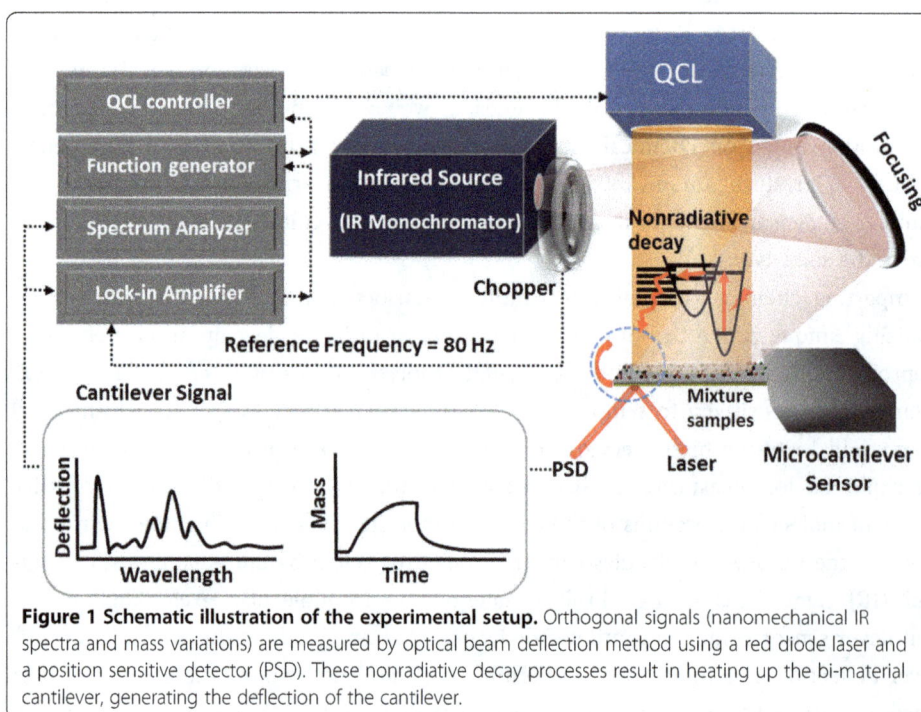

Figure 1 Schematic illustration of the experimental setup. Orthogonal signals (nanomechanical IR spectra and mass variations) are measured by optical beam deflection method using a red diode laser and a position sensitive detector (PSD). These nonradiative decay processes result in heating up the bi-material cantilever, generating the deflection of the cantilever.

absorption of IR energy are directly related to the vibrational modes of the adsorbed molecules as well as to the heat capacity and thermal conductivity of the cantilever beam. The observed peak amplitudes of nanomechanical IR spectra are proportional to the amount of the adsorbed molecules, the impinging power of IR radiation, the absorption mode, and the thermomechanical sensitivity of the bi-material cantilever. Since the PCDS is based on nonradiative decay, it does not rely on Beer-Lambert's principle. Unlike, techniques based on Beer-Lambert's principle, such as Fourier transform infrared (FTIR) and conventional IR spectroscopy, the signal-to-noise ratio of PCDS increases with the power of the illumination source. Therefore, by increasing the power of the IR light source, it is possible to increase the sensitivity of the technique [21].

Since the resonance frequency of the cantilever can be monitored before and after molecular adsorption, the adsorbed mass can be simultaneously determined using equation 3. Therefore, within a certain dynamic range, the peak amplitudes of the nanomechanical IR spectra can be normalized by adsorbed mass and these can serve as basis for quantitative nanomechanical IR spectral analysis to determine the adsorbed mass of each target molecule in a mixture since the IR spectrum of a mixture is a linear superposition of individual spectra.

Since IR spectra are different for different molecules, high chemical selectivity can be achieved in detection. This technique can be used for selective detection and quantification of the binary and ternary mixtures of explosive molecules such as trinitrotoluene (TNT), cyclotrimethylene trinitramine (RDX), and pentaerythritol tetranitrate (PETN)) [21]. Figure 2 shows the normalized nanomechanical IR absorption spectra of explosive molecules. Figure 2a presents the normalized IR spectra of 1:1 binary mixtures such as TNT&RDX (blue), PETN&TNT (cyan), and RDX&PETN (magenta). The nanomechanical IR spectra of the individual explosives were taken separately as references and agreed quite well with our previous report [20]. Figure 2b shows the normalized nanomechanical IR absorption spectra of individual explosive molecules, TNT (black), PETN (red), and RDX (green), using a conventional monochromator. The signal amplitude is in the range of mV/ng. However, the amplitude of normalized nanomechanical IR absorption spectra for the same samples using the QCL in Figure 2c were three orders of magnitude higher than that of spectra using a monochromator in Figure 2b. The sensitivity of PCDS was improved by employing a broadly tunable QCL as a powerful IR source [21]. Several distinct peaks and shoulders appeared in the binary mixture spectrum since the mixture spectrum is a linear superposition of individual spectra. Comparing these peaks with those of individual TNT, RDX, and PETN spectra, it is apparent that the peaks at 6.49 and 7.46 μm are from TNT, the peaks at 6.38 and 7.27 μm are from RDX, and the peaks at 6.06 and 7.82 μm are from PETN molecules [26–30]. PCDS can detect and distinguish differences between such closely related molecular species and between other interfering compounds and target molecules while sensing in mixture samples.

Selective and sensitive detection has direct relevance in many industrial applications such as chemical process and environmental monitoring [22,31]. For example, naphtha - a complex hydrocarbon – is routinely used for extraction of bitumen from oil sands, needs to be monitored on a routine basis in oil sands processing. Since the process of stripping naphtha from bitumen is not perfect, they easily lost into tailing streams during bitumen extraction process. However, the detection of industrial solvent and

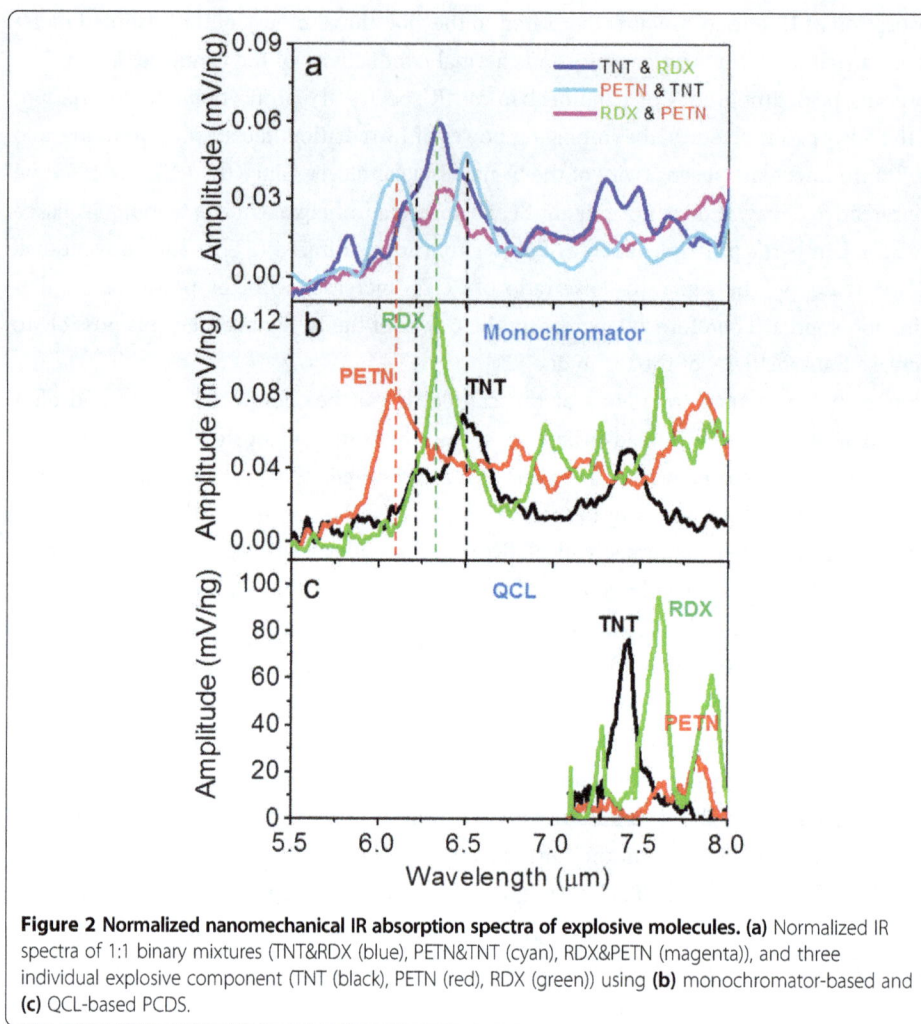

Figure 2 Normalized nanomechanical IR absorption spectra of explosive molecules. (a) Normalized IR spectra of 1:1 binary mixtures (TNT&RDX (blue), PETN&TNT (cyan), RDX&PETN (magenta)), and three individual explosive component (TNT (black), PETN (red), RDX (green)) using **(b)** monochromator-based and **(c)** QCL-based PCDS.

chemicals such as high volatile organic compounds poses many challenges as a result of non-specific interactions with surfaces and interferences from identical groups in other compounds. By monitoring multiple peaks of adsorbed molecules in PCDS, it is possible to overcome the selectivity and sensitivity challenges of currently used techniques. Figure 3 demonstrates selective detection of naphtha and bitumen using PCDS. The IR spectra of bitumen and naphtha reveal that they are largely composed of the aliphatic compounds. The common bands correspond to C–H asymmetrical stretching of $-CH_2$ and $-CH_3$ in range of 3.35 to 3.5 μm. The two bands at 6.8 and 7.2 μm are associated with C–H deformation in $-CH_2$ and $-CH_3$, respectively. Carbon rings in a cyclic compound and CH out of plane deformations in vinyl compounds show peaks in the range of 9.7–10.3 μm [22,31]. However, we can clearly distinguish them with some characteristic peaks, which can be attributed to different ratio of each component and some minor mixture components having various functional groups such as carboxylic acids (C–O, C–O–H) and ring vibrations (C = C, C–C).

The PCDS technique also offers high selectivity in detection of biological samples. Despite intrinsic selectivity in the binding of antigen-antibody or DNA hybridization, the microfabricated biosensors incur several problems such as non-uniform

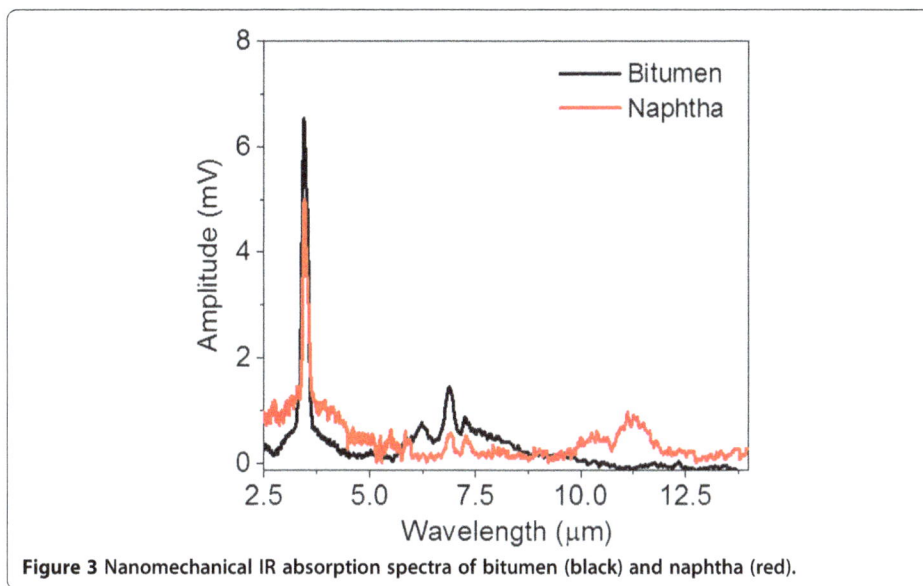

Figure 3 Nanomechanical IR absorption spectra of bitumen (black) and naphtha (red).

immobilization, reduced activity of receptors, and non-specific binding of unwanted molecules. Additionally, it is not easy to rapidly detect biological target molecules due to the intensive sample delivery process and reaction time involved in the receptor-target interactions. To overcome these issues, the PCDS technique can be used as an effective label-free and receptor-free biosensing concept. Figure 4 shows the normalized nanomechanical IR absorption spectra of the ssDNA of 20 bases of guanine using a conventional monochromator (Figure 4a). Figure 4b shows the same ssDNA spectrum obtained using a QCL which can scan from 5.68 μm to 6.39 μm. In Figure 4, the peaks between 5.68 and 6.39 μm result from the stretching vibration of the carbonyl group and ring bonds in guanine. The peaks of guanine are from carbonyl $C = O$ displacement (5.91 μm) and ring bond $C = N$ stretching (6.1, 6.25 μm) [32,33]. The sensitivity of QCL-based PCDS was found to be superior to that of monochromator-based PCDS as a result of using a high power IR source. These experiments show that QCL-based PCDS has the great potential for rapid identification and quantification of biological analytes. Although this technique is not able to discern different DNA sequences, it is highly useful as a rapid screening method for potential variation (mutation) in a given DNA strand with a known-sequence as well as detection of conjugation and byproduct of biomolecular interactions [32,34,35]. This technique may also find forensic applications where the need is to discriminate DNA rather than sequencing.

The relative intensities of the PCDS peaks were different than those observed with conventional IR spectrum or FTIR method. Some peaks that are not very prominent in conventional IR spectra appear to have higher intensity in the PCDS spectra. This may be directly related to the efficiency of nonradiative decay of these excited states. The position of absorption peaks were in general agreement with those observed with conventional IR absorption spectroscopies. We did not observe any appreciable shift in the position of the PCDS peaks. Since the energy resolution of the PCDS depends on the energy resolution of the IR sources, any small shift induced by the presence of the substrate is below the resolution limit of the current device.

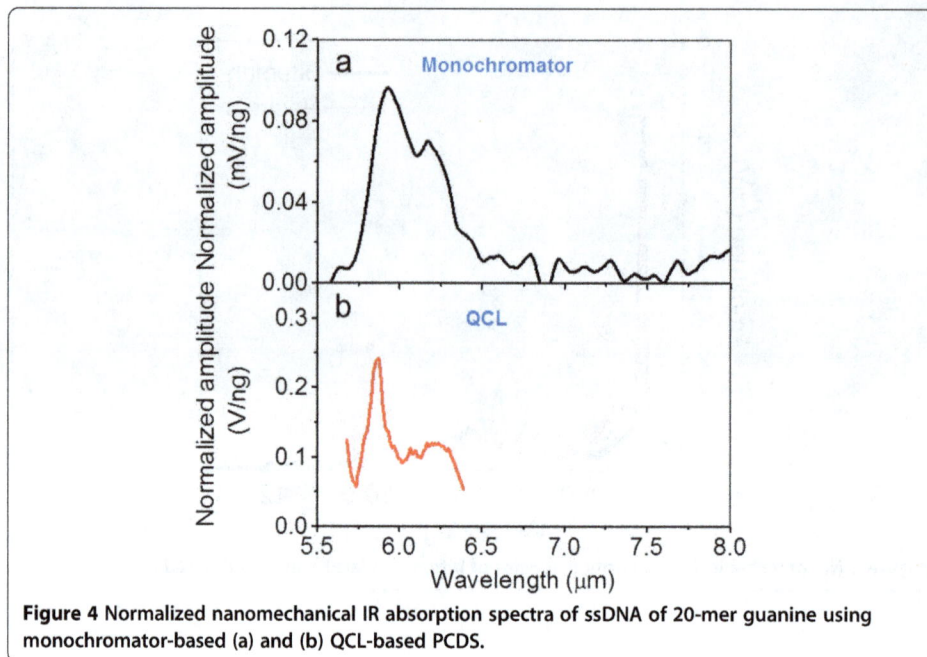

Figure 4 Normalized nanomechanical IR absorption spectra of ssDNA of 20-mer guanine using monochromator-based (a) and (b) QCL-based PCDS.

Advantages of the PCDS technique

The PCDS does not use any chemical receptors or interfaces for molecular recognition. The cantilever used in this technique has no surface functional groups. The molecules can be deposited on the surface with no chemical interaction between the analyte and the substrate. Even placing a drop of solution and drying it under ambient conditions is sufficient to obtain PCDS signals.

- High Selectivity: The signal generation is based on resonance excitation of various molecular bonds in the analyte. Molecular vibrations are extremely selective. In addition, in a complex mixture, the resultant signal is a linear combination of molecular vibrations from various constituents. Therefore, using a simple superposition principle it is possible to analyze the signal for the target analyte.
- High Sensitivity: The PCDS combines the extreme high thermomechanical sensitivity of a bi-material cantilever beam with the selectivity of mid-IR molecular spectroscopy. Our earlier results have shown mass sensitivity of 10 pg on the cantilever (corresponding to a few monolayers of analyte) under ambient conditions.
- Quantitative: The magnitude of the IR absorption-induced cantilever bending is proportional to the amount of analyte on the cantilever. This can be further quantified by measuring the resonance frequency of the cantilever (mass loading).
- Real-time Monitoring: Since the heat generation due to non-radiative decay is instantaneous, the PCDS technique is capable of real-time monitoring of adsorbed analytes.
- Small Size: The first generation device can be packaged to be smaller than a shoe box (including IR source, power supplies, display unit, etc.) making it a portable device. A second generation device can be made even smaller (the size of a cell phone) by optimizing packaging.

- Low Cost: Once produced in bulk, the cost of the device will be a few thousand dollars.
- Versatility: The same device can be used for different analytes.

Conclusions

PCDS provides simple technique to impart chemical selectivity onto microcantilever sensors. The demonstrated sensitivity and selectivity of this approach offers new possibilities for receptor-free sensing of a wide range of materials beyond what is currently possible using conventional techniques. These devices have the obvious advantages of requiring sub-nanogram samples, quick detection time and the potential to be inexpensive. It is possible to increase the sensitivity of PCDS by optimizing the bi-material cantilever parameters as well as increasing the power of the illuminating IR source. For example, by selecting different metals and optimizing the thickness of the coating it is possible to make a bi-material cantilever very sensitive to thermal changes. It is also possible to fabricate cantilevers with an optimized spring constant for increased thermomechanical bending or to pattern the cantilever surface for increased adsorption. Additionally, sensitivity can be increased by restricting the heat flow from the cantilever onto the base of the cantilever by making the contact area smaller.

Method

Maximizing the thermomechanical sensitivity of a bi-material cantilever

A microcantilever which is made bi-material by depositing a thin layer of metal on one of its side becomes an excellent, miniature thermal sensor, capable of detecting extremely small temperature changes at room temperature (there is no need to cool the device). The deflection of the cantilever tip due to the bi-material effect can be found from the following equation [36]:

$$z = -\frac{3}{4}(\alpha_1 - \alpha_2)\frac{t_1 + t_2}{t_2^2 K}\frac{l^3}{(\lambda_1 t_1 + \lambda_2 t_2)w}P. \tag{5}$$

where z is the cantilever deflection at the tip, α_1 and α_2 are the coefficients of thermal expansion for the two layers, l is the length of the cantilever, t_1 and t_2 are the layer thicknesses, λ_1 and λ_2 are the thermal conductivities, w is the width of the cantilever, and P is the total power absorbed by the cantilever. The subscripts 1 and 2 represent metal and cantilever substrate, respectively. The parameter K stands for the expression:

$$K = 4 + 6\left(\frac{t_1}{t_2}\right) + 4\left(\frac{t_1}{t_2}\right)^2 + \frac{E_1}{E_2}\left(\frac{t_1}{t_2}\right)^3 + \frac{E_2}{E_1}\left(\frac{t_2}{t_1}\right). \tag{6}$$

In the PCDS technique, the cantilever with adsorbed chemical species is scanned using successive pulses of monochromatic IR radiation. The amplitude of cantilever deflection as a function of IR wavelength resembles the IR absorption spectrum of the adsorbed molecules. The limit of detection of this technique is determined by the power of IR source and the thermomechanical sensitivity of the cantilever, which can be optimized by properly designing the bi-material aspect of the cantilever.

Instrumentation

Preparation of bi-material cantilevers

The experiments were carried out using commercially available silicon microcantilevers. Rectangular silicon cantilevers (CSC12-E) were obtained from MikroMasch USA (San Jose, CA). The dimension of each cantilever was 350 μm in length, 35 μm in width, and 1 μm in thickness. Another rectangular silicon cantilevers (Octosensis) were obtained from Micromotive GmbH Germany. A silicon chip contained an array of eight cantilevers having typical dimensions of 500 μm in length, 90 μm in width, and 1.0 ± 0.3 μm in thickness. The microcantilevers were cleaned by rinsing with acetone, ethanol, and treating them with UV ozone and then coated with 10 nm of chromium (adhesion layer) followed by 200 nm of gold using an e-beam evaporator at the deposition rate of 0.2 Å/s. A bare Octosensis silicon cantilever was used for bitumen experiments. Bitumen was diluted with toluene and coated on top of the silicon cantilever using a micro glass capillary. The bitumen-coated cantilever was dried in a vacuum oven at 150°C overnight to remove any low molecular weight solvent mixed in with the bitumen. Three standard explosive samples (TNT, RDX, and PETN) were purchased from AccuStandard, Inc. (New Haven, CT) and used without further purification. As indicated by the manufacturer, the standard concentration of each explosive is 1 mg/mL. The binary mixtures (by volume of standard sample solution) of explosive molecules were deposited on CSC12-E microcantilevers using micro glass capillaries. Naphtha was deposited on the gold side of the Octosensis cantilever by drop casting. The used chemicals (acetone, ethanol, toluene, and naphtha) were purchased from Fisher Scientific and used without further purification. We also procured synthetic single-stranded DNA (ssDNA) of 20 bases of guanine (G20) from Oligos Etc. The initial concentration of the ssDNA solution was 10 μM in deionized water. DNAs were deposited on the cantilever by inserting the cantilever into a micro glass capillary containing the DNA solution and slowly taking the cantilever out of the capillary.

PCDS setup

The sample-coated cantilever was mounted on a stainless steel cantilever holder and then positioned in the head unit of a Multi-Mode atomic force microscope (AFM) (Bruker, Santa Barbara, CA). The nanomechanical deflection and resonance frequency of the cantilever were measured using the optical beam deflection method with a red laser diode and a position sensitive detector (PSD). The IR radiation from the monochromator (Foxboro Miran 1A-CVF) was mechanically chopped at a frequency of 80 Hz and focused onto the cantilever by a spherical mirror. The filter wheel of the monochromator scans from 2.5 μm to 14.5 μm (i.e., 4000 cm^{-1} to 690 cm^{-1} in wavenumber) and has a spectral resolution of approximately 0.05 μm at 3 μm, 0.12 μm at 6 μm, and 0.25 μm at 11 μm according to the manufacturer. The 200 kHz pulsed IR radiation with a 10% duty cycle from a ÜT-8 QCL (Daylight Solutions, San Diego, CA) was electrically modulated at 80 Hz, using a DS345 function generator (Stanford Research Systems, Sunnyvale, CA) and directed to the cantilever for explosive sensing experiments. The IR wavelength was scanned from 7.1 μm to 8.73 μm (1408 cm^{-1} to 1145 cm^{-1}) with a step size of 5 nm. For DNA sensing experiments, 100 kHz pulsed IR radiation with a 5% duty cycle from a QCL (Daylight Solutions 6 μm laser) was electrically modulated at 80 Hz and directed to the DNA-coated cantilever. The IR wavelength was scanned from 5.68 μm to 6.39 μm (1760 cm^{-1} to 1565 cm^{-1}) with a step size of 5 nm. The resonance frequencies of the cantilever were measured before and after sample deposition using an SR760 spectrum analyzer (Stanford Research Systems, Sunnyvale, CA). The nanomechanical IR deflection signals before and after sample deposition were taken using an SR850 lock-in amplifier (Stanford Research Systems, Sunnyvale, CA) and the differential signals were presented as the nanomechanical IR spectra.

Abbreviations
IR: Infrared; PCDS: Photothermal cantilever deflection spectroscopy; QCL: Quantum cascade laser; FTIR: Fourier transform infrared; TNT: Trinitrotoluene; RDX: Cyclotrimethylene trinitramine; PETN: Pentaerythritol tetranitrate.

Competing interests
The authors declare that they have no competing interests.

Authors' contributions
SK & DL designed, carried out the experiments, collected and analyzed the data, and wrote the manuscript; TT conceived the experiments, analyzed results, and wrote the manuscript. All authors discussed the results and commented on the manuscript. All authors read and approved the final manuscript.

Authors' information
Seonghwan Kim is an Assistant Professor in the Department of Mechanical and Manufacturing Engineering at the University of Calgary. He earned his B.S. and M.S. degree in aerospace engineering at Seoul National University in 1998 and 2000, respectively, and received his Ph.D. in mechanical engineering at the University of Tennessee, Knoxville in 2008.
Dongkyu Lee is a Postdoctoral fellow in the Department of Chemical and Materials Engineering, University of Alberta, Edmonton. He received his Ph.D. in chemical engineering from Pohang University of Science and Technology, Pohang, Korea in 2011.
Thomas Thundat is a Canada Excellence Research Chair Professor at the University of Alberta, Edmonton. He received his Ph.D. in physics from the State University of New York at Albany in 1987.

Acknowledgements
This work is supported by Canada Excellence Research Chairs (CERC) Program. D.L. also would like to acknowledge the partial support from Basic Science Research Program through the National Research Foundation of Korea (NRF) funded by the Ministry of Education, Science and Technology (2012R1A6A3A03040416). SK also acknowledges the partial support from the Schulich School of Engineering in the University of Calgary.

Author details
[1]Department of Mechanical and Manufacturing Engineering, University of Calgary, T2N 1N4 Calgary, AB, Canada.
[2]Department of Chemical and Materials Engineering, University of Alberta, T6G 2V4 Edmonton, AB, Canada.

References
1. Thundat T, Warmack RJ, Chen GY, Allison DP: **Thermal and ambient-induced deflections of scanning force microscope cantilevers.** *Appl Phys Lett* 1994, **64**:2894–2896.
2. Gimzewski JK, Gerber C, Meyer E, Schlittler RR: **Observation of a chemical reaction using a micromechanical sensor.** *Chem Phys Lett* 1994, **217**:589–594.
3. Baller MK, Lang HP, Fritz J, Gerber C, Gimzewski JK, Drechsler U, Rothuizen H, Despont M, Vettiger P, Battiston FM, Ramseyer JP, Fornaro P, Meyer E, Güntherodt HJ: **A cantilever array-based artificial nose.** *Ultramicroscopy* 2000, **82**:1–9.
4. Fritz J, Baller MK, Lang HP, Rothuizen H, Vettiger P, Meyer E, Güntherodt HJ, Gerber C, Gimzewski JK: **Translating biomolecular recognition into nanomechanics.** *Science* 2000, **288**:316–318.
5. Chen GY, Thundat T, Wachter EA, Warmack RJ: **Adsorption-induced surface stress and its effects on resonance frequency of microcantilevers.** *J Appl Phys* 1995, **77**:3618–3622.
6. Thundat T, Chen GY, Warmack RJ, Allison DP, Wachter EA: **Vapor detection using resonating microcantilevers.** *Anal Chem* 1995, **67**:519–521.
7. Cherian S, Thundat T: **Determination of adsorption-induced variation in the spring constant of a microcantilever.** *Appl Phys Lett* 2002, **80**:2219–2221.
8. Wachter EA, Thundat T: **Micromechanical sensors for chemical and physical measurements.** *Rev Sci Instrum* 1995, **66**:3662–3667.
9. Hsieh MD, Zellers ET: **Limits of recognition for simple vapor mixtures determined with a microsensor array.** *Anal Chem* 2004, **76**:1885–1895.
10. Then D, Vidic A, Ziegler C: **A highly sensitive self-oscillating cantilever array for the quantitative and qualitative analysis of organic vapor mixtures.** *Sens Actuators B* 2006, **117**:1–9.
11. Zhao W, Pinnaduwage LA, Leis JW, Gehl AC, Allman SL, Shepp A, Mahmud KK: **Identification and quantification of components in ternary vapor mixtures using a microelectromechanical-system-based electronic nose.** *J Appl Phys* 2008, **103**:104902.

12. Jin C, Kurzawski P, Hierlemann A, Zellers ET: Evaluation of multitransducer arrays for the determination of organic vapor mixtures. *Anal Chem* 2008, **80:**227–236.

13. Jin C, Zellers ET: Limits of recognition for binary and ternary vapor mixtures determined with multitransducer arrays. *Anal Chem* 2008, **80:**7283–7293.

14. Barnes JR, Stephenson RJ, Welland ME, Gerber C, Gimzewski JK: Photothermal spectroscopy with femtojoule sensitivity using a micromechanical device. *Nature* 1994, **372:**79–81.

15. Wachter EA, Thundat T, Oden PI, Warmack RJ, Datskos PG, Sharp SL: Remote optical detection using microcantilevers. *Rev Sci Instrum* 1996, **67:**3434–3439.

16. Li G, Burggraf LW, Baker WP: Photothermal spectroscopy using multilayer cantilever for chemical detection. *Appl Phys Lett* 2000, **76:**1122–1124.

17. Datskos PG, Rajic S, Sepaniak MJ, Lavrik N, Tipple CA, Senesac LR, Datskou I: Chemical detection based on adsorption-induced and photoinduced stresses in microelectromechanical systems devices. *J Vac Sci Technol B* 2001, **19:**1173–1179.

18. Arakawa ET, Lavrik NV, Rajic S, Datskos PG: Detection and differentiation of biological species using microcalorimetric spectroscopy. *Ultramicroscopy* 2003, **97:**459–465.

19. Wig A, Arakawa ET, Passian A, Ferrell TL, Thundat T: Photothermal spectroscopy of Bacillus anthracis and Bacillus cereus with microcantilevers. *Sens Actuators B* 2006, **114:**206–211.

20. Krause AR, Van Neste C, Senesac L, Thundat T, Finot E: Trace explosive detection using photothermal deflection spectroscopy. *J Appl Phys* 2008, **103:**094906.

21. Kim S, Lee D, Liu X, Van Neste C, Jeon S, Thundat T: Molecular recognition using receptor-free nanomechanical infrared spectroscopy based on a quantum cascade laser. *Sci Rep* 2013, **3:**1111.

22. Bagheri M, Chae I, Lee D, Kim S, Thundat T: Selective detection of physisorbed hydrocarbons using photothermal cantilever deflection spectroscopy. *Sens Actuators B* 2014, **191:**765–769.

23. Lee D, Kim S, Van Neste CW, Lee M, Jeon S, Thundat T: Photoacoustic spectroscopy of surface adsorbed molecules using a nanostructured coupled resonator array. *Nanotechnology* 2014, **25:**035501.

24. Khan MF, Kim S, Lee D, Schmid S, Boisen A, Thundat T: Nanomechanical identification of liquid reagents in a microfluidic channel. *Lab Chip* 2014, **14:**1302–1307.

25. Lee D, Kim S, Chae I, Jeon S, Thundat T: Nanowell-patterned TiO2 microcantilevers for calorimetric chemical sensing. *Appl Phys Lett* 2014, **104:**141903.

26. Pristera F, Halik M, Castelli A, Fredericks W: Analysis of Explosives Using Infrared Spectroscopy. *Anal Chem* 1960, **32:**495–508.

27. Lewis IR, Daniel NW Jr, Griffiths PR: Interpretation of raman spectra of nitro-containing explosive materials. part i: group frequency and structural class membership. *Appl Spectrosc* 1997, **51:**1854–1867.

28. Makashir PS, Kurian EM: Spectroscopic and thermal studies on 2,4,6-trinitro toluene (TNT). *J Therm Anal Cal* 1999, **55:**173–185.

29. Makashir PS, Kurian EM: Spectroscopic and thermal studies on Pentaerythritol Tetranitrate (PETN). *Propell Explos Pyrot* 1999, **24:**260–265.

30. Beal RW, Brill TB: Vibrational Behavior of the –NO2 Group in Energetic Compounds. *Appl Spectrosc* 2005, **59:**1194–1202.

31. Yoon S, Son J, Lee W, Lee H, Lee CW: Prediction of bitumen content in oil sand based on FT-IR measurement. *J Ind Eng Chem* 2009, **15:**370–374.

32. Dovbeshko GI, Gridina NY, Kruglova EB, Pashchuk OP: FTIR spectroscopy studies of nucleic acid damage. *Talanta* 2000, **53:**233–246.

33. Kim S, Lee D, Thundat R, Bagheri M, Jeon S, Thundat T: Photothermal cantilever deflection spectroscopy. *ECS Trans* 2013, **50:**459–464.

34. Mateo-Marti E, Briones C, Roman E, Briand E, Pradier CM, Martin-Gago JA: Self-assembled monolayers of peptide nucleic acids on gold surfaces: a spectroscopic study. *Langmuir* 2005, **21:**9510–9517.

35. Liao W, Wei F, Liu D, Qian MX, Yuana G, Zhaoa XS: FTIR-ATR detection of proteins and small molecules through DNA conjugation. *Sens Actuators B* 2006, **114:**445–450.

36. Lai J, Perazzo T, Shi Z, Majumdar A: Optimization and performance of high-resolution micro-optomechanical thermal sensors. *Sens Actuators A* 1997, **58:**113–119.

6

Novel method for state selective determination of electron-impact-excitation cross sections from 0° to 180°

Marvin Weyland[1,2]*, Xueguang Ren[1,2], Thomas Pflüger[1,2], Woon Yong Baek[1], Klaus Bartschat[3], Oleg Zatsarinny[3], Dmitry V Fursa[4], Igor Bray[4], Hans Rabus[1] and Alexander Dorn[2]

*Correspondence:
weyland@mpi-hd.mpg.de
[1] Physikalisch-Technische
Bundesanstalt, Bundesallee 100,
38116 Braunschweig, Germany
[2] Max-Planck-Institut für Kernphysik,
Saupfercheckweg 1, 69117
Heidelberg, Germany
Full list of author information is
available at the end of the article

Abstract

We use an improved target recoil momentum spectroscopy setup to determine differential cross sections for excited metastable state production in atoms and molecules by electron impact and show its capabilities for an atomic helium target. A crossed beam setup with a supersonic helium jet and a pulsed electron beam at energies close to the excitation threshold of 19.82 eV was used. Measuring the recoil momentum vector of the target instead of the momentum of the scattered electron removes common restrictions to the accessible scattering angles while the microchannel plate detector ensures a high counting efficiency. Using a photoemission electron source we reach an energy resolution of about 200 meV at 1 μA peak current. Results are compared with simulations using theoretical convergent-close-coupling (CCC), R-matrix with pseudo-states (RMPS) and B-spline R-matrix (BSR) calculations and show good agreement.

Keywords: Inelastic scattering; Electron impact excitation

Background

In traditional electron impact experiments, the scattered electron is measured using a movable detector. The scattering angles are scanned by changing the position of the detector [1]. This technique allows measurement within a limited angular range, as the electron detector would not be able to measure the scattered electrons in the backward and forward directions due to the interference of the spectrometer with the incoming or outgoing projectile beam. Detecting the recoil momentum of the excited target instead of the scattered electron has the advantage that the projectile beam has no influence on the measurement. This principle has been shown by Murray and Hammond [2,3] for electron impact excitation at intermediate projectile energies. Our improved setup allows for state selective measurements close to the threshold energy by using a time- and position-sensitive microchannel plate detector, where events from all electron scattering angles can be registered simultaneously and by using a photo-emission electron source.

Electron impact excitation is the simplest inelastic scattering process to study experimentally, as it has only two free particles before and after the collision. To satisfy momentum conservation, the projectile can only transfer its momentum to the target.

Because of the large mass of the target compared to the mass of the projectile electron, the corresponding transfer of kinetic energy to the target is negligible. The electron with the initial energy E_0^e loses the energy E_{exc} needed to excite a certain state in the target and remains with the energy E_1^e after the collision:

$$E_1^e = E_0^e - E_{exc}. \tag{1}$$

The momentum $\mathbf{q}$, which is transferred to the target in the process, is

$$\mathbf{q} = \mathbf{p}_0^e - \mathbf{p}_1^e, \tag{2}$$

where $\mathbf{p}_0^e$ and $\mathbf{p}_1^e$ are the momenta of the projectile electron before and after the collision, respectively. The absolute values of these momenta depend only on the corresponding electron energies:

$$|\mathbf{p}_0^e| = \sqrt{2m_e E_0^e} \tag{3}$$

$$|\mathbf{p}_1^e| = \sqrt{2m_e \left(E_0^e - E_{exc} \right)}, \tag{4}$$

where m_e is the electron mass. All relevant momenta are depicted in Figure 1, with the incoming electron beam moving along the positive y-direction and the target gas jet moving along the positive z-direction. Scattered excited atoms in metastable states are detected by the microchannel plate (MCP) detector indicated as the light gray disc. Before the collision, the target has the initial momentum $\mathbf{p}_0^t$ in the gas jet. As a result of the momentum transfer in the collision, its final momentum is

$$\mathbf{p}_1^t = \mathbf{p}_0^t + \mathbf{q} = \mathbf{p}_0^t + \mathbf{p}_0^e - \mathbf{p}_1^e. \tag{5}$$

For a given projectile momentum $\mathbf{p}_0^e$ and a particular excited state energy E_{exc}, the absolute value of the final electron momentum $|\mathbf{p}_1^e|$ is fixed. The target momenta therefore lie on a sphere with origin $\left(\mathbf{p}_0^t + \mathbf{p}_0^e \right)$ and radius $|\mathbf{p}_1^e|$ for all possible scattering angles. In Figure 1, momentum transfer is shown only in the x-y plane for simplicity: therefore, all events that correspond to the excitation of the same state have target momenta on the dotted dark gray circle while the momentum of the outgoing electron lies on the full circle. The radius of this circle is proportional to the amount of electron momentum after the collision, putting energetically lower lying states - which correspond to a higher excess energy - on larger radii.

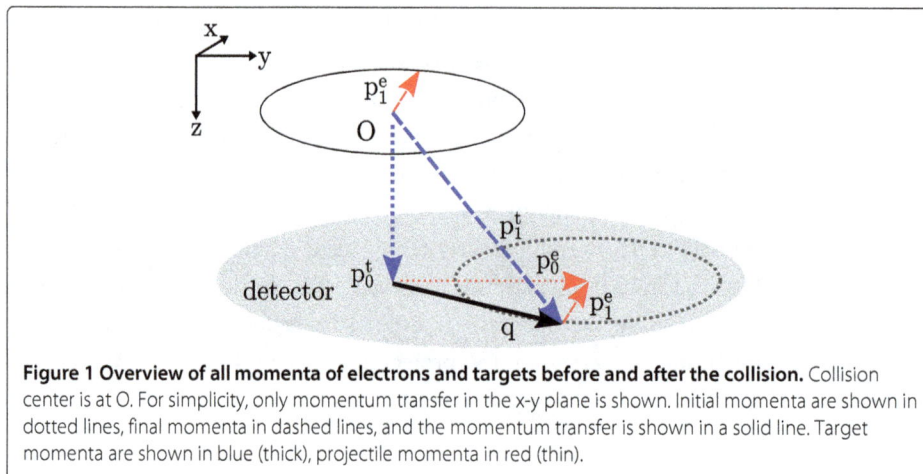

Figure 1 Overview of all momenta of electrons and targets before and after the collision. Collision center is at O. For simplicity, only momentum transfer in the x-y plane is shown. Initial momenta are shown in dotted lines, final momenta in dashed lines, and the momentum transfer is shown in a solid line. Target momenta are shown in blue (thick), projectile momenta in red (thin).

Momentum in x- and y-direction can be calculated from the impact position of the excited target on the detector and its time-of-flight t_f:

$$q_x = m_t \frac{x}{t_f} \qquad q_y = m_t \frac{y}{t_f}. \tag{6}$$

For these directions perpendicular to the gas jet, a high momentum resolution of 0.068 a.u.[a] can be reached. Momentum transfer in z-direction can be calculated by

$$q_z = p_{0,z}^t - \frac{m_t d}{t_f}, \tag{7}$$

where d is the distance between the interaction region and the detector plane. Unfortunately, our measurements showed that the initial longitudinal momentum spread in the supersonic jet due to its finite temperature is about 0.75 a.u. and therefore of the same order of magnitude as the transferred momentum, making the momentum transfer in z-direction inaccessible for all practical purposes. The rotational symmetry of the electron scattering cross section around the y-axis, however, allows for reconstruction of the complete differential cross section, as will be described in Section 'Data analysis'.

Helium has long been a benchmark target for the investigation of electron-atom collisions due to its simple structure. Many experiments have been conducted to measure electron impact excitation cross sections [2,4-10] and theories were developed to explain the experiments [11-13]. Total cross sections are known very precisely from experiment [8] and agree very well with theoretical calculations [12], but measurement of differential scattering cross sections has for a long time been restricted to an intermediate range of projectile scattering angles, usually ranging from 10° to 130° [7,10].

One way of accessing electron scattering angles around 180° is the use of a magnetic angle changer [14], in which a strong localized magnetic field in the interaction region changes the ejection direction of the electrons, depending on their energy. The few experiments that have been conducted to measure inelastic electron scattering on helium using this technique [15-17] show discrepancies between experiment and theory especially at high scattering angles. Detection of the scattered helium atoms, as performed with the instrument described here, is therefore a sensible addition to the existing methods.

In earlier experiments, the momentum transfer to the scattered helium was used to determine electron scattering angles by Zajonc et al. [9]. They only measured the time-of-flight of metastable helium atoms, which varied due to the momentum transfer in the flight direction of the helium. This method gave access to all scattering angles, but was unable to distinguish between different excited states and was therefore only applicable in the energy range from 19.8 eV to 20.6 eV, where just the 2^3S state can be excited. Murray and Hammond [18] used a rotatable detector setup to measure the deflection angle of helium atoms after electron impact excitation. In their arrangement the incoming electron beam crosses a gas jet perpendicularly and electron scattering within the plane determined by both beams is studied. Thus, scattered electron momentum determination relies on the measurement of the atomic recoil momentum along the incoming electron beam and along the gas jet. Since the resolution along the latter direction, due to the thermal velocity spread, is limited to about 0.76 a.u., a larger projectile excess energy was chosen such that different electron scattering angles result in more strongly varying recoil momenta. Also, excitation to the different accessible states was not resolved but summed differential cross sections were obtained.

Using a microchannel-plate detector with time of flight resolution and x-y-position resolution for detection of the recoiling metastable excited atoms, we obtain 4π-acceptance for electron scattering. In the plane perpendicular to the gas jet we obtain strongly improved momentum resolution, discriminating excitation of different states at low energies above the excitation threshold.

Our new setup is not limited to a helium target, but rather it can be used with all light targets, for which the deflection from their original direction after the collision is large enough to be resolved.

Results and discussion

The recorded momenta associated with the excitation of metastable states provide diverse information. On the one hand, the excited atom yield and, thus, the total metastable state excitation cross section as a function of impact energy are obtained. On the other hand, excited state resolved cross sections can be extracted as, e.g., excitation functions and cross sections differential in the projectile scattering angle.

In Figure 2, the total yield of excited helium atoms is depicted as a function of projectile energy. The metastable state production starts at the excitation threshold of the lowest 2^3S state and shows a characteristic shape due to negative ion resonances and additional excited states whose energies are marked in the diagram. The features of this curve are well known and described, e.g., by Brunt et al. [8]. In the energy range shown, only the 2^1P state at 21.2 eV decays to the ground state before detection and is therefore ignored. The red solid line in the diagram represents the theoretical cross section obtained from an RMPS calculation after convolution with a Gaussian describing the energy spread of the projectile beam. Expected helium yield curves at higher and lower energy resolution are shown for comparison as well. Energy spread and an energy offset were obtained by fitting the convolved theoretical curve to the experimental data. The FWHM of the energy spread varied according to the operating conditions of the electron gun between 150 meV and 250 meV, and is 200 meV in Figure 2. The energy offset arises due to contact potentials in the cathode and in the interaction region.

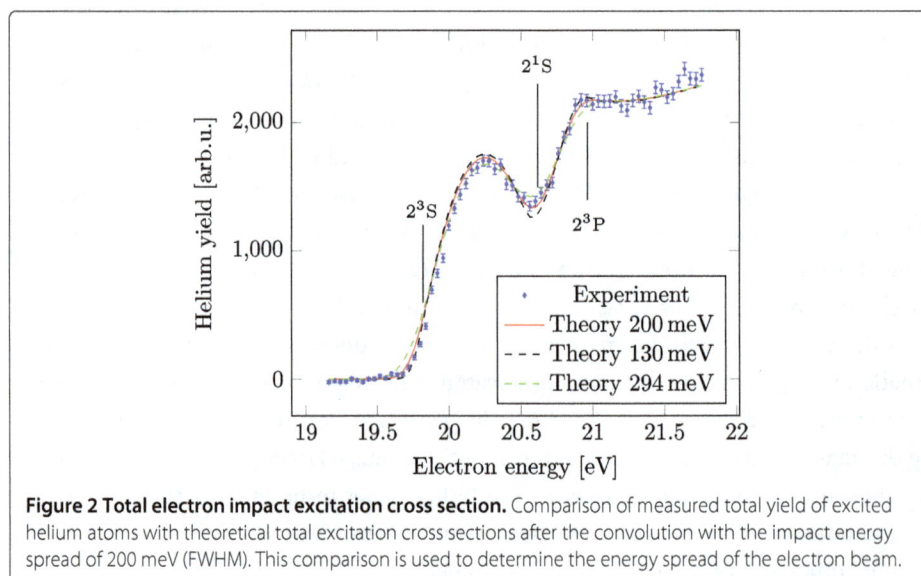

Figure 2 Total electron impact excitation cross section. Comparison of measured total yield of excited helium atoms with theoretical total excitation cross sections after the convolution with the impact energy spread of 200 meV (FWHM). This comparison is used to determine the energy spread of the electron beam.

Figure 3(a) shows the experimental cross section as a function of scattered electron momentum $\mathbf{p}_1^e$ for an impact energy of $E_0 = 22.0$ eV. It shows the projection of all momenta onto the detector plane. In Figure 3(b), the out-of-plane scattering contribution was restricted by including only events close to the velocity distribution maximum as described in Section 'Data analysis'. The various features that correspond to different excited states can be separated better than in the projection of all events. Theoretical cross sections displayed in Figure 3(c) and 3(d) are CCC and RMPS calculations, respectively. In these plots, scattered electron momenta for the excitation of a particular atomic state lie on a circle with a certain radius matching the corresponding excess energy. Circles for the three accessible states 2^3S (solid line), 2^1S (dashed line) and 2^3P (dotted line) are shown in the plots. Scattering leading to excitation of the 2^3S state can be separated from the higher lying 2^1S and 2^3P states. These, on the other hand, are fairly closely spaced with energies of 20.616 eV ($|\mathbf{p}_1^e| = 0.319$ a.u.) and 20.964 eV ($|\mathbf{p}_1^e| = 0.276$ a.u.) and cannot be resolved due to the combined uncertainties of impact energy and target size. The cross section for the excitation of the 2^3S state exhibits maxima at roughly $\pm$ 90° scattering angle and for backward scattering. The excitation cross sections for 2^1S and 2^3P states show a strong and relatively sharp maximum in the forward direction, while the maximum in the backward direction extends over a larger angular range. Experimental results are in excellent agreement with the theories.

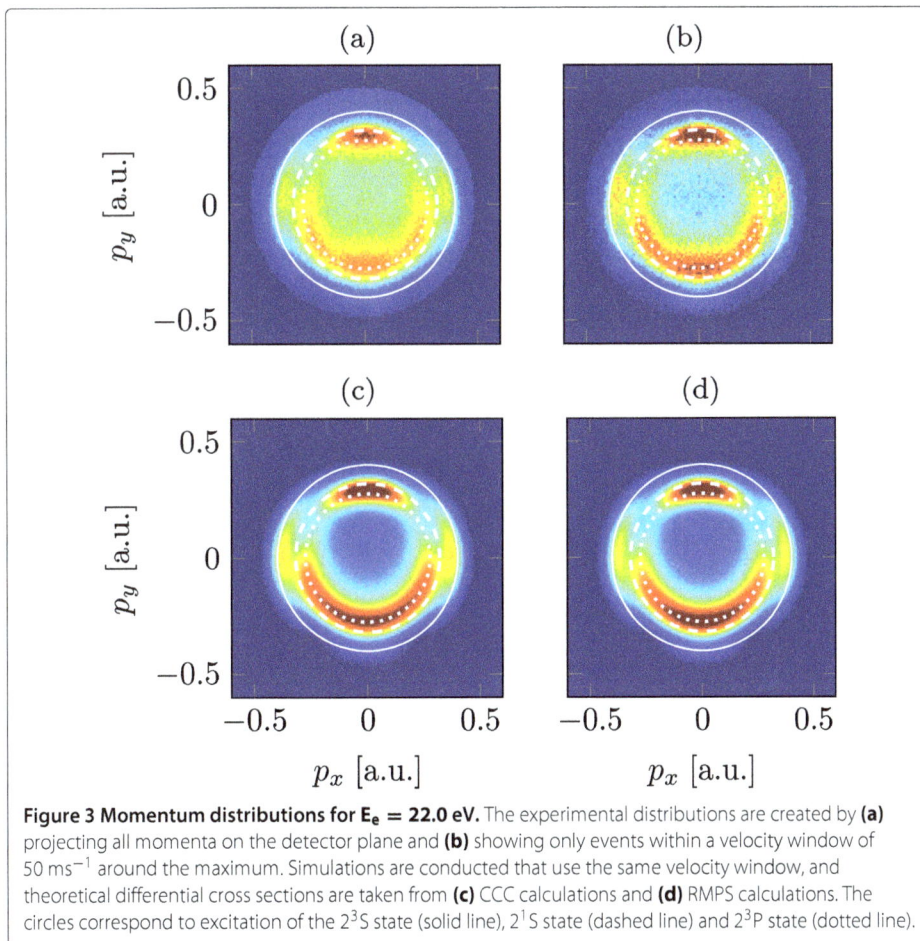

Figure 3 Momentum distributions for $E_e = 22.0$ eV. The experimental distributions are created by **(a)** projecting all momenta on the detector plane and **(b)** showing only events within a velocity window of 50 ms^{-1} around the maximum. Simulations are conducted that use the same velocity window, and theoretical differential cross sections are taken from **(c)** CCC calculations and **(d)** RMPS calculations. The circles correspond to excitation of the 2^3S state (solid line), 2^1S state (dashed line) and 2^3P state (dotted line).

In Figure 4, the differential cross sections for the excitation of 2^1S and 2^3P states are shown. The data were obtained by integrating the yield of metastable excited helium atoms over the energy range (1.2 ± 0.05) eV of the scattered electrons for both the experimental data and the theoretical CCC, RMPS and BSR data. The total cross section was not determined. Therefore, theoretical and experimental data sets were normalized to yield the same integrated cross section. Figure 5 depicts the data for (2.2 ± 0.05) eV excess energy, obtained in the same way and corresponding to the excitation of 2^3S state.

There is no significant difference between CCC, RMPS and BSR calculations in the differential cross section for 2^3S state excitation, and the experiments are in good agreement with all theories. For the 2^1S and 2^3P state excitation, however, the CCC calculations show lower yield in the broad backward scattering peak and a more pronounced peak at about 110°, showing a reminiscent feature of the 2^3S state excitation (cf. Figure 4). However, all theories are in good agreement with the experimental results within our measurement uncertainty.

Finally, Figure 6 displays results from the data analysis using the inverse Abel transformation. In Figure 3(a), the experimental data are shown for the projection of all recorded momenta onto the detector plane (x-y plane). Applying the inverse Abel transformation to this data, the pure x-y plane cross section $\sigma(p_x, p_y; p_z = 0)$ displayed in Figure 6(a) is obtained. A simulation of the experiment using the RMPS calculation is shown in Figure 6(b). It is evident from the comparison with Figure 3 that the inverse Abel transformation provides a better resolution of the different states.

Conclusions

Our setup allows for the measurement of electron impact excitation to metastable states in light targets, resolving all scattering angles as well as different excited states. We demonstrated the setup's ability by using a helium target. Differential cross sections for the excitation of the 2^1S and 2^3P state in helium close to the excitation threshold energy are shown for the first time at all scattering angles, including the scattering angles in the

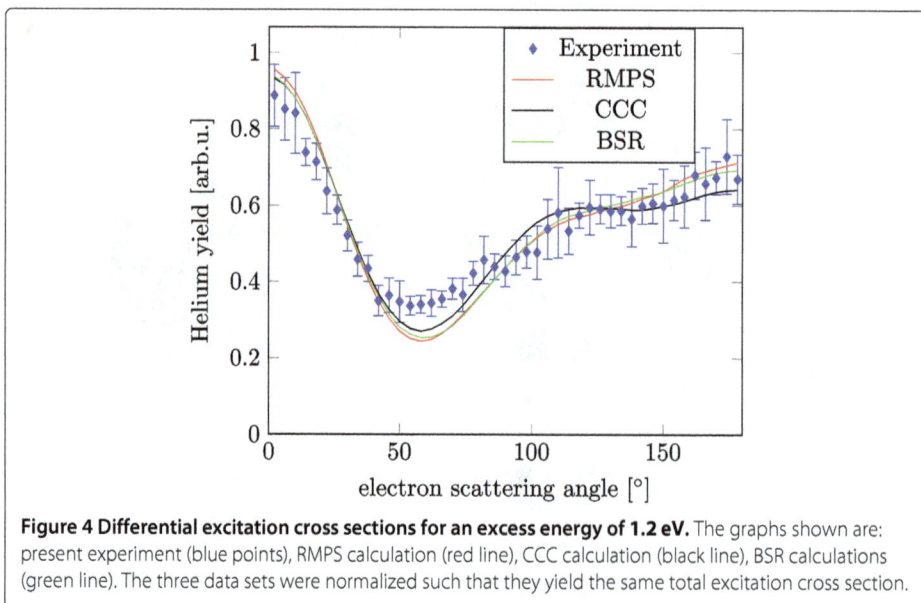

Figure 4 Differential excitation cross sections for an excess energy of 1.2 eV. The graphs shown are: present experiment (blue points), RMPS calculation (red line), CCC calculation (black line), BSR calculations (green line). The three data sets were normalized such that they yield the same total excitation cross section.

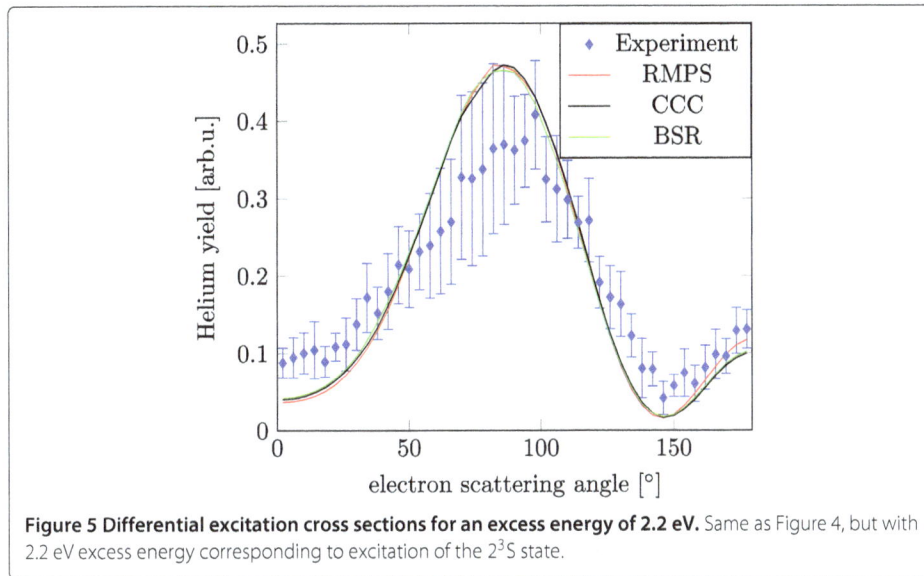

Figure 5 Differential excitation cross sections for an excess energy of 2.2 eV. Same as Figure 4, but with 2.2 eV excess energy corresponding to excitation of the 2^3S state.

vicinity of 0° and 180°. This instrument provides an additional tool to check theoretical predictions of electron impact excitation cross sections.

Methods

Setup

The experimental setup (cf. Figure 7) consists of an electron beam from a pulsed photoemission electron source, crossed with a continuous supersonic gas jet, a potentially variable interaction region and a time and position-sensitive MCP detector.

We have created a supersonic jet of the target gas by expansion through a 30 µm nozzle at a backing pressure of 5 bar. The jet is collimated by two skimmers 250 µm and 400 µm in diameter. Each skimmer is set in a differentially pumped stage, maintaining pressures of 2×10^{-3} mbar in the first jet stage, 5×10^{-6} mbar in the second jet stage, and 5×10^{-8} mbar in the main chamber during operation. The longitudinal jet temperature

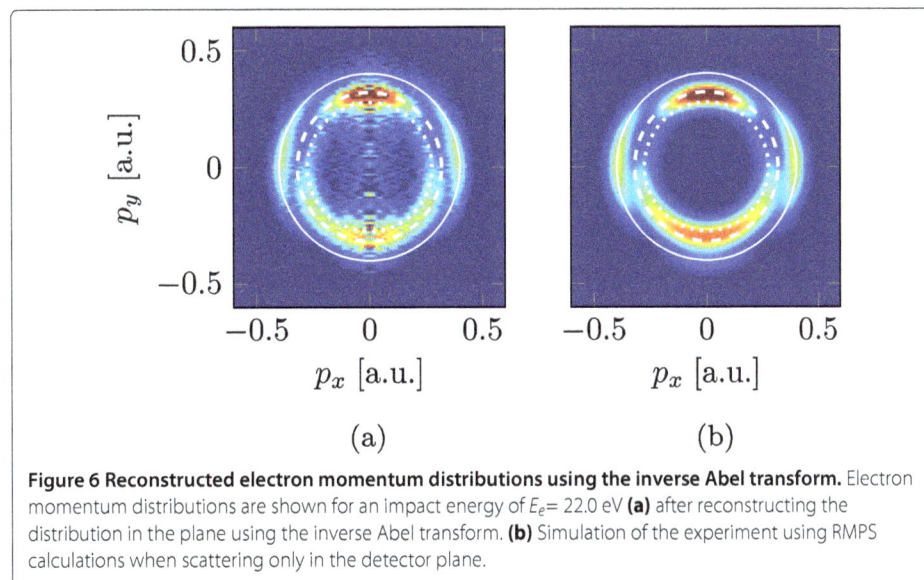

Figure 6 Reconstructed electron momentum distributions using the inverse Abel transform. Electron momentum distributions are shown for an impact energy of E_e= 22.0 eV **(a)** after reconstructing the distribution in the plane using the inverse Abel transform. **(b)** Simulation of the experiment using RMPS calculations when scattering only in the detector plane.

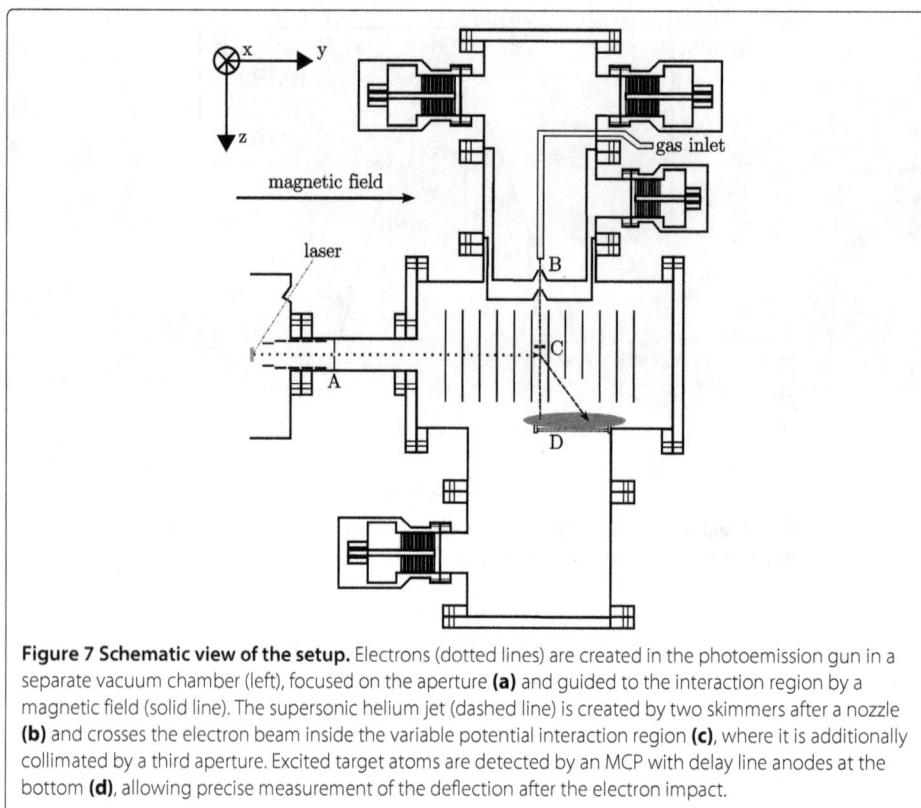

Figure 7 Schematic view of the setup. Electrons (dotted lines) are created in the photoemission gun in a separate vacuum chamber (left), focused on the aperture **(a)** and guided to the interaction region by a magnetic field (solid line). The supersonic helium jet (dashed line) is created by two skimmers after a nozzle **(b)** and crosses the electron beam inside the variable potential interaction region **(c)**, where it is additionally collimated by a third aperture. Excited target atoms are detected by an MCP with delay line anodes at the bottom **(d)**, allowing precise measurement of the deflection after the electron impact.

is 3 K. With this setup a beam diameter of about 2 mm and a target number density of 1.5×10^{11} cm^{-3} can be obtained in the region of interaction with the electron beam. To increase resolution, the beam is additionally collimated by inserting a 500 μm aperture a few millimeters above the interaction point, which is held at the same potential as the surrounding region at all times. The beam is dumped directly into the chamber without an additional dump stage.

The electron beam is produced in a photoemission electron gun [19]. A GaAs cathode is coated with a monolayer of cesium and oxygen to obtain negative electron affinity (NEA) conditions. In a NEA semiconductor, the conduction band minimum is above vacuum energy. Illuminating the surface with a laser excites electrons to the conduction band and can induce electron emission by tunneling through the thin surface barrier. The electron gun has been described in detail in the work by Schröter et al. [20]. To keep the surface stable, ultra-high vacuum is needed. The electron gun is therefore placed in a separately pumped chamber, which can be sealed off completely from the measurement vacuum chamber by a gate valve. This so-called gun chamber was heated to 220°C for one week to reach a base pressure below 1×10^{-11} mbar at room temperature afterwards. During the measurements, the two chambers are connected by an aperture 1 mm in diameter, which allows a pressure of 1×10^{-10} mbar to be maintained in the electron gun chamber.

The pulsed electron beam is produced by a laser pulse, illuminating the cathode. Previous studies have shown that multi-mode lasers can induce a high energy spread in the electron beam, as do high current densities [21]. Kolac et al. [22] explain this large energy spread from multi-mode lasers with high temporal intensity fluctuations which lead to an inhomogeneous space charge distribution in the emission current.

In our setup, the cathode is illuminated by a single-longitudinal-mode laser with a constant temporal intensity profile. The laser has a wavelength of 671 nm corresponding to a photon energy of 1.84 eV, which is high enough to populate the conduction band in the used crystal that has a band gap of 1.42 eV [23]. When operated in continuous-wave mode at a laser intensity of 25 mW, currents of several μA are reached. The pulsed electron beam is produced by guiding the incident laser through an acousto-optic modulator and coupling the light of the first diffraction maximum into an optic fiber, where it is guided to the focussing optics outside the vacuum chamber. Using the acousto-optic modulator allows switching the laser within 50 ns and obtaining extinction ratios better than 10^{-3}. The laser is focused to a diameter of 1 mm at the cathode, which restrains the electron source to an area of 1.4 mm × 1.0 mm, due to the laser incidence angle of 45°. Additional attenuation filters and an aperture are used to regulate the laser beam intensity such as to compensate for a decreasing quantum efficiency of the cathode over time.

The electrons are focused by four electrostatic lenses and two pairs of deflection plates onto the 1 mm aperture, which separates the electron gun chamber from the detection chamber. Additional steering to the interaction point is supplied by a pair of Helmholtz coils, creating a magnetic field of 1.2 mT pointing from the center of the aperture to the center of the interaction region, thereby confining the electron beam motion perpendicular to the direction of the magnetic field. Fine control of the beam position in x- and z-direction is possible with two additional pairs of coils that create a low magnetic field up to 0.1 mT perpendicular to the direction of the electrons.

The excitation process takes place in an electric-field-free region that is provided by a number of electrically connected metal rings. The distance between two adjacent metal rings is 10 mm and the symmetry axis of the rings is coincident with the electron beam direction. The rings are set to a variable voltage to change the electron impact energy. The electric-field-free region between the two central rings is accessible to the electron beam and the target jet. Rings blocking the flight path of the scattered target atoms are cut on the lower end to clear the path to the detector. The electron gun, interaction region boundary rings and the magnetic field are aligned precisely to keep the projectile beam on the axis of the apparatus when the interaction potential is changed. Any change in overlap of the electron beam and the jet leads to changing count rates. In the aligned setup, this effect was negligible for scanning amplitudes up to 2 V. The scanning voltage is read by an analog-to-digital converter (Hytec ADC 521).

After the inelastic electron-atom scattering, the excited atoms impinge on an MCP detector 80 mm in diameter with a position-sensitive delay line anode, located about 110 mm below the interaction region. Here they are efficiently detected since their internal energy gives rise to electron release at the detector surface. It is therefore necessary to shield the MCP from other electrons. An additional grid in front of the detector, set to -200 V, and the presence of the magnetic field eliminate background due to charged particles. The MCP is mounted at 32 mm offset of the target jet in positive y-direction to account for momentum transfer to the atoms and to detect excited atoms for all projectile scattering angles at excess energies up to 30 eV. Time-of-flight information and delay line signals are read by a time-to-digital converter (LeCroy 3377 TDC) operated in common stop mode and gated to accept only events in the expected time-of-flight range from 55 μs to 87 μs. Data from the TDC and ADC are read into a personal computer by a CAMAC system and are analyzed in offline mode. To correct for varying detection efficiency at different positions of the delay line anodes, the detector was homogeneously illuminated using an ^{241}Am α-source. The efficiency map obtained this way was used to weigh counts in the electron-scattering measurements.

Experiments on helium

Measurements were conducted at an 11 kHz gun pulse repetition rate and with electron pulse lengths ranging from 50 ns to 100 ns, which are needed to precisely measure the time-of-flight of the excited atoms. Time-of-flight and the position on the MCP detector are then used to calculate the projection of the momentum transfer $\mathbf{q}$ on the detector plane. The pulse length is also considerably shorter than the width of the time-of-flight distribution of the excited helium atoms that originate from the thermal velocity spread in the jet and the momentum transfer along the jet direction. The initial velocity distribution of the jet is therefore assumed to be equal to the measured velocity distribution at the excitation threshold. As can be seen from Eq. (4), the complete momentum of the electron is transferred to the helium, if $E_0^e = E_{\mathrm{exc}}$. In this case, no momentum spread arises due to different scattering angles. The velocity distribution for this case is shown in Figure 8. The mean jet velocity is 1624 ms^{-1} and the distribution has a full width at half maximum of 220 ms^{-1}, corresponding to a longitudinal jet temperature of 3.0 K [24,25]. In the simulation discussed in the next section, these jet characteristics are used as one source of the momentum spread.

The cathode potential of the electron gun is fixed to -22.9 V. Different impact energies are created by changing the potential of the interaction region while keeping all electron gun settings. For impact energy scans, a triangular scan with a 1 Hz repetition rate is carried out on the voltage applied to the interaction region boundary, usually providing the energy scan range from 19 eV to 23 eV. Energy scans are used to measure the electron-impact-energy dependence of the total cross section for metastable state excitation that is used to calibrate the impact energy scale based on RMPS calculations. From the measured energy dependence of the cross section the width of the electron energy distribution is

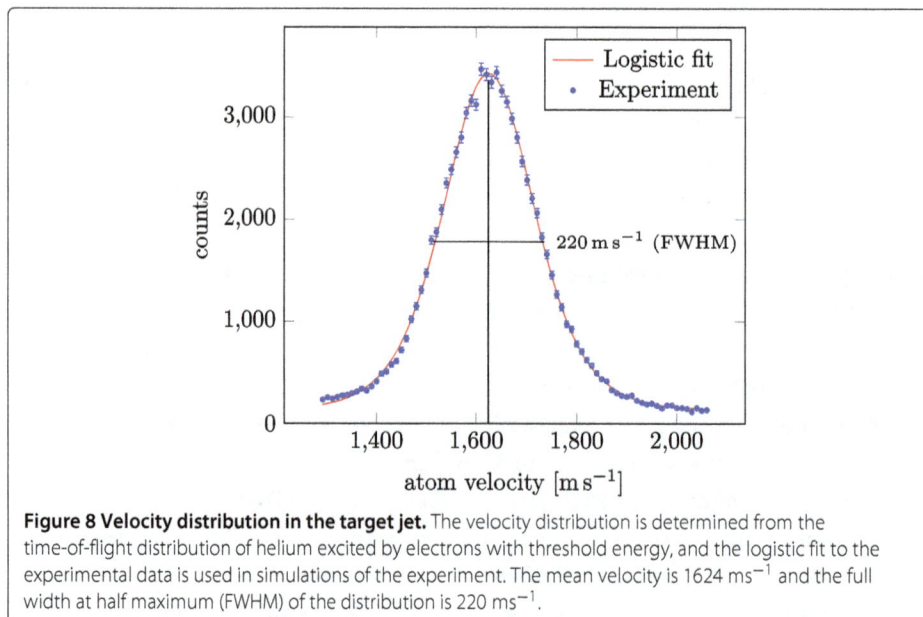

Figure 8 Velocity distribution in the target jet. The velocity distribution is determined from the time-of-flight distribution of helium excited by electrons with threshold energy, and the logistic fit to the experimental data is used in simulations of the experiment. The mean velocity is 1624 ms^{-1} and the full width at half maximum (FWHM) of the distribution is 220 ms^{-1}.

determined by comparison with theoretical cross-sectional data [12] that are convoluted with Gaussian distributions of different standard deviation.

Differential cross sections were measured with constant impact energy and count rates of around 300 counts per second were achieved. The counts were accumulated over an acquisition period of 10 to 24 hours. The discussed exemplary cross sections were obtained for an impact energy E_0^e of 22.0 eV. At this energy, theoretical calculations predict only minor changes in differential scattering cross sections within several 100 meV, making the measurement less sensitive to the finite energy spread of the electron beam.

Flight times in this setup are about 70 μs, making the measurement insensitive to fast decaying states like the 2^1P state in helium, which has a radiative lifetime of 0.6 ns [26] and decays to the undetected ground state. The three lowest excited states 2^3S, 2^1S and 2^3P are metastable because radiative dipole transitions are forbidden. They have sufficiently long lifetimes of about 8000 s (2^3S) and 20 ms (2^1S) and are detected before a significant amount of the excited metastable atoms decays [27,28]. The 2^3P state radiatively decays with a lifetime of about 100 ns [8,26] to the 2^3S state, which is then detected.

The 2^1P state produces photons of 21.2 eV when it decays. These photons can be detected by the MCP, but are not recognized, because they arrive before the start of the TDC gate and are therefore ignored.

Data analysis

In the experiment, the momentum transfer in the detector plane is measured very accurately. Error propagation analysis for Eq. (6) with typical values results in an uncertainty of $\Delta q_{x,y} = 0.068$ a.u. that originates mainly from the transversal size of the gas jet. Analogously, the uncertainty of the momentum in the jet direction given by Eq. (7) is an order of magnitude higher: $\Delta q_z = 0.8$ a.u., which is almost exclusively due to the initial velocity distribution in the gas jet. Because the highest transferred momentum is 0.4 a.u. at 2.2 eV excess energy, no useful information about the momentum transfer in z-direction is available.

Therefore, in the analysis, projections of the momenta on the x-y plane are used. Two methods of data analysis are applied. First, restricting analysis to events near the peak of the time-of-flight distribution is used to suppress large q_z-values. Second, the fact that the scattering has rotational symmetry around the y-axis allows the use of an inverse Abel transform [29,30] to reconstruct the full 3D-scattering cross section from the x-y projection, based on the axial symmetry of the scattering. Both sets of results are compared to the predictions from simulations.

The first approach offered the advantage that only little post-processing was necessary. Scattering out of the detector plane was partly suppressed by using only events with velocities in a 50 ms^{-1} window around the maximum of the velocity distribution (cf. Figure 8). In this case, only out-of-plane scattering events with a vertical momentum transfer up to $|q_z| = 0.17$ a.u. contribute to the momentum transfer maps. On the other hand, atoms from the wings of the velocity distribution can contribute if they experience a vertical momentum transfer such that their resulting time-of-flight is within the acceptance interval. This contribution is smaller due to the lower number of atoms in the wings of the velocity distribution. Owing to this condition, the obtained momentum transfer maps show pronounced in-plane scattering features.

The second way to analyze the data was to reconstruct the momentum transfer in the detector plane by exploiting the rotational symmetry around the electron impact direction. Using all events we obtained the projection of all momenta on the detector plane. The projection of a cylindrically symmetrical object on a plane that contains the symmetry axis is an Abel transformation [31]. In the experiment, the symmetry axis was along the electron impact direction in y-direction and the projection plane was the detection plane, i.e. the x-y plane. The original distribution was reconstructed using the inverse Abel transform [30]. Here, the results are independent of the velocity distribution in the jet and show better momentum resolution in x- and y-direction. These results are compared to a simulation where scattering was confined to the detector plane, thus eliminating the jet velocity distribution from the calculations.

To allow for a comparison of the experimental results with theoretical cross sections, the expected momentum transfer maps were simulated by a Monte Carlo-type program. This program takes into account the differential cross sections provided by R-matrix with pseudo-states (RMPS), B-spline R-matrix (BSR), and convergent-close-coupling (CCC) calculations [13], as well as the following experimental factors: target initial velocity, collision position, and collision energy for every event. Finally, the momentum of the excited target atom is mapped as a projection on the x-y plane. The time-of-flight is calculated from the initial velocity of the atom and the momentum change during the collision. Events produced in this program are filtered by their time-of-flight using the same velocity condition as was used in the experiment. Thus, out-of-plane scattering is taken into account in the same way as in the experiments and the measured momentum distributions are reproduced. The finite size of the jet has been modeled as an additional momentum spread which is specific to the setup geometry. All random variables are assumed to have Gaussian or logistic distributions and parameters of the distributions are taken from measured properties where possible.

The method of using only events with a certain velocity strongly improves the visibility of details in the momentum distributions. At the same time, the result depends on experimental parameters like electron beam resolution and all jet properties. A comparison with theoretical cross sections is difficult, as all parameters enter in the simulation as well.

The momentum distributions that were reconstructed using the inverse Abel transformation were compared to a similar simulation. Scattering angles were restricted to the x-y plane and no velocity condition was applied to reconstruct the expected momentum transfer maps for this case.

Endnote

[a]atomic unit of momentum, 1 a.u. $= 1.993 \times 10^{-24}\,\mathrm{kg\,m\,s^{-1}}$.

Competing interests
The authors declare that they have no competing interests.

Authors' contributions
MW carried out the measurements, interpreted the data and drafted the manuscript. KB, OZ, DVF and IB provided theoretical calculations for comparison with the experiments. TP, AD and XR participated in setting up the experiment and in discussing the results. WB and HR participated in the design of the experiment and helped to draft the manuscript. All authors read and approved the final manuscript.

Acknowledgements
This work was supported, in part, by the United States National Science Foundation under grants No. PHY-1068140, PHY-1212450, and PHY-14230245, as well as the XSEDE allocation PHY-090031 (KB and OZ). Support of the Australian Research Council and iVEC via the The Pawsey Centre's advanced computing resources is gratefully acknowledged as well (DVF and IB).

Author details
[1]Physikalisch-Technische Bundesanstalt, Bundesallee 100, 38116 Braunschweig, Germany. [2]Max-Planck-Institut für Kernphysik, Saupfercheckweg 1, 69117 Heidelberg, Germany. [3]Department of Physics and Astronomy, Drake University, Des Moines, IA 50311, USA. [4]ARC Centre for Antimatter-Matter Studies, Curtin University, Perth, WA 6845, Australia.

References
1. Hall RI, Joyez G, Mazeau J, Reinhardt J, Schermann C: **Electron impact differential and integral cross sections for excitation of the n = 2 states of helium at 29.2 eV, 39.2 eV and 48.2 eV.** *J Phys Paris* 1973, **34:**827–843.
2. Murray AJ, Hammond P: **Laser probing of metastable atoms and molecules deflected by electron impact.** *Phys Rev Lett* 1999, **82:**4799–4802.
3. Murray AJ, Hammond P: **Studies of electron-Excited targets using recoil momentum spectroscopy with laser probing of the excited state.** *Adv Atom Mol Opt Phy* 2001, **47:**163–204.
4. Schulz GJ, Fox RE: **Excitation of metastable levels in helium near threshold.** *Phys Rev* 1957, **106:**1179–1181.
5. Holt HK, Krotkov R: **Excitation of n = 2 states in helium by electron bombardment.** *Phys Rev* 1966, **144**(1):82–93.
6. Dugan JLG, Richards HL, Muschlitz EEJr: **Excitation of the metastable states of helium by electron impact.** *J Chem Phys* 1967, **46**(1):346–351.
7. Pichou F, Huetz A, Joyez G, Landau M, Mazeau J: **Electron impact excitation of helium: absolute differential cross sections of the n = 2 and 3^3S states from threshold to 3-6 eV above.** *J Phys B* 1976, **9**(6):933–944.
8. Brunt JNH, King GC, Read FH: **A study of resonance structure in helium using metastable excitation by electron impact with high energy resolution.** *J Phys B* 1977, **10**(3):433–448.
9. Zajonc AG, Pearl JC, Zorn JC: **Differential cross section for electron impact excitation of metastable helium.** *Phys Rev A* 1978, **1**(4):1408–1414.
10. Hoshino M, Kato H, Tanaka H, Bray I, Fursa DV, Buckman SJ, Ingólfsson O, Brunger MJ: **Benchmark differential cross sections for electron impact excitation of the n = 2 states in helium at near-ionization-threshold energies.** *J Phys B* 2009, **42:**145202.
11. Fursa D, Bray I: **Calculation of electron-helium scattering.** *Phys Rev A* 1995, **52**(2):1279–1297.
12. Bartschat K: **Electron-impact excitation of helium from the 1^1S and 2^3S states.** *J Phys B: At Mol Opt Phys* 1998, **31:**469–476.
13. Lange M, Matsumoto J, Lower J, Buckman S, Zatsarinny O, Bartschat K, Bray I, Fursa D: **Benchmark experiment and theory for near-threshold excitation of helium by electron impact.** *J Phys B* 2006, **39**(20):4179–4190.
14. Read FH, Channing JM: **Production and optical properties of an unscreened but localized magnetic field.** *Rev Sci Instrum* 1996, **67**(6):2372–2377.
15. Cubric D, Mercer DJL, Channing JM, King GC, Read FH: **A study of inelastic electron scattering in He covering the complete angular range from 0 to 180°.** *J Phys B* 1999, **32:**45–50.
16. Allan M: **Excitation of the 2^3S state of helium by electron impact from threshold to 24 eV: measurements with the 'magnetic angle changer'.** *J Phys B* 2000, **33:**L215–L220.
17. Ward R, Cubric D, Bowring N, King GC, Read FH, Fursa DV, Bray I, Zatsarinny O, Bartschat K: **Differential cross sections for electron impact excitation of the n = 2 states of helium at intermediate energies (80, 100 and 120 eV) measured across the complete angular scattering range (0 to 180°).** *J Phys B* 2011, **44**(4):045209.
18. Murray AJ, Hammond P: **A novel spectrometer combining laser and electron excitation and deflection of atoms and molecules.** *Rev Sci Instrum* 1999, **70:**1939–1950.
19. Feigerle CS, Pierce DT, Seiler A, Celotta RJ: **Intense source of monochromatic electrons: photoemission from GaAs.** *Appl Phys Lett* 1984, **44**(9):866–868.
20. Schröter CD, Rudenko A, Dorn A, Moshammer R, Ullrich J: **Status of the pulsed photoelectron source for atomic and molecular collision experiments.** *Nucl Instrum Meth A* 2005, **536**(3):312–318.
21. Groves T, Hammond DL, Kuo H: **Electron-beam broadening effects caused by discreteness of space charge.** *J Vac Sci Technol* 1979, **16**(6):1680–1685.
22. Kolac U, Donath M, Ertl K, Liebl H, Dose V: **High-performance GaAs polarized electron source for use in inverse photoemission spectroscopy.** *Rev Sci Instrum* 1988, **59**(9):1933.
23. Blakemore JS: **Semiconducting and other major properties of gallium arsenide.** *J Appl Phys* 1982, **53**(10):R123–R181.
24. Toennies JP, Winkelmann K: **Theoretical studies of highly expanded free jets: influence of quantum effects and a realistic intermolecular potential.** *J Chem Phys* 1977, **66**(9):3965.
25. Haberland H, Buck U, Tolle M: **Velocity distribution of supersonic nozzle beams.** *Rev Sci Instrum* 1985, **56**(9):1712–1716.
26. Gabriel AH, Heddle DWO: **Excitation processes in helium.** *P Roy Soc A* 1960, **258**(1292):124–145.
27. Moos HW, Woofworth JR: **Observation of the forbidden $2^3S_1 \rightarrow 1^1S_0$ spontaneous emission line from helium and measurement of the transition rate.** *Phys Rev Lett* 1973, **30**(17):775–778.
28. Van Dyck RS, Johnson CEJr, Shugart HA: **Radiative lifetime of the metastable 2^1S_0 state of helium.** *Phys Rev Lett* 1970, **25**(20):1403–1405.
29. Pretzler G: **A new method for numerical Abel-inversion.** *Z Naturforsch* 1991, **46a:**639–641.
30. Killer C: **Abel inversion algorithm** [http://www.mathworks.com/matlabcentral/fileexchange/43639-abel-inversion-algorithm]. Published 2013-09-26, Updated 2014-02-18.
31. Bracewell RN: *The Fourier Transform and Its Applications.* Singapore: McGraw-Hill Higher Education; 2000.

7

Design strategies for controlling damping in micromechanical and nanomechanical resonators

Surabhi Joshi, Sherman Hung and Srikar Vengallatore[*]

* Correspondence:
srikar.vengallatore@mcgill.ca
Department of Mechanical
Engineering, McGill University, 817
Sherbrooke Street West, Montreal,
Quebec H3A 0C3, Canada

Abstract

Damping is a critical design parameter for miniaturized mechanical resonators used in microelectromechanical systems (MEMS), nanoelectromechanical systems (NEMS), optomechanical systems, and atomic force microscopy for a large and diverse set of applications ranging from sensing, timing, and signal processing to precision measurements for fundamental studies of materials science and quantum mechanics. This paper presents an overview of recent advances in damping from the viewpoint of device design. The primary goal is to collect and organize methods, tools, and techniques for the rational and effective control of linear damping in miniaturized mechanical resonators. After reviewing some fundamental links between dynamics and dissipation for systems with small linear damping, we explore the space of design and operating parameters for micromechanical and nanomechanical resonators; classify the mechanisms of dissipation into *fluid–structure interactions* (viscous damping, squeezed-film damping, and acoustic radiation), *boundary damping* (stress-wave radiation, microsliding, and viscoelasticity), and *material damping* (thermoelastic damping, dissipation mediated by phonons and electrons, and internal friction due to crystallographic defects); discuss strategies for minimizing each source using a combination of models for dissipation and measurements of material properties; and formulate design principles for low-loss micromechanical and nanomechanical resonators.

Keywords: Dissipation; Damping; Nanomechanical sensing; MEMS; NEMS; Optomechanical systems; Quality factor; Atomic force microscopy; Structural design; Optimization

Introduction

Although damping has been studied for well over a hundred years, the rational design and control of structural damping has seemed a distant goal to many generations of engineers. By way of illustration, we quote from two excellent articles published in the 1990s: "All structures exhibit vibration damping, but despite a large literature on the subject damping remains one of the least well-understood aspects of general vibration analysis" [1], and damping in microcantilevers "is not readily susceptible to engineering analysis" [2].

During the past 15 years, however, there has been a remarkable resurgence of interest in structural damping, especially at small length scales (1 nm to 100 μm), motivated by a host of emerging technologies that include microelectromechanical systems (MEMS), nanoelectromechanical systems (NEMS), nanomechanical sensors, and optomechanical systems. This paper presents an overview of major advances in this field from the viewpoint of device design. Our goal is to collect methods and techniques that can provide designers with guidelines and tools for analyzing, controlling, and minimizing damping in miniaturized resonators.

To this end, the next section reviews the fundamental relationships between dissipation and structural dynamics for systems with small linear damping. Subsequently, we explore the space of design and operating parameters for miniaturized mechanical resonators; review the major mechanisms of dissipation by classifying them into three categories: *fluid–structure interactions, boundary damping*, and *material damping*; and discuss a set of strategies for controlling each source of damping by combining models for dissipation with measurements of material properties. The concluding section integrates the various strategies in the form of design principles and presents a case-study to illustrate an intriguing trade-off between functionality and performance for bilayered resonators.

Foundations of damping

Let us consider a single mode of oscillations of a linear mechanical resonator. Damping refers to *dissipation* (that is, the conversion of useful mechanical energy into disordered thermal energy) in the oscillating structure, and to the effects of dissipation on *structural dynamics* (see, for instance, [3-9]). Dissipation is quantified using two dimensionless measures: (i) the specific damping capacity, $\Psi = (\Delta W / W)$; and (ii) the loss factor, $\eta = (\Delta W / 2\pi W)$. Here, ΔW is the energy dissipated during one cycle of vibration, and W is the maximum elastic strain energy during the vibration cycle. It is difficult to measure energy dissipation or entropy generation directly; hence, damping is estimated by monitoring structural dynamics using such techniques as harmonic excitation, free decay, thermomechanical noise, and wave propagation [5]. Harmonic excitation is associated with two dimensionless measures of damping: the loss angle, ϕ, by which the stress (σ) leads the strain (ε); and the quality factor, $Q \equiv (\omega_n / \Delta\omega)$, where $\Delta\omega$ is the half-power bandwidth of the resonance peak and ω_n is the angular natural frequency. Alternately, the quality factor can be estimated by fitting the resonance peak in the thermomechanical noise spectrum [10-12]. The free decay technique quantifies damping in terms of the logarithmic decrement, δ, and wave propagation techniques measure the attenuation, $\hat{\alpha}$, of the amplitude of elastic waves with wavelength λ [9].

In general, the relationships between the various measures of damping can be quite complicated. However, when damping is linear and small [5,8],

$$\tan\phi = Q^{-1} = \frac{\delta}{\pi} = \frac{\hat{\alpha}\lambda}{\pi} = \frac{\Psi}{2\pi} = \eta = \frac{\Delta W}{2\pi W} \qquad (1)$$

Eq. (1) is a cornerstone of the subject of damping. The various relationships are widely used to: (i) quantify damping in miniaturized mechanical structures; (ii) make connections between dissipation and structural dynamics; (iii) compare measurements from different techniques; (iv) compare theoretical predictions with experimental measurements; and (v) design micromechanical and nanomechanical resonators. Indeed, the quality factor is frequently used to estimate damping even when the identities and frequency dependence of the dominant mechanisms of dissipation are not known. In this context, it is worth highlighting the fact that Eq. (1) is *neither exact nor universal*.

Each relationship is an approximation derived from three simple models for linear damping. The first is the classical system consisting of a mass, spring, and viscous dashpot. Viscous damping has its origins in Rayleigh's dissipation function and assumes that the damping force is proportional to the instantaneous velocity of the mass [3]. The second model uses the concept of a complex spring, $k^* = k_1 + i\, k_2$ with $\tan \phi = (k_2/k_1)$. The real and imaginary parts of k^* are called the storage modulus and loss modulus, respectively [9]. The third model is the standard anelastic solid with a constitutive relationship of the form $\sigma + \tau_\sigma \dot{\sigma} = E(\varepsilon + \tau_\varepsilon \dot{\varepsilon})$, where E is the elastic modulus, and τ_σ and τ_ε are material parameters [4,5]. Thus, anelasticity represents an extension of elasticity (as embodied in Hooke's law, $\sigma = E\,\varepsilon$) to incorporate rate effects and relaxation [5]. For these elementary models of viscous, viscoelastic, and anelastic behavior, the approximations in Eq. (1) are within 1% of the exact value for small linear damping with $\phi < 0.01$ [5].

Devices, applications, and design parameters

The first microelectromechanical oscillator was demonstrated in 1967 in the form of a device called the *resonant gate transistor* (RGT) [13]. The RGT consisted of a gold microcantilever (oscillating at 5 kHz with a quality factor of 500) fabricated on a silicon integrated circuit for signal processing applications. Even at this early stage, the benefits of miniaturization for attaining high frequencies and quality factors were well recognized. Soon thereafter, Newell [14] highlighted the value of integrating mechanical structures with microelectronic circuits using microfabrication techniques, and discussed the effects of scale on certain material and structural properties (including fatigue, thermomechanical noise, and viscous air damping). These ideas lay dormant for a decade before exploding into a creative burst of activity that continues unabated to the present day. The invention of the atomic force microscope in 1986 [15], and the rapid evolution of MEMS technologies during the 1990s, provided the motivation and processing capabilities for exploiting miniaturized mechanical resonators for sensing, signal processing, timing, vibration energy harvesting, and precision measurements for fundamental studies in diverse areas of science and engineering [16,17].

In this paper, we focus on one design parameter, namely, the level of damping in miniaturized mechanical resonators. Controlling damping is a ubiquitous and vital requirement because miniaturized resonators require high quality factors ($Q > 10^4$) or ultrahigh quality factors ($Q > 10^6$) for optimal performance [16-19]. The former is typical of MEMS used for signal processing and inertial sensing [16], and the latter of miniaturized resonators used for precision measurements [6]. Unfortunately, controlling damping is a formidable task because numerous mechanisms of dissipation operate in miniaturized resonators. The magnitude of damping is controlled by a large number of variables which include *structural dimensions* (1 nm to 100 µm); *type of structure* (beams, plates, shells, membranes, hollow resonators, fluid-filled micropipes, coupled arrays); *class of material* (metal, alloy, ceramic, glass, elastomer, polymer, nanocomposite); *chemistry* (dopants, impurities); *defects* (type, density, distribution, mobility), *microstructure* (amorphous, polycrystalline, nanocrystalline, single-crystal); *residual stress*; *surface chemistry* and *topography*; *mode of vibration* (flexural, torsional, longitudinal, disc-mode, wineglass-mode, bulk mode); *frequency of oscillation* (100 Hz to 10 GHz); *temperature* (10^{-3} K to 10^3 K); *pressure* (ultrahigh vacuum, atmospheric pressure, immersion in viscous liquid); *amplitude of oscillation*; nature and structure of *boundaries* (structural, thermal, fluidic);

Table 1 Factors that influence damping in miniaturized mechanical resonators

Operation	Structure	Materials	Processing
Mode	Shape	Chemistry	Deposition
Frequency	Size	Alloying	Patterning
Temperature	Architecture	Residual stress	Etching
Pressure	Supporting frame	Microstructure	Bonding
Transduction		Defects	Annealing
		Surfaces	Interconnection
		Interfaces	Packaging

supports, *clamps*, *interconnection* and *packaging*; and *processing techniques* used for micromachining and nanofabrication. Table 1 classifies the factors that control damping in miniaturized mechanical resonators. Damping in *oscillators* (that is, a combination of mechanical resonators and an external source of energy for sustaining time-harmonic oscillations) is influenced by an additional set of variables associated with the external source operating in different energy domains [16].

Review of damping

The mechanisms of linear damping can be classified in several different ways. Table 2 shows a scheme based on the spatial origin of dissipation [1]; thus, damping is divided into three categories: *boundary damping*, *fluid–structure interactions* (FSI), and *material damping*. The superscripts indicate the current state of knowledge. The squares (■) denote mechanisms for which detailed models are available for computing dissipation. Hence, design guidelines can be formulated for controlling viscous damping, acoustic radiation, squeezed-film damping, elastic wave radiation, and thermoelastic damping using closed-form expressions or by recourse to numerical analysis using finite-element and finite-difference methods. The section sign (§) denotes mechanisms for which first-order models are available. The challenging task of modeling higher-order phonon-phonon, phonon-surface, phonon-defect, and phonon–electron interactions is an area of current research. Finally, the triangles (▲) denote mechanisms that cannot yet be predicted from first principles. Measurements of material properties are usually required to guide the selection of materials, structures, and processes for controlling internal friction, microsliding, and viscoelasticity.

To gain a qualitative understanding of the various mechanisms, let us consider a generic description of micromechanical and nanomechanical resonators. Typically, the resonators are attached to relatively large *supporting frames* to facilitate handling and

Table 2 Classification of linear damping mechanisms

Boundary damping		Fluid–structure interactions		Material damping	
Elastic wave radiation■	[20-39]	Viscous damping■	[40-42]	Thermoelastic damping■	[43-59]
Microsliding▲	[60-63]	Squeeze-film damping■	[16,64]	Phonon-phonon interactions§	[65-67]
Viscoelasticity▲	[9]	Acoustic radiation■	[68]	Phonon–electron interactions§	[6,65]
		Internal flow■	[69,70]	Internal Friction▲	[4,5,71-84]

interconnection, and then *packaged* by mounting the supporting frame inside test stations, instruments, or customized packages. Illustrative examples include commercial silicon microcantilevers (1 to 10 μm thick and 0.1 to 0.5 mm long) used for atomic force microscopy (AFM), and silicon nitride nanomembranes (30 to 200 nm thick with lateral dimensions ranging from 10 μm to 1 mm) used in optomechanical systems [85,86]. In both cases, the devices are attached to silicon frames with thickness of a few hundred microns and lateral dimensions of a few millimeters. The attachment can be monolithic (exemplified by bulk-micromachined silicon resonators) or engendered by the adhesion of thin films to the supporting frame in the case of surface-micromachined devices [16].

The materials, methods, and designs used for packaging vary widely depending on the device and application. For example, silicon microcantilevers are loaded into AFMs using clamps, spring-loaded clips, or adhesive gels, but other devices require custom-designed packages for electrical connections, thermal interfaces, microfluidic manifolds, or vacuum packaging to maintain the resonator at low pressure [16]. The dissipation due to all components associated with the supporting frame and package is collectively called boundary damping. There are three main mechanisms of boundary damping: (i) *support losses* or *anchor losses* due to the radiation of elastic waves (or stress waves) from the resonator into the supporting frame [20-35]; (ii) *microsliding* at the interfaces between the resonator and supporting frame, and between the supporting frame and package [61]; and (iii) *viscoelasticity* in the gels and adhesives used to bond the supporting frame to the package [9]. (The term *clamping loss* is frequently encountered in the literature. Depending upon the context, it can refer either to stress-wave radiation or to microsliding).

The application dictates the fluidic environment of the resonator. When a miniaturized mechanical structure oscillates in a chamber containing a gas or liquid, energy is dissipated during every cycle of vibration due to the conversion of ordered structural energy into the thermal energy of the molecules of the fluid. Fluid–structure interactions (FSI) have been studied extensively using experiment, theory, and computation because they are a major source of damping. Indeed, the immense literature on FSI in microscale and nanoscale resonators deserves a dedicated full-fledged review in its own right. A detailed understanding of viscous damping, squeezed-film damping, and acoustic radiation has been obtained for numerous devices including plates, membranes, beams, torsional resonators, and hollow resonators containing internal microfluidic channels. The dependence of damping on material properties, fluidic properties, geometry, size, confinement, mode, frequency, pressure, and temperature has been captured well by models (see, for example, [14,16,40-42,64,68-70]).

Material damping refers to all dissipative mechanisms that operate within the volume (*bulk*), at the *free surfaces*, and at the internal *interfaces* of the resonator. Composite architectures are ubiquitous in MEMS and NEMS technologies; hence, several types of interfaces (layer boundaries, grain boundaries, and phase boundaries) can be encountered in miniaturized resonators. The creation of interfaces is often unintentional and, in some cases, undesirable. For example, the free surfaces of silicon, titanium, and aluminum are invariably covered with ultrathin coatings of native oxide under typical processing and operating conditions. Similarly, internal oxide layers and interfaces can be created due to unintentional oxidation during deposition [87].

Material damping can be divided into two categories: *fundamental damping* and *internal friction* [6]. The former is a set of mechanisms (thermoelastic damping, phonon-phonon interactions, and phonon–electron interactions) that set the ultimate lower limit on damping. These mechanisms have their origin in the electronic, atomic, and molecular structure of materials, and operate even in the idealized limit of perfectly engineered devices (for example, high-quality single-crystal materials with negligible defect density). Internal friction refers to damping caused by the irreversible motion of crystallographic defects (for example, vacancies, divacancies, interstitial atoms, substitutional atoms, surface adatoms, edge dislocations, screw dislocations, grain boundaries, phase boundaries, layer boundaries, and precipitates) [4], and each type of defect can give rise to several mechanisms of dissipation [5].

The standard anelastic solid provides a conceptually simple picture of material dissipation as a relaxation process. When damping is governed by a single relaxation time τ, the frequency dependence of material damping is a Debye peak given by [4]

$$Q^{-1}_{material} = \Delta \frac{\omega \tau}{1 + (\omega \tau)^2} \qquad (2)$$

where Δ is called the relaxation strength. Table 3 lists the relaxation strength and time for the two main sources of fundamental dissipation–*thermoelastic damping* due to irreversible heat conduction and *Akhiezer damping* due to phonon-phonon interactions–for an isotropic, homogeneous beam of thickness h. In this table, E is the Young's modulus, ρ is the density, α is the coefficient of thermal expansion, T is the absolute temperature, C is the specific heat per unit volume, k is the thermal conductivity, v is the velocity of sound, and γ is Grüneisen's constant. A preliminary estimate of fundamental losses can be obtained using the nominal properties listed in Table 4 for common metals and ceramics used in MEMS and NEMS. For detailed calculations, it is necessary to account for changes in properties due to size effects, microstructure, crystallographic anisotropy, and processing conditions.

Techniques for measuring internal friction in thin films were developed over 30 years ago [72], but many classes of materials, structures, and defects still remain largely unexplored. The sparse literature on this topic can be classified into two broad categories. The first set focuses on the effects of temperature on dissipation in an effort to measure relaxation peaks and study the underlying defect interactions (see, for example, [72-77]). The second set explores the effects of processing conditions (including post-deposition annealing), residual stress, and frequency on internal friction (see, for example, [78-83]). In general, the dissipation spectra rarely exhibit the Debye peak predicted by Eq. (2); instead, internal friction is often a weak monotonic function of frequency [9]. Explanations for this behavior range from a distribution of activation energies and relaxation times [5,9], to a hierarchically constrained sequence of serial relaxation processes in which the fast degrees of freedom (involving the motion of single atoms) must relax

Table 3 Simple expressions for estimating fundamental material losses using Eq. (2) [5,6]

Mechanism	Relaxation strength	Relaxation time
Thermoelastic damping (TED)	$\Delta_{TED} = \frac{E \alpha^2 T}{C}$	$\tau_{TED} = \frac{C h^2}{\pi^2 k}$
Akhiezer damping	$\Delta_{Akhiezer} = \frac{C T \gamma^2}{\rho v^2}$	$\tau_{Akhiezer} = \frac{3k}{C v^2}$

Table 4 Nominal properties of common metals and ceramics at 300 K [98]

Material	E (GPa)	ρ (kg/m³)	k (W/m/K)	α (K⁻¹)	C (J/m³/K)	v (m/s)	γ
Aluminum	70	2.7×10^3	220	24.0×10^{-6}	2.4×10^6	5.1×10^3	1.7
Copper	120	8.9×10^3	400	20.0×10^{-6}	3.8×10^6	3.6×10^3	1.9
Gold	82	1.9×10^4	320	15.0×10^{-6}	2.5×10^6	2.1×10^3	2.9
Nickel	210	8.9×10^3	92	13.0×10^{-6}	3.9×10^6	4.8×10^3	1.8
Silicon	160	2.3×10^3	150	2.6×10^{-6}	1.6×10^6	8.3×10^3	0.4
Silicon carbide	400	3.2×10^3	70	3.0×10^{-6}	3.0×10^6	1.1×10^4	1.1
Silicon oxide	70	2.2×10^3	1	0.5×10^{-6}	1.5×10^6	5.6×10^3	1.9
Silicon nitride	250	3.2×10^3	8	3.0×10^{-6}	3.0×10^6	8.8×10^3	1.2
Silver	76	1.1×10^4	430	18.0×10^{-6}	3.0×10^6	2.7×10^3	2.4

before the slower degrees of freedom involving the coordinated motion of groups of atoms [84].

Strategies for controlling damping

Fluid–structure interactions (FSI)

Fluid–structure interactions typically limit the quality factor of miniaturized mechanical resonators to low values ranging from 1 to 10^3. The most effective strategy for eliminating FSI is to operate the device at low pressure. For values of the Knudsen number (Kn) below 0.01, the gas is effectively a viscous continuum and damping is independent of pressure [14]. As the pressure is reduced, the flow transitions to the free molecular regime (FMR) for Kn > 10, and the damping is given by [14]

$$Q_{\text{FMR}}^{-1} = \sqrt{\frac{32M}{\pi R T}} \frac{P}{h \rho \omega} \tag{3}$$

where M is the molar weight of the gas, R is the universal gas constant, and P is the gas pressure. When the pressure is reduced further, fluidic damping becomes negligible at a critical value which is a function of the size, shape, and mode of the resonator. The critical pressure has been measured to range from 0.1 Pa to 10^3 Pa for miniaturized mechanical resonators [14,29,40-42].

Elastic wave radiation

Elastic waves are generated when the resonator applies time-harmonic forces and moments at the point of attachment to the support. This phenomenon is effectively a source of damping because the ordered mechanical energy of the resonator is transferred to the substrate in the form of stress waves, and eventually dissipated in the substrate. By assuming that the radiated energy does not reflect back into the resonator, damping can be estimated by performing calculations solely in the elastic domain without specifying the detailed mechanisms by which energy is dissipated in the substrate [20-25]. One of the first models for support loss is an analysis by Jimbo and Itao in 1968 of a two-dimensional system consisting of an isotropic, homogeneous, linear elastic cantilever of length L, width w, and thickness h that is attached to an elastic half-space. For this idealized geometry, the support loss is proportional to $(h/L)^3$ [20]. When the finite size of the supporting frame is taken into account, support losses in typical bulk-micromachined cantilevers are given by

$$Q^{-1}_{\text{support}} = 0.95 \frac{w}{L} \frac{h^2}{h_s^2} \qquad (\text{for } h_s \ll \lambda)$$

$$Q^{-1}_{\text{support}} = 0.31 \frac{w}{L} \left(\frac{h}{L}\right)^4 \qquad (\text{for } h_s \gg \lambda)$$

(4)

where h_s is the thickness of the support and λ is the wavelength of the elastic wave propagating in the support [22].

Support losses can be reduced by using analytical and numerical models for stress-wave radiation to guide the selection of materials, structures, dimensions, modes, and frequencies. Alternately, the generation and propagation of elastic waves can be disrupted by contacting the resonator at its nodal points using anchors or tethers [26-31] and incorporating phononic band-gap structures [32-35], acoustic reflectors [36,37], and vibration isolators [38,39]. In each case, well-established models from vibrations and elasticity are available to guide design.

Microsliding and viscoelasticity

The other sources of boundary damping face challenges that are common to many aspects of thin-film adhesion and packaging of MEMS [16]. The variety of designs, processes, techniques, and materials makes it difficult to develop general guidelines for a large class of devices and applications. Models for adhesion and friction can provide useful qualitative insights into the underlying mechanisms but improving thin-film adhesion and reducing microsliding remains an art. Nevertheless, several general strategies can be formulated: (i) minimize deformations and strains at the interface between the supporting frame and package: for example, by employing the anti-symmetric modes of dual-cantilever beams [60] or double-paddle oscillators [62,63], and acoustically isolating the resonator from the supporting frame and package; (ii) use precision-machined clamps and avoid spring-loaded clips, gels, polymer-based adhesives, and die bonds for packaging; and (iii) ensure good adhesion between the resonator and supporting frame by activating the substrate before depositing structural thin films, employing adhesion promoters in the form of ultrathin films of Ti or Cr, and using ion-beam assisted deposition techniques [88].

Thermoelastic damping

Thermoelastic damping (TED) is a rare example of a dissipative mechanism that can be computed accurately from first principles [4]. Consider a beam or a plate that is subjected to time-harmonic bending forces or moments. The elastic stresses in the structure give rise to elastic strains (which are in phase with the stress) and thermal strains due to thermoelastic coupling. In general, the thermal strains are not in phase with the driving elastic stresses; hence, energy is dissipated during every cycle of vibration. Alternately, TED can be viewed as the result of oscillating temperature gradients generated by oscillating stress gradients in thermoelastic solids. Heat conduction across finite temperature gradients leads to entropy generation and energy dissipation [43-48].

The analysis of TED is conceptually straightforward for heat conduction in the diffusive regime (that is, the mean free path of thermal phonons is much less than the characteristic length scale of the resonator [46]). The formula in Table 3 is due to a celebrated analysis by Zener in 1937. More recently, Zener's analysis of homogenous, isotropic beams has been extended in multiple directions and there are now over 350 significant publications on this topic. A selection of the literature includes models to account for the effects of structural boundaries [47,48]; polycrystalline grain structure [49,50]; layered composite architecture [51-53]; electrostatic actuation [54]; and geometry (plates [54,55], slotted, channeled, and hollow beams [56-58], double-paddle oscillators [63], and bulk-mode, ring-mode, and disc-mode resonators [59]). Using these models, the design space can be explored to formulate detailed guidelines for selecting geometries, structures, modes, materials, and frequencies to minimize thermoelastic damping.

Internal friction

The magnitude of internal friction is governed by the type, distribution, density, and mobility of defects, and by the interactions between different classes of defects. Quantifying these details, especially in micrometer and nanometer scale structures, requires extensive experimental studies using a suite of microscopic and spectroscopic techniques (see, for example, [82,87,89]). Even when such information is available, it is difficult to model the dynamics of defects over multiple scales of length, time, and energy. Therefore, in common with many other material properties, design guidelines are derived by measuring internal friction, characterizing the microstructure, and formulating *process-structure–property relationships* [5,9,71].

There are three main strategies for controlling dissipation due to internal friction. The first is to control the type, distribution, and density of defects. High-quality single-crystals can be used for resonators but fabrication challenges currently limit this option to a small set of materials (for example, commercial silicon wafers and epitaxial thin films grown using molecular beam epitaxy). More practically, defects can be controlled by an appropriate selection of the deposition technique, processing conditions, and post-deposition heat treatment. For example, when sputter-deposited aluminum films were annealed at 450°C for 1 hour in an inert atmosphere, the average grain size increased from 100 nm to 390 nm, and the room-temperature values of internal friction reduced from 0.05 to 0.02 [82].

The second strategy is to reduce the mobility of defects by materials selection and materials design. Defect mobility is a function of several variables including atomic bonding, crystal structure, microstructure, melting temperature, operating temperature, and frequency. The lattice self-diffusivity at the melting point is $\sim 10^{-16}$ m^2/s for covalently bonded diamond-cubic materials, $\sim 5 \times 10^{-14}$ m^2/s for oxides, and $\sim 5 \times 10^{-13}$ m^2/s for face-centered cubic metals [90]. All else being equal, defects are less mobile in brittle ceramics (Si, SiO_2, quartz, SiC, TiO_2, Al_2O_3, and Ta_2O_5) than in common metals and alloys [6]. Thus, for precision measurements requiring ultrahigh quality factors, multilayer stacks of dielectric films consisting of alternate layers of amorphous silica (SiO_2) and amorphous tantala (Ta_2O_5) are preferred over metallic thin films for optical coatings [17]. Furthermore, small quantities of alloying additions can have a disproportionately large influence on defect mobility. For instance, damping in aluminum alloy Al 5056 is an order of magnitude less than that in pure aluminum over a broad range of temperatures and frequencies [6,74].

Finally, internal friction can be altered by changing the operating temperature but the implementation of this strategy must take into account the non-monotonic behavior of dissipation. For example, internal friction in bulk quartz crystals increased by over two orders of magnitude (from 10^{-7} to 4×10^{-5}) upon cooling from 300 K to 50 K, and then reduced precipitously to a remarkably low value of 2×10^{-10} when cooled further to 2 K [91].

Stress-engineered resonators

Micromechanical and nanomechanical resonators are commonly subjected to a pre-stress (or residual stress) originating from intrinsic stresses generated during the growth of thin film materials [92] and differential expansion caused by thermal excursions during processing [16]. The pre-stress can affect several properties (stiffness, natural frequencies, mode shapes, linearity, and damping) and reliability (fracture under tension, buckling under compression, inelastic deformation, and stress relaxation by creep [93]). Therefore, the control and mitigation of residual stresses is a major consideration in the design of surface-micromachining processes [16].

In some cases, however, large tensile residual stresses can be used to reduce damping, as exemplified by high-Q devices fabricated using amorphous silicon nitride films grown by low-pressure chemical vapor deposition (see, for example, [83,86,94-96]). Indeed, quality factors of 50 million have been obtained at room temperature with highly-stressed (~1 GPa) nanomembrane resonators [94]. In general, the effects of the tensile pre-stress are governed by the relative magnitudes of two factors: (i) increase in the elastic stored energy (both extensional and flexural), and (ii) stress-induced changes in dissipation. The former can be obtained by analyzing the elastic deformation of the resonator (see, for example, [95-97] for expressions of the quality factors for stretched-string resonators), but the latter has been analyzed explicitly only for a few mechanisms. Notable examples include thermoelastic damping in plates [54] and membranes [94]; in both cases, the magnitude of TED can be reduced by applying an in-plane tensile pre-stress.

Design principles

Miniaturized mechanical resonators are enabling a remarkably large and diverse set of applications ranging from sensing, timing, and signal processing to scanning probe microscopy and precision measurements for fundamental studies in several fields of engineering and science. New concepts continue to emerge; devices are growing in sophistication, performance, and functionality; and there is a growing emphasis on transitioning from proofs-of-concept to full-fledged commercialization. All these trends have created an urgent need for developing effective and rational methods for design, analysis, and optimization.

When microresonator technologies began to emerge in the 1980s, damping was controlled using iterative cycles of fabrication, testing, and analysis at the device level [2]. Unfortunately, the design space is much too vast and complex to be probed efficiently in this manner, especially considering the time and resources required for integrating new materials and structures into micromachining and nanofabrication process flows [16]. It is therefore useful and timely to develop methods that account for damping at every stage of the design cycle. In this section, we present an approach that is based on identifying the mechanisms of damping, and then developing strategies for controlling each mechanism with a judicious combination of models for dissipation and measurements of material properties.

Using the standard concept-embodiment-detail description of the design process [98], consider a typical problem that begins by translating market needs and application requirements into a set of device specifications, which are then used to generate a set of preliminary designs. At this stage, a careful estimation of fundamental material losses can establish the ultimate limits of dissipation and serve as a criterion for ranking the various designs. The device can approach the ultimate limits only to the extent that all other sources of dissipation (FSI, boundary damping, and internal friction) are eliminated. The optimal solutions are to: (i) eliminate FSI by operating at a sufficiently low pressure (ranging from 0.1 Pa to 10^3 Pa, depending on the size, shape, and mode of the resonator); (ii) eliminate boundary damping by decoupling the resonator from the supports and package by using nodal supports, phononic bandgaps, and acoustic isolators; and (iii) eliminate internal friction by using high-quality single-crystals or materials with low defect mobility. In the literature, we can now find a small, but steadily growing, set of devices that can approach the ultimate limits of damping (see, for example, [6,17,28,63,80]).

In many cases, however, the optimal solutions cannot be implemented because the design is constrained by the application, functionality, and actuation, and by the limitations of processing and packaging technologies. As an example, consider the class of bilayered metal-coated ceramic resonators that are widely used in MEMS and NEMS. The ceramic structure can be designed to oscillate with low damping approaching the fundamental limits. The coating adds valuable functionality by enhancing the optical reflectivity and electrical conductivity, but severely degrades the quality factor and performance of the device due to the relatively large internal friction in polycrystalline metallic films. Thus, the design problem is to identify an optimal trade-off between performance and functionality.

When dissipation is dominated by internal friction (IF) in the metallic film, Eq. (1) can be used to express the inverse quality factor of the bilayer as

$$Q_{\text{bilayer}}^{-1} = \frac{\Delta W_{\text{bilayer}}}{2\pi W_{\text{bilayer}}} \simeq \frac{\Delta W_{\text{IF, film}}}{2\pi W_{\text{bilayer}}} = \left(\frac{W_{\text{film}}}{W_{\text{bilayer}}}\right)\frac{\Delta W_{\text{IF, film}}}{2\pi W_{\text{film}}} = \left(\frac{W_{\text{film}}}{W_{\text{bilayer}}}\right)Q_{\text{IF, film}}^{-1} \quad (5)$$

where $(W_{\text{film}}/W_{\text{bilayer}})$ is obtained by computing the elastic deformation of the film and bilayer. Eq. (5) suggests two distinct strategies for increasing the quality factor. The first is to control internal friction in the film using process-structure–property relationships. For sputtered aluminum films (which are widely used as coatings for numerous applications including commercial probes for atomic force microscopy), internal friction can be reduced by post-deposition annealing to increase grain size [82], reducing film thickness [80], and alloying aluminum with small amounts (5%) of magnesium [6,74]. Alternately, aluminum can be replaced with gold [80] or multilayer dielectric stacks (for example, alternate layers of silica and tantala [17]). The second strategy is to minimize the ratio of elastic strain energies by confining the metallic coating to regions of low deformation and strain [99,100]. For microcantilevers oscillating in the fundamental bending mode, internal friction due to metallization can be made negligible by coating only the tip of the beam [99].

Competing interests

The authors declare that they have no competing interests.

Authors' contributions

SJ analyzed the relationships between dissipation and dynamics, formulated the case study, and contributed to the sections on fluid–structure interactions. SH contributed to the sections on boundary damping. SV contributed to the sections on material damping and design methodologies. The manuscript was edited by SV, and then read and approved by all authors.

Acknowledgements

This paper is based on experience gained over several years of work on damping. During that period, we benefited greatly from valuable discussions with a number of generous colleagues around the world. We owe a debt of gratitude to all of them. Financial support from the Canada Research Chairs program and the Natural Sciences and Engineering Research Council of Canada is gratefully acknowledged.

References

1. Woodhouse J: **Linear damping models for structural vibration.** *J Sound Vib* 1998, **215**:547–569.
2. Sidles JA, Garbini JL, Bruland KJ, Rugar D, Züger O, Hoen S, Yannoni CS: **Magnetic resonance force microscopy.** *Rev Modern Phys* 1995, **67**:249–265.
3. Strutt JW: *Theory of Sound.* New York: Dover; 1945.
4. Zener CM: *Elasticity and Anelasticity of Metals.* Chicago: University of Chicago Press; 1948.
5. Nowick AS, Berry BS: *Anelastic Relaxation in Crystalline Solids.* New York: Academic Press; 1972.
6. Braginsky VB, Mitrofanov VP, Panov VI: *Systems with Small Dissipation.* Chicago: Chicago University Press; 1985.
7. Nashif AD, Jones DIG, Henderson JP: *Vibration Damping.* New York: Wiley; 1985.
8. Graesser J, Wong CR: **The relationship between traditional damping measures for materials with high damping capacity: a review.** In *M3D: Mechanics and Mechanisms of Material Damping.* Edited by Kinra VK, Wolfenden A. Philadelphia: American Society for Testing and Materials; 1992:316–343.
9. Lakes R: *Viscoelastic Materials.* Cambridge: Cambridge University Press; 2009.
10. Saulson PR: **Thermal noise in mechanical experiments.** *Phys Rev D* 1990, **42**:2437–2445.
11. Sader JE, Hughes BD, Sanelli JA, Bieske EJ: **Effect of multiplicative noise on least-squares parameter estimation with applications to the atomic force microscope.** *Rev Sci Instrum* 2012, **83**:055106.
12. Kuter-Arnebeck O, Labuda A, Joshi S, Das K, Vengallatore S: **Estimating damping in microresonators by measuring thermomechanical noise using laser Doppler vibrometry.** *J Microelectromech Syst* 2014, in press.
13. Nathonson HC, Newell WE, Wickstrom RA, Davis JR: **The resonant gate transistor.** *IEEE Trans Elec Dev* 1967, **ED-14**:117–133.
14. Newell WE: **Miniaturization of tuning forks.** *Science* 1968, **161**:1320–1326.
15. Binnig G, Quate CF, Gerber C: **Atomic force microscope.** *Phys Rev Lett* 1986, **56**:930–933.
16. Senturia SD: *Microsystem Design.* Boston: Kluwer; 2001.
17. Harry G, Bodiya TP, DeSalvo R: *Optical Coatings and Thermal Noise in Precision Measurement.* Cambridge: Cambridge University Press; 2012.
18. Gabrielson TB: **Mechanical-thermal noise in micromachined acoustic and vibration sensors.** *IEEE Trans Elec Dev* 1993, **40**:903–909.
19. Labuda A, Bates JR, Grütter PH: **The noise of coated cantilevers.** *Nanotechnology* 2012, **23**:025503.
20. Jimbo Y, Itao K: **Energy loss of a cantilever vibrator.** *J Horological Institute of Japan* 1968, **47**:1–15.
21. Hao Z, Erbil A, Ayazi F: **An analytical model for support loss in micromachined beam resonators with in-plane flexural vibrations.** *Sens Actuators A* 2003, **109**:156–164.
22. Photiadis DM, Judge JA: **Attachment losses of high Q oscillators.** *Appl Phys Lett* 2004, **85**:482–484.
23. Park YH, Park KC: **High-fidelity modeling of MEMS resonators–Part I: Anchor loss mechanisms through substrate.** *J Microelectromech Syst* 2004, **13**:238–247.
24. Wilson-Rae I, Barton RA, Verbridge SS, Southworth DR, Ilic B, Craighead HG, Parpia JM: **High-Q nanomechanics via destructive interference of elastic waves.** *Phys Rev Lett* 2011, **106**:047205.
25. Bindel DS, Govindjee S: **Elastic PMLs for resonator anchor loss simulation.** *Int J Numer Meth Eng* 2005, **64**:789–818.
26. Wang K, Wong AC, Nguyen CTC: **VHF free-free beam high-Q micromechanical resonators.** *J Microelectromech Syst* 2000, **9**:347–360.
27. Ferguson AT, Li L, Nagaraj VT, Balachandran B, Piekarski B, DeVoe DL: **Modeling and design of composite free-free beam piezoelectric resonators.** *Sens Actuators A* 2005, **118**:63–69.
28. Anetsberger G, Rivière R, Schliesser A, Arcizet O, Kippenberg TJ: **Ultralow-dissipation optomechanical resonators on a chip.** *Nature Photonics* 2008, **2**:627–633.
29. Khine L, Palaniapan M: **High-Q bulk-mode SOI square resonators with straight-beam anchors.** *J Micromech Microeng* 2009, **19**:015017.
30. Lee JEY, Yan J, Seshia AA: **Study of lateral mode SOI-MEMS resonators for reduced anchor loss.** *J Micromech Microeng* 2011, **21**:045010.
31. Cole GD, Wilson-Rae I, Werbach K, Vanner MR, Aspelmeyer M: **Phonon-tunnelling dissipation in mechanical resonators.** *Nature Commun* 2011, **2**:231.
32. Mohammadi S, Adibi A: **Waveguide-based phononic crystal micro/nanomechanical high-Q resonators.** *J Microelectromech Syst* 2012, **21**:379–384.
33. Hsu FC, Hsu JC, Huang TC, Wang CH, Chang P: **Reducing support loss in micromechanical ring resonators using phononic band-gap structures.** *J Phys D* 2011, **44**:375101.

34. Yu PL, Cicak K, Kampel NS, Tsaturyan Y, Purdy TP, Simmonds RW, Regal CA: A phononic bandgap shield for high-Q membrane microresonators. *Appl Phys Lett* 2014, **104**:023510.

35. Tsaturyan Y, Barg A, Simonsen A, Villanueva LG, Schmid S, Schliesser A, Polzik ES: Demonstration of suppressed phonon tunneling losses in phononic bandgap shielded resonators for high-Q optomechanics. *Optics Express* 2014, **22**:6810–6821.

36. Pandey M, Reichenbach RB, Zehnder AT, Lal A, Craighead HG: Reducing anchor loss in MEMS resonators using mesa isolation. *J Microelectromech Syst* 2009, **18**:836–844.

37. Harrington BP, Abdolvand R: In-plane acoustic reflectors for reducing effective anchor loss in lateral–extensional MEMS resonators. *J Micromech Microeng* 2011, **21**:085021.

38. Yoon SW, Lee S, Perkins NC, Najafi K: Analysis and wafer-level design of a high-order silicon vibration isolator for resonating MEMS devices. *J Micromech Microeng* 2011, **21**:015017.

39. Le Foulgoc B, Bourouina T, Le Traon O, Bosseboeuf A, Marty F, Breluzeau C, Grandchamp JP, Masson S: Highly decoupled single-crystal silicon resonators: An approach for the intrinsic quality factor. *J Micromech Microeng* 2006, **16**:S45–S53.

40. Blom FR, Bouwstra S, Elwenspoek M, Fluitman JHJ: Dependence of the quality factor of micromachined silicon beam resonators on pressure and geometry. *J Vac Sci Technol B* 1992, **10**:19–26.

41. Pandey AK, Pratap R, Chau FS: Effect of pressure on fluid damping in MEMS torsional resonators with flow ranging from continuum to molecular regime. *Exp Mech* 2008, **48**:91–106.

42. Svitelskiy O, Sauer V, Liu N, Cheng KM, Finley E, Freeman MR, Hiebert WK: Pressurized fluid damping of nanoelectromechanical systems. *Phys Rev Lett* 2009, **103**:244501.

43. Zener C: Internal friction in solids I: Theory of internal friction in reeds. *Phys Rev* 1937, **52**:230–235.

44. Zener C: Internal friction in solids II: General theory of thermoelastic internal friction. *Phys Rev* 1938, **53**:90–99.

45. Kinra VK, Milligan KB: A second-law analysis of thermoelastic damping. *J Appl Mech* 1994, **61**:71–76.

46. Lifshitz R, Roukes ML: Thermoelastic damping in micro- and nanomechanical systems. *Phys Rev B* 2000, **61**:5600–5609.

47. Prabhakar S, Vengallatore S: Theory for thermoelastic damping in micromechanical resonators with two-dimensional heat conduction. *J Microelectromech Syst* 2008, **17**:494–502.

48. Prabhakar S, Païdoussis MP, Vengallatore S: Analysis of frequency shifts due to thermoelastic coupling in flexural-mode micromechanical and nanomechanical resonators. *J Sound Vib* 2009, **323**:385–396.

49. Randall RH, Rose FC, Zener C: Intercrystalline thermal currents as a source of internal friction. *Phys Rev* 1939, **56**:343–348.

50. Srikar VT, Senturia SD: Thermoelastic damping in fine-grained polysilicon flexural beam resonators. *J Microelectromech Syst* 2002, **11**:499–504.

51. Bishop JE, Kinra VK: Elastothermodynamic damping in laminated composites. *Int J Solids Struct* 1997, **34**:1075–1092.

52. Vengallatore S: Analysis of thermoelastic damping in laminated composite micromechanical beam resonators. *J Micromech Microeng* 2005, **15**:2398–2404.

53. Nourmohammadi Z, Prabhakar S, Vengallatore S: Thermoelastic damping in layered microresonators: critical frequencies, peak values, and rule of mixture. *J Microelectromech Syst* 2013, **22**:747–754.

54. Nayfeh AH, Younis MI: Modeling and simulations of thermoelastic damping in microplates. *J Micromech Microeng* 2004, **14**:1711–1717.

55. Norris AN: Dynamics of thermoelastic thin plates: a comparison of four theories. *J Thermal Stresses* 2006, **29**:169–195.

56. Duwel A, Candler RN, Kenny TW, Varghese M: Engineering MEMS resonators with low thermoelastic damping. *J Microelectromech Syst* 2006, **15**:1437–1445.

57. Prabhakar S, Vengallatore S: Thermoelastic damping in slotted and hollow microresonators. *J Microelectromech Syst* 2009, **18**:725–735.

58. Abdolvand R, Johari H, Ho GK, Erbil A, Ayazi F: Quality factor in trench-refilled polysilicon beam resonators. *J Microelectromech Syst* 2006, **15**:471–478.

59. Chandorkar SA, Candler RN, Duwel A, Melamud R, Agarwal M, Goodson KE, Kenny TW: Multimode thermoelastic dissipation. *J Appl Phys* 2009, **105**:043505.

60. Baur J, Kulik A: Optimal sample shape for internal friction measurements using a dual cantilevered beam. *J Appl Phys* 1985, **58**:1489–1492.

61. Nouira H, Foltête E, Ait Brik B, Hirsinger L, Ballandras S: Experimental characterization and modeling of microsliding on a small cantilever quartz beam. *J Sound Vib* 2008, **317**:30–49.

62. Kleiman RN, Kaminsky GK, Reppy JD, Pindak R, Bishop DJ: Single crystal silicon high-Q torsional oscillators. *Rev Sci Instrum* 1985, **56**:2088–2091.

63. Borrielli A, Bonaldi M, Serra E, Bagolini A, Conti L: Wideband mechanical response of a high-Q silicon double-paddle oscillator. *J Micromech Microeng* 2011, **21**:065019.

64. Bao M, Yang H: Squeeze film air damping in MEMS. *Sens Act A* 2007, **136**:3–27.

65. Cleland AN: *Foundations of Nanomechanics*. Berlin: Springer; 2003.

66. Kiselev AA, Iafrate GJ: Phonon dynamics and phonon assisted losses in Euler-Bernoulli nanobeams. *Phys Rev B* 2008, **77**:205436.

67. Kunal K, Aluru NR: Akhiezer damping in nanostructures. *Phys Rev B* 2011, **84**:245450.

68. Vishwakarma SD, Pandey AK, Parpia JM, Southworth DR, Craighead HG, Pratap R: Evaluation of mode dependent fluid damping in a high frequency drumhead microresonator. *J Microelectromech Syst* 2014, **23**:334–346.

69. Rinaldi S, Prabhakar S, Vengallatore S, Païdoussis MP: Dynamics of microscale pipes containing internal fluid flow: damping, frequency shift and stability. *J Sound Vib* 2010, **329**:1081–1088.

70. Sader JE, Burg TP, Manalis SR: Energy dissipation in microfluidic beam resonators. *J Fluid Mech* 2010, **650**:215–250.

71. Blanter MS, Golovin IS, Neuhauser H, Sinning HR: *Internal Friction in Metallic Materials*. Berlin: Springer; 2007.

72. Berry BS, Pritchet WC: Extended capabilities of a vibrating-reed internal friction apparatus. *Rev Sci Instrum* 1983, **54**:254–256.

73. Prieler M, Bohn HG, Schilling W, Trinkaus H: **Grain boundary sliding in thin substrate-bonded Al films.** *J Alloys Compd* 1994, **211–212**:424–427.
74. Liu X, Thompson E, White BE, Pohl RO: **Low-temperature internal friction in metal films and in plastically deformed bulk aluminum.** *Phys Rev B* 1999, **59**:11767–11776.
75. Uozumi K, Honda H, Kinbara A: **Internal friction of vacuum-deposited silver films.** *J Appl Phys* 1978, **49**:249–252.
76. Zhu AW, Bohn HG, Schilling W: **Internal friction associated with grain boundary diffusion in thin gold films.** *Phil Mag A* 1995, **72**:805–812.
77. Choi DH, Kim H, Nix WD: **Anelasticity and damping of thin aluminum films on silicon substrates.** *J Microelectromech Syst* 2004, **13**:230–237.
78. Vengallatore S: **Gorsky damping in nanomechanical structures.** *Scripta Mater* 2005, **52**:1265–1268.
79. Ono T, Esashi M: **Effect of ion attachment on mechanical dissipation of a resonator.** *Appl Phys Lett* 2005, **87**:044105.
80. Sosale G, Prabhakar S, Frechette L, Vengallatore S: **A microcantilever platform for measuring internal friction in thin films using thermoelastic damping for calibration.** *J Microelectromech Syst* 2011, **20**:764–773.
81. Paolino P, Bellon L: **Frequency dependence of viscous and viscoelastic dissipation in coated micro-cantilevers from noise measurement.** *Nanotechnology* 2009, **20**:405705.
82. Sosale G, Almecija D, Das K, Vengallatore S: **Mechanical spectroscopy of nanocrystalline aluminum films: effects of frequency and grain size on internal friction.** *Nanotechnology* 2012, **23**:155701.
83. Yu PL, Purdy TP, Regal CA: **Control of material damping in high-Q membrane microresonators.** *Phys Rev Lett* 2012, **108**:083603.
84. Palmer RG, Stein DL, Abrahams E, Anderson PW: **Models of hierarchically constrained dynamics for glassy relaxation.** *Phys Rev Lett* 1984, **53**:958–961.
85. Yasumura KY, Stowe TD, Chow EM, Pfafman T, Kenny TW, Stipe BC, Rugar D: **Quality factors in micron- and submicron-thick cantilevers.** *J Microelectromech Syst* 2000, **9**:117–125.
86. Zwickl BM, Shanks WE, Jayich AM, Yang C, Jayich ACB, Thompson JD, Harris JGE: **High quality mechanical and optical properties of commercial silicon nitride membranes.** *Appl Phys Lett* 2008, **92**:103125.
87. Stoffels S, Autizi E, van Hoof R, Severi S, Puers R, Witvrouw A, Tilmans HAC: **Physical loss mechanisms for resonant acoustical waves in boron doped poly-SiGe deposited with hydrogen dilution.** *J Appl Phys* 2010, **108**:084517.
88. Ohring M: *Materials Science of Thin Films.* San Diego: Academic Press; 2002.
89. Lee Z, Ophus C, Fischer LM, Nelson-Fitzpatrick N, Westra KL, Evoy S, Radmilovic V, Dahmen U, Mitlin D: **Metallic NEMS components fabricated from nanocomposite Al-Mo films.** *Nanotechnology* 2006, **17**:3063–3070.
90. Brown AM, Ashby MF: **Correlations for diffusion constants.** *Acta Metall* 1980, **28**:1085–1101.
91. Smagin AG: **A quartz resonator for a frequency of 1 MHz with a Q-value of 4.2 x 10^9 at a temperature of 2 K.** *Cryogenics* 1975, **15**:483–485.
92. Spaepen F: **Interfaces and stresses in thin films.** *Acta Mater* 2000, **48**:31–42.
93. Lee HJ, Cornella G, Bravman JC: **Stress relaxation of free-standing aluminum beams for microelectromechanical systems applications.** *Appl Phys Lett* 2000, **76**:3415–3417.
94. Chakram S, Patil YS, Chang L, Vengalattore M: **Dissipation in ultrahigh quality factor SiN membrane resonators.** *Phys Rev Lett* 2014, **112**:127201.
95. Unterreithmeier QP, Faust T, Kotthaus JP: **Damping of nanomechanical resonators.** *Phys Rev Lett* 2010, **105**:027205.
96. Schmid S, Jensen KD, Nielsen KH, Boisen A: **Damping mechanisms in high-Q micro and nanomechanical string resonators.** *Phys Rev B* 2011, **84**:165307.
97. Berry BS: **Damping mechanisms in thin-layer materials.** In *M3D: Mechanics and Mechanisms of Material Damping.* Edited by Kinra VK, Wolfenden A. Philadelphia: American Society for Testing and Materials; 1992:28–44.
98. Ashby MF: *Materials Selection in Mechanical Design.* Oxford: Butterworth-Heinemann; 2011.
99. Sosale G, Das K, Frechette L, Vengallatore S: **Controlling damping and quality factors of silicon microcantilevers by selective metallization.** *J Micromech Microeng* 2011, **21**:105010.
100. Serra E, Cataliotti FS, Marin F, Marino F, Pontin A, Prodi GA, Bonaldi M: **Inhomogeneous mechanical losses in micro-oscillators with high reflectivity coating.** *J Appl Phys* 2012, **111**:113109.

8

Characterization of a gold coated cantilever surface for biosensing applications

Ann-Lauriene Haag[1*], Yoshihiko Nagai[2], R Bruce Lennox[3] and Peter Grütter[1]

* Correspondence:
haagal@physics.mcgill.ca
[1]Department of Physics, McGill University, 3600 Rue University, Montreal, QC H3A 2T8, Canada
Full list of author information is available at the end of the article

Abstract

Cantilever based sensors are a promising tool for a very diverse spectrum of biological sensors. They have been used for the detection of proteins, DNA, antigens, bacteria viruses and many other biologically relevant targets. Although cantilever sensing has been described for over 20 years, there are still no viable commercial cantilever-based sensing products on the market. Several reasons can be found for this – a lack of detailed understanding of the origin of signals being an important one. As a consequence application-relevant issues such as shelf life and robust protocols distinguishing targets from false responses have received very little attention. Here, we will discuss a cantilever sensing platform combined with an electrochemical system. The detected surface stress signal is modulated by applying a square wave potential to a gold coated cantilever. The square wave potential induces adsorption and desorption onto the gold electrode surface as well as possible structural changes of the target and probe molecules on the cantilever surface resulting in a measurable surface stress change. What sets this approach apart from regular cantilever sensing is that the quantification and identification of observed signals due to target-probe interactions are not only a function of stress value (i.e. amplitude), but also of the temporal evolution of the stress response as a function of the rate and magnitude of the applied potential change, and the limits of the potential change.

This paper will discuss three issues that play an important role in future successful applications of cantilever-based sensing. First, we will discuss what is required to achieve a large surface stress signal to improve sensitivity. Second, a mechanism to achieve an optimal probe density is described that improves the signal-to-noise ratio and response times of the sensor. Lastly, lifetime and long term measurements are discussed.

Keywords: Surface stress; Cantilever sensing; Biosensor; Oligonucleotide; Electrochemistry

Introduction

Nanomechanical structures can be used for label-free and low-cost biosensors that offer high sensitivities. In recent years, several nano and micromechanical structures have been described as possible biosensor platforms, such as nanomechanical cantilevers [1-4], resonators [5,6], and optomechanical structures [7]. The most common detection principles due to biological binding effects are changes in surface stress [8,9] and mass [10,11].

Here we focus on a cantilever sensing platform that detects changes in surface stress. In our platform, a cantilever is coated with a gold layer that serves two purposes. First, this gold layer is used as a support structure of probe molecules bound to the surface typically using thiol linkers; this in principle gives the sensor specificity [12]. What is often not considered is the second role of this gold layer, as it can act as a very sensitive transducer that is located within nanometers of the probe molecules that sense the biological binding events [9,13]. In our system, the surface potential of the gold coated cantilever is controlled and changed over time to induce changes of the surface coverage of the adsorbing ion. Changing the presence of any ionic or charged species near the surface leads to a large change of surface stress. This is based on the well established fact that surface stress is directly proportional to the surface charge density [14]. Surface concentration changes of charged species can be induced by applying an electrochemical potential which generates conformational changes of probe molecules.

Our approach to increasing the dynamic range of the stress signal is to drive the adsorption and desorption of ions to the cantilever surface, thus inducing a large measurable and characteristic surface stress change [15]. This movement of ions can be modulated as a function of time, allowing signal averaging techniques to be used. If clean gold surfaces are used, the resultant reproducible time dependent stress signals include information on the target-probe system, such as ion diffusion times and polymer dynamics. This information can be used for biochemical sensors or in fundamental studies (e.g. for the investigation of the folding dynamics of proteins). Reliable signal and thus target identification can be based on recognition of the complex time dependent stress patterns in addition to the information given by signal amplitudes.

In our experiments, we change the presence of ions near the surface by combining a conventional cantilever stress sensing system with a standard three-electrode electrochemical system. All experiments are performed in buffer solution with the cantilever acting as a working electrode (WE), a platinum wire as the counter electrode (CE) and a Ag/AgCl (sat. KCl) electrode as the reference electrode (RE). The electrodes are connected through a potentiostat allowing a voltage to be applied between the working (cantilever) and the reference electrode thus measuring the current flowing between the working and the counter electrode (voltammetry) [16]. Upon application of a square wave potential to the gold coated cantilever between +/− 200 mV, chloride ions that are present in solution will ad-/desorb on the surface which leads to a change in surface charge density [17] and therefore to a change in surface stress [14]. In our system the stress-induced bending of the cantilever is measured by optical beam deflection methods and translated into a quantitative surface stress signal by using Stoney's formula [18] and appropriate calibrations [19,20].

The electrochemical aspect of our sensor system serves two distinct purposes. First, it is used to clean and electrochemically characterize the surface of the gold coated cantilever. Secondly, it is used to apply a controlled, time dependent potential to the cantilever to induce repetitive surface stress changes. This first point is very important, as surface stress and surface stress changes are driven by surface charge density, which is a function of the cleanliness of a system [9]. Recalling that a clean metallic surface typically takes about 1 microsecond to be contaminated in air by absorbable organic molecules – hence the need for ultra high vacuum conditions (UHV) to investigate surface phenomena. Electrochemistry allows a systematic cleaning and characterization of surfaces in solution. Note that compared to the concentration of rest gas ('contaminations') in UHV, solutions are very seldom as pure – clean solution to background contaminations would need to be at a level of 1 part in 10^{13} to achieve similar lifetimes of clean surfaces in solution as in UHV.

An important insight is that the surface stress change on the cantilever is proportional to the available and accessible gold surface area. This can be used to measure and optimize the concentration of probes on a cantilever that leads to a decrease in the available gold surface area due to the target molecules covering part of the gold surface (Figure 1) [9]. On a clean gold coated surface, a large number of ions can interact with the surface resulting in a large surface stress change signal. If part of the surface is covered by molecules, in this case thiolated single-stranded oligonucleotide, fewer ions can access the surface leading to a smaller surface stress change. Covering the surface complete densely with a monolayer of molecules will hinder the ions to reach the surface and no large change in surface stress is observed. Residual (much smaller) stress changes can be due to steric hindrance and other effects. Our group has previously shown surface stress changes for aptamer functionalized gold surfaces. The aptamer undergoes a conformation change into a more compact state upon binding to its cognate ligand therefore increasing the available gold area, i.e. increased surface stress, compared to its relaxed state [15].

In this paper we will discuss three issues that most nanomechanical based sensors are facing and present protocols to improve these issues. This will be the foundation for possible applications of biosensors applicable to real-life samples containing not only the analyte of interest, but many background 'contaminants'. These three challenges are: 1. How can large surface stress signals be achieved? Any contaminants on the surface will reduce the available surface area and therefore lead to smaller surface stress change. We will present an electrochemical cleaning protocol that results in a clean gold surface, leading to a large and quantitatively reproducible surface stress signal when surface charge densities change. 2. How can the signal-to-noise ratio be improved? An optimal probe density is required for good signal-to-noise ratios of the surface stress change. This is achieved by using a multi-step functionalization protocol recently described by Nagai et al. [15]. 3. How can sensors with long term stability and realistic shelf life be manufactured? For device applications one needs to know how the sensors perform during long term measurements in the analyte of interest. We will present long term measurements of our cantilever platform in solution and discuss our observations.

Results and discussion
Electrochemical cleaning

We have tested different cleaning procedures by quantifying surface cleanliness electrochemically *in situ* and using surface science techniques *ex situ*. Based on results

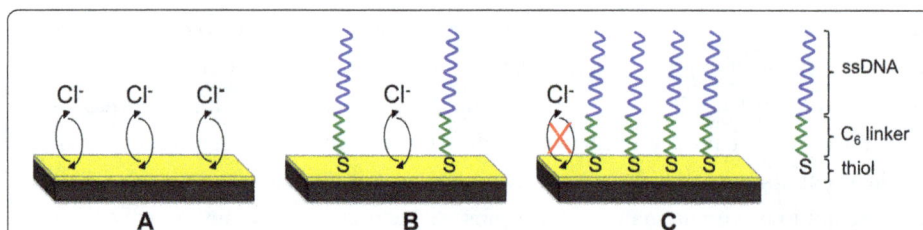

Figure 1 Potential induced adsorption/desorption of chloride ions on different functionalized gold surfaces. Schematics describe the relationship between surface stress change and available surface area. **A**, a bare and clean gold surface is shown. Chloride ions can freely interact with the whole surface and large surface stress changes are measured. Upon binding of some single-stranded oligonucleotide to the surface **B**, fewer ions can react with the surface leading to a smaller surface stress change. Once the layer is densely packed on the surface **C**, the surface stress change vanishes as no ions can reach the surface anymore.

published by Fischer et al. [21], we tested the following cleaning protocols: 1. Electrochemical sweep of the gold coated cantilever in 50 mM potassium hydroxide (KOH) from −0.2 to 1.2 V (vs. Ag/AgCl (sat. KCl)), 2. Electrochemical sweep in 50 mM potassium perchlorate ($KClO_4$) from −0.8 to 1.4 V (vs. Ag/AgCl (sat. KCl)) and 3. Piranha solution (Three parts concentrated sulfuric acid (H_2SO_4) and one part hydrogen peroxide (H_2O_2); note that great caution is necessary when using piranha solution) treatment of the cantilever for 5 min. We find that the $KClO_4$ and KOH- mediated processes result in the cleanest surface as monitored using *ex situ* X-ray photoelectron spectroscopy (XPS) and *in situ* cyclic voltammetry. The atomic percent surface composition measured from XPS results of the survey scan for the piranha cleaning method results in 40.2% gold, 33.6% carbon and 26.1% oxygen. A huge improvement of these value, i.e. higher gold percentage can be seen for the KOH sweep as well as the $KClO_4$ sweep. For KOH the composition is 65.3% gold, 30.0% carbon and 4.74% oxygen, which is very comparable to the values for the $KClO_4$ sweep with 61.7% gold, 33.8% carbon and 4.5% oxygen. The $KClO_4$ sweep is chosen to be the primary cleaning step for all further experiments, as this is a standard media for electrochemical cleaning of gold. Additionally, perchlorate has a very small affinity for gold and will not adsorb onto the gold surface.

In detail, with this method the cantilever is electrochemically cleaned in 50 mM $KClO_4$ by sweeping the applied potential in solution from −0.8 to 1.4 V (vs. Ag/AgCl (sat. KCl)). Sweeping the potential from −0.8 to 1.4 V oxidizes the gold surface, whereas the reverse step, going from 1.4 to −0.8 V reduces the resulting gold oxide. This cleaning step serves to intrinsically remove contaminants from the surface. The cyclic voltammogram (current vs. potential, CV) scan is performed at 20 mV/sec and is continuously repeated until a reproducible CV of gold is achieved, indicating the removal of any contaminates. In Figure 2A, a CV from a bare clean gold evaporated cantilever is shown. Multiple peaks that overlap are observed between 0.9-1.2 V, which correspond to the oxidation of gold. A significant sharp reduction peak is observed at 0.35 V. Figure 2B shows a cleaning process in action. The CV spectrum has two additional peaks at 0.35 V and 0.15 V consistent with chloride on the gold surface. Over the course of six full potential cycles, the chloride redox peaks vanish, leaving a clean bare gold surface.

To further verify the effectiveness of the chosen cleaning protocol, X-ray photoelectron spectroscopy (K-Alpha X-Ray Photoelectron Spectrometer system, Thermo Scientific, USA) was performed on evaporated gold surface on a silicon substrate. In Figure 3, the individual high-resolution spectra of Au4f, C1s and O1s are shown for two different samples: (A) gold sample that has not been cleaned, and (B) an electrochemically cleaned gold sample. Both samples are made from a piece of a silicon wafer with a thermally evaporated 2 nm titanium adhesion layer followed by thermal deposition of 100 nm gold. The samples were stored under ambient condition for 1 week. Prior to the experiment, one sample is rinsed with MilliQ water and blow-dried using a nitrogen stream (uncleaned), the other sample is electrochemically cleaned using 50 mM $KClO_4$ and dried with nitrogen (cleaned). The atomic percent surface composition of the uncleaned sample is measured to be 43.7% gold, 41.2% carbon and 9.4% oxygen based on the survey scan. In Figure 3B, the electrochemically cleaned sample is shown. The time between the cleaning and measuring the sample with XPS was about

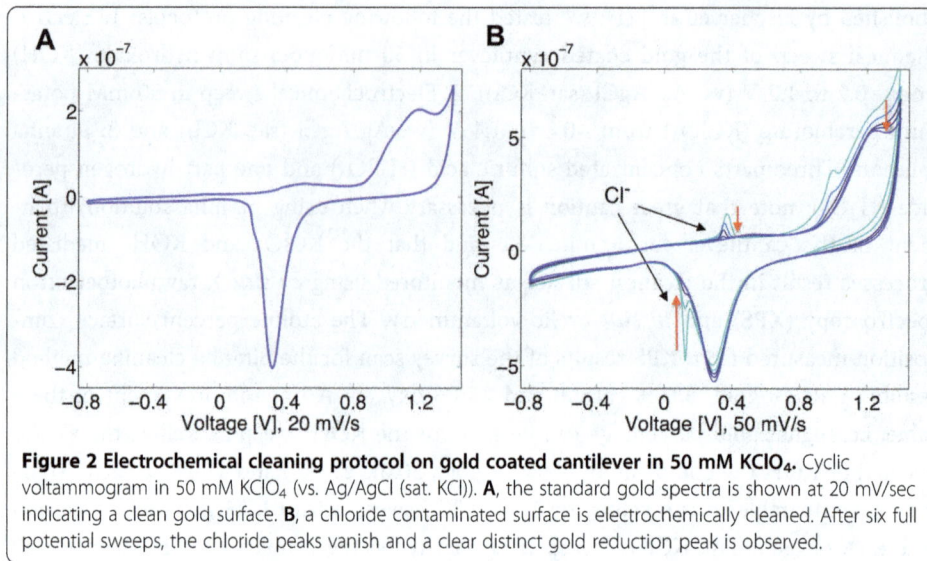

Figure 2 Electrochemical cleaning protocol on gold coated cantilever in 50 mM KClO₄. Cyclic voltammogram in 50 mM KClO₄ (vs. Ag/AgCl (sat. KCl)). **A**, the standard gold spectra is shown at 20 mV/sec indicating a clean gold surface. **B**, a chloride contaminated surface is electrochemically cleaned. After six full potential sweeps, the chloride peaks vanish and a clear distinct gold reduction peak is observed.

30 min. The overall composition of the surface was 61.7% gold, 33.8% carbon and 4.5% oxygen. Compared to the dirty sample, a relative increase of about 40% was observed for the gold peak and a relative decrease of 20% and 50% for the carbon and oxygen peak contamination was observed. Higher measured levels of gold on the surface means there is less contamination surface, demonstrating the effectiveness of the electrochemical cleaning protocol. The remaining oxygen and carbon peaks result from exposing the sample to air for 30 min prior to measuring the surface composition and cannot be avoided. This was verified by sputter cleaning a gold sample in UHV until a clean Auger spectrum was acquired, then exposing it to air for 20 minutes. The surface

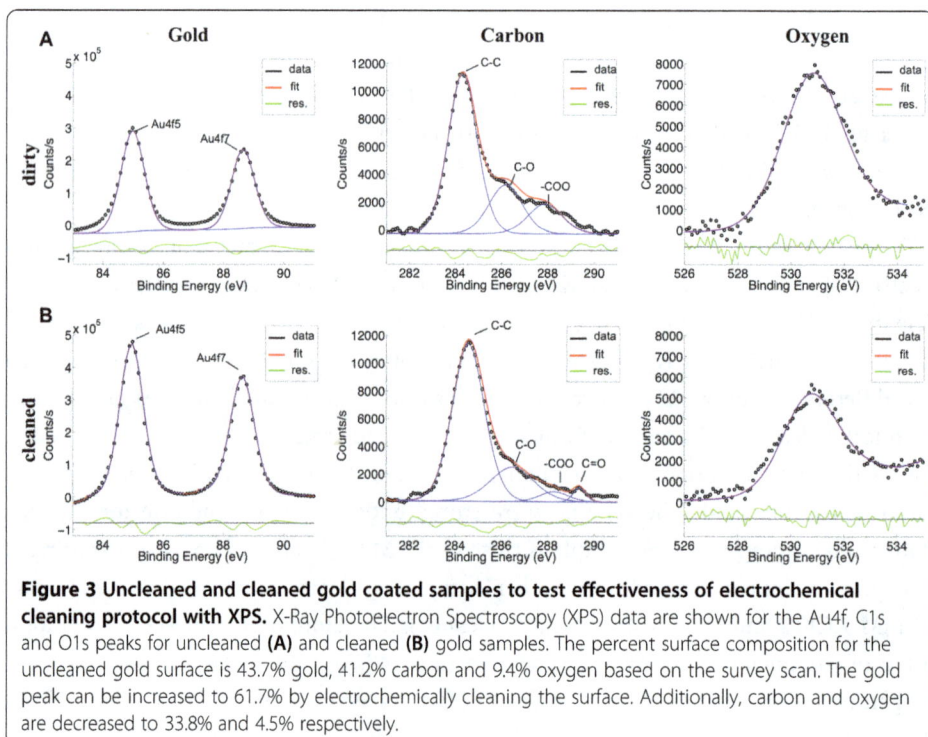

Figure 3 Uncleaned and cleaned gold coated samples to test effectiveness of electrochemical cleaning protocol with XPS. X-Ray Photoelectron Spectroscopy (XPS) data are shown for the Au4f, C1s and O1s peaks for uncleaned **(A)** and cleaned **(B)** gold samples. The percent surface composition for the uncleaned gold surface is 43.7% gold, 41.2% carbon and 9.4% oxygen based on the survey scan. The gold peak can be increased to 61.7% by electrochemically cleaning the surface. Additionally, carbon and oxygen are decreased to 33.8% and 4.5% respectively.

composition measured by the survey scan resulted in 66.1% gold, 31.8% carbon and 2.2% oxygen.

An important feature of the intrinsic electrochemical cleaning protocol is the ability to revert the sensor surface to its base state in situ by removing the oligonucleotide functionalization layer. A cantilever that is functionalized with 10 μM 25-mer thiolated oligonucleotide was measured with XPS and the % at composition for gold, nitrogen and phosphorus is 11.1%, 11.64% and 5.97%. A clear phosphorus peak in the XPS spectra indicates successful oligonucleotide functionalization, see Figure 4A. Subsequently, a sample functionalized under the same conditions is electrochemically cleaned with 50 mM KClO$_4$ to remove the oligonucleotide. XPS of the oligonucleotide functionalized and electrochemically cleaned sample is shown in Figure 4B. The % at composition for gold, nitrogen and phosphorus is measured as 45.86%, 9.73% and 0% (not measurable). The clear removal of the phosphorous peak indicated the removal of the oligonucleotide functionalization layer from the surface. Subsequently, a more thorough electrochemical cleaning can be done.

Surface stress measurements

A large signal-to-noise ratio can be achieved by precisely controlling the probe density on the surface. If the surface is completely covered with the molecule of interest, ions cannot interact with the surface, and only very small surface stress changes are measured. Additionally, a complete coverage of the surface by probe molecules is detrimental to a fast response time of the sensor, as many target molecules will not be able to interact with the probe molecules. Therefore, it is crucial to achieve an optimal probe density to maximize the signal-to-noise ratio of the measurements. Our group has developed a multistep oligonucleotide functionalization protocol that enables for systematic control of the functionalization density and thus leads to high quality sensor functionalization with good reproducibility's. After the electrochemical cantilever

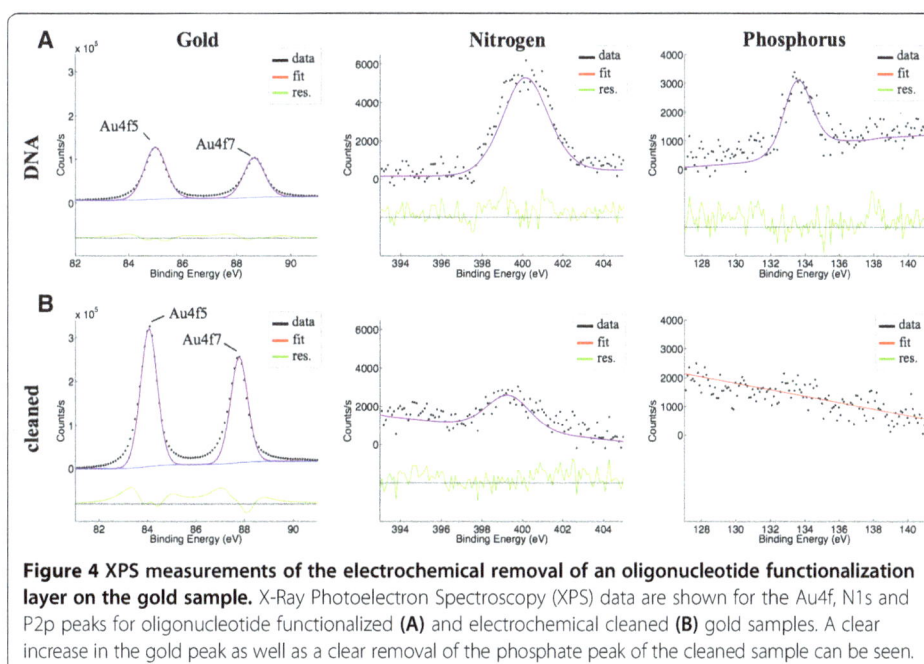

Figure 4 XPS measurements of the electrochemical removal of an oligonucleotide functionalization layer on the gold sample. X-Ray Photoelectron Spectroscopy (XPS) data are shown for the Au4f, N1s and P2p peaks for oligonucleotide functionalized **(A)** and electrochemical cleaned **(B)** gold samples. A clear increase in the gold peak as well as a clear removal of the phosphate peak of the cleaned sample can be seen.

cleaning described above, a surface stress pattern is recorded for 30 min during application of a square wave potential between +/− 200 mV with a 10 min period. Afterwards, the cantilever is incubated in a 10 μM thiolated single stranded oligonucleotide solution for 5 min and another surface stress pattern is recorded. This step is repeated until the desired coverage is achieved. This can be controlled by first monitoring the decrease in surface stress amplitude due to the increased probe coverage and therefore a decreased availability in gold area and then evaluating the surface stress pattern change due to comparative adsorption of chloride ions in solution and the negatively charged oligonucleotide phosphate backbone [22]. The effective density of the oligonucleotide layer can be determined by using 12-ferrocenyl-1-dodecanethiol ($Fc(CH_2)_{12}SH$) to label unfunctionalized areas of the gold surface. The net area associated with unfunctionalized gold is determined from the integrated area of the electrochemically active ferrocene label. This process was previously shown by Nagai et al. [15], details are described below.

From an applications point of view achieving a reproducible sensor response is highly desirable. In our system this translates into the necessity of achieving a reproducible probe surface coverage. The surface probe density can be characterized by measuring the chloride-induced stress changes of the cantilever (all experiments are performed in Tris–HCl 10 mM NaCl 50 mM pH 7.4 buffer (TN buffer)). To drive adsorption and desorption of chloride ions to the cantilever gold surface, a square-wave potential is switched between −200 and +200 mV, with a 10 min period. As a result of the square wave potential, the cantilever will undergo characteristic bending due to the induced surface stress change. In Figure 5, the surface stress change patterns for a gold surface that is clean (blue), partially functionalized with single stranded thiolated oligonucleotide (red) and 6-mercapto-1-hexanol (MCH) (green) versus time are shown. These three cases demonstrate the relationship between surface stress change and available gold surface area very well. The surface stress change for clean gold results in a large

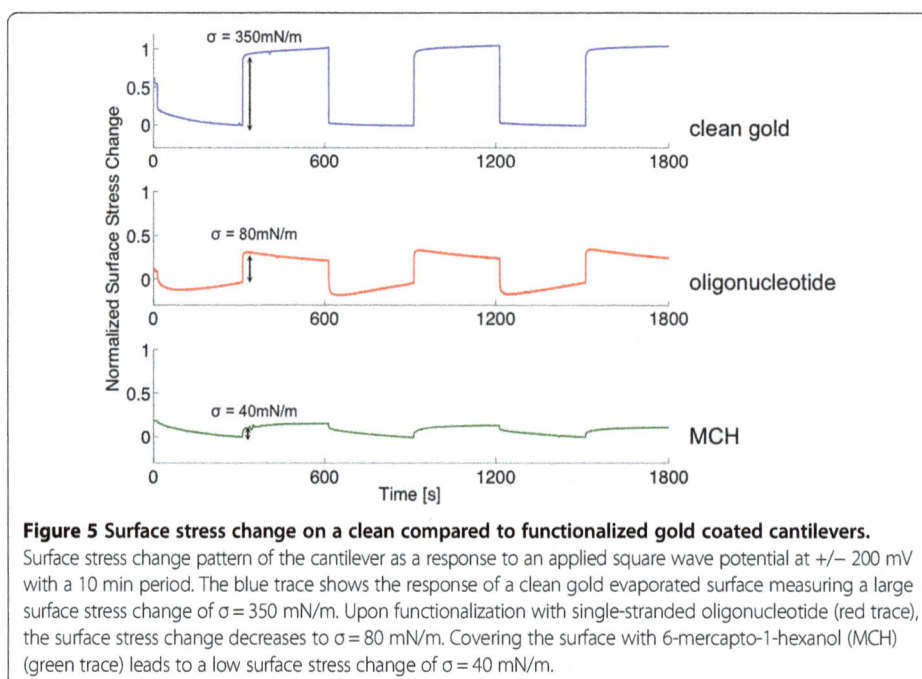

Figure 5 Surface stress change on a clean compared to functionalized gold coated cantilevers.
Surface stress change pattern of the cantilever as a response to an applied square wave potential at +/− 200 mV with a 10 min period. The blue trace shows the response of a clean gold evaporated surface measuring a large surface stress change of σ = 350 mN/m. Upon functionalization with single-stranded oligonucleotide (red trace), the surface stress change decreases to σ = 80 mN/m. Covering the surface with 6-mercapto-1-hexanol (MCH) (green trace) leads to a low surface stress change of σ = 40 mN/m.

signal with an amplitude of $\sigma = 350$ mN/m. Furthermore, it shows a response pattern that is in phase and similar in shape to the applied square wave potential. This is because chloride ions are essentially immediately driven to/from the surface; surface charge density (leading to a change in surface stress) thus follows the profile of the applied potential.

Upon functionalization of the cantilever gold surface with 25-mer thiolated single stranded oligonucleotide with a packing density of roughly 9% as described in [15], the surface stress amplitude decreases and the response pattern starts to deviate from that of the applied square potential. The trace shows an upward slope trend for regions where +200 mV is applied and a downward slope for regions where −200 mV is applied. The amplitude decreases from 350 mN/m to 80 mN/m compared to the clean gold surface. This supports the fact that if less gold surface is available for the chloride ions to adsorb to, the smaller the surface stress change will be. The oligonucleotide covers a part of the surface and makes it less accessible for chloride ions to adsorb/desorb at the applied potentials. The change in shape of the response curve is due to changes in the structure of single-stranded oligonucleotide when potential is applied and results from interactions between the charged oligonucleotide phosphate backbone and the gold surface and hydration shell dynamics. Quantitative modeling of these various phenomena and their interplay is presently being investigated. What is clear at this stage is that the detailed shape of the stress response curve allows determination of the probe surface coverage.

To further demonstrate this principle, another sample was functionalized with MCH, a short thiolated C_6 linker for 5 min. MCH binds strongly to the gold surface resulting in a densely packed layer. Ion adsorption is blocked and the capacitance of the electrode is reduced [23]. The surface stress change pattern shows an even stronger deviation from the shape of the applied potential and a further decrease in stress amplitude to values such as 40 mN/m. Note that the pattern change provides a potentially more robust signal to detect hybridization than the amplitude, which can vary depending on the number and type of defects in the self-assembled monolayer [24].

In summary, we point out two key observations. This first is that the amplitude of the surface stress change at the switching potential decreases, the more the gold surface is blocked by any molecules, allowing fewer chloride ions to ad-/desorb onto the surface. The absolute value of the amplitude is a function of the initial gold cleanliness. Contaminants in the solution that competitively bind to the clean gold surface (e.g. bromide in a chloride solution) and potential-induced conformational changes in the probe molecules also affect the signal amplitude. Reproducible large absolute signal values can be achieved by suitable gold cleaning protocols as described above. Note that the average absolute surface stress value is 280 mN with a reproducibility of 40%.

The second key point is that the surface stress pattern is characteristic of the nature of the surface bound molecules (either of the probe functionalization layer or the probe-target complex). The former can be used to (re-)generate well defined probe functionalization layers *in situ*. The latter allows for the determination of the presence of target molecules as recently demonstrated by Nagai et al. [15], who reported on the pattern shape change due to oligonucleotide hybridization and aptamer-protein interactions of optimized sensing layers. The change in the pattern can be attributed to the negatively charged phosphate backbone of the DNA and mechanical property changes upon hybridization. A negative applied potential will repel the DNA from the surface leaving the DNA in a standing up position [25]. Over time, a double layer will build up that screens the DNA charge which results in a relaxation of the DNA into its neutral state. This is reflected in the slope change in the surface stress change pattern observed at −200 mV. At positive potentials, the DNA is attracted to the gold surface and is lying down. The relaxation of this position is visible in the slope change of the surface stress change pattern. In passing, we note that by varying the temporal period of the applied potential, conformation dynamics can be studied, potentially allowing label free fundamental experimental insights into important topics such as protein folding.

Long-term measurements

An important question is how long the sensor response remains stable. Because cantilever based sensors are very sensitive, their response is expected to drift and change as a function of time, concentration of contaminants, etc. We have performed long-term stability recordings in the TN buffer used for our oligonucleotide and aptamer protein measurements. The cantilever was electrochemically cleaned in 50 mM $KClO_4$, rinsed with MilliQ water and placed in buffer. No oligonucleotide functionalization was performed. A square wave potential between +/− 200 mV with a period of 10 min is applied to the cantilever for 14 hrs (84 cycles). The surface stress change of the cantilever was recorded over time, as shown in Figure 6. Overall, the surface stress amplitude decreases from 150 mN/m to 100 mN/m after 4 hrs, to 60 mN/m after 10 hrs, and finally less than 50 mN/m after 14 hrs. Additionally, a pattern change is observed. The slope of the pattern at +200 mV changes from a positive to a negative trend after 8 hrs. A similar change is observed for the slope at −200 mV changing from a negative to a positive trend after 11 hrs. A zoom into three different regions of the curve after 1, 9.5 and 13 hrs is shown in the Figure 6. The overall decrease in amplitude is attributed to chemisorbed ions on the surface leading to a decrease in the available gold area. A CV was recorded in 50 mM $KClO_4$ before and after the long-term measurement. The charge increases by 85%, indicating an increase in the capacitance, assuming that the active area has remained constant. The experiment starts by applying a potential

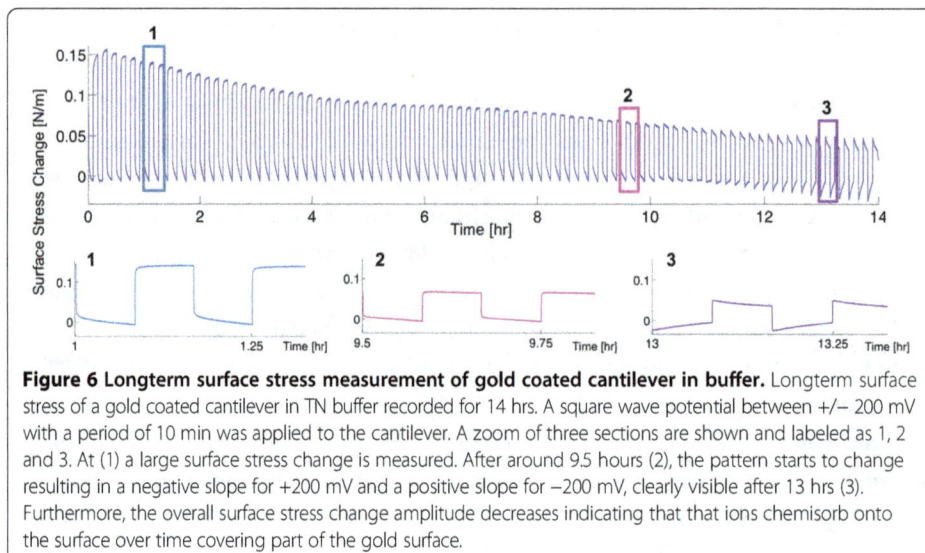

Figure 6 Longterm surface stress measurement of gold coated cantilever in buffer. Longterm surface stress of a gold coated cantilever in TN buffer recorded for 14 hrs. A square wave potential between +/− 200 mV with a period of 10 min was applied to the cantilever. A zoom of three sections are shown and labeled as 1, 2 and 3. At (1) a large surface stress change is measured. After around 9.5 hours (2), the pattern starts to change resulting in a negative slope for +200 mV and a positive slope for −200 mV, clearly visible after 13 hrs (3). Furthermore, the overall surface stress change amplitude decreases indicating that that ions chemisorb onto the surface over time covering part of the gold surface.

of −200 mV and ends 14 hours later at +200 mV after 84 cycles. Generally, one observes that the sensor remains stable for 10 hrs in TN buffer before competitive chemisorption takes place leading to an unreliable measurement. This is likely a function of the purity of the buffer components and water used to prepare the buffer solution. With the intrinsic electrochemical cleaning set-up, the cantilever can be restored to its original state, i.e. a clean gold surface *in situ*.

Conclusion

We have addressed and discussed three issues that are fundamental to successful cantilever biosensor integration and relevant for many other sensor platforms and applications. First, in sensors where the signal relies on surface charge changes such as in chemFETS or cantilever based sensors, a clean chemical functionalization layer support surface is crucial in order to obtain large signals. Here we report an electrochemical cleaning method of the gold surface often used to support the thiolated probe molecules by sweeping a potential that is applied to the sensor surface between −0.8 and 1.4 V in 50 mM $KClO_4$ until a reproducible cyclic voltammogram is obtained. XPS data verifies the effectiveness of this cleaning method. The advantage of this method is that the sensor can be cleaned intrinsically without the use of any harsh chemicals that might harm the sensor integration environment. This cleaning step will remove the functionalization layer of the sensor restoring it to its original state. Second, we demonstrate how to achieve a high signal-to-noise ratio by carefully controlling the probe coverage of the sensor. A multistep functionalization protocol is described relying on characteristic changes in the stress response as a function of probe density to *in situ* electrochemical stimulation. This systematically provides a higher quality layer and a better control of the surface coverage, leading to higher signal-to-noise ratios and to a reproducible, predictable sensor response. Surface stress change measurements on a clean gold surface, oligonucleotide and MCH modified gold cantilever surfaces are described. In all experiments described here, a square wave potential between +/− 200 mV is applied to the cantilever in TN buffer. These experiments confirm that the surface stress change is proportional to the available gold area. Additionally, the surface stress change pattern gives detailed information about conformational changes on the surface upon applying a potential. Lastly, long term stability measurements are shown in buffer indicating the sensor lifetime to be about 10 hrs. The origin of this limitation is currently being investigated. After 10 hrs, a electrochemical cleaning step is necessary to recover the surface to its initial high sensitivity state.

Methods
Oligonucleotide preparation
All experiments were performed using a 25-mer thiolated single stranded oligonucleotide with a sequence of 5′-HS-SC6- TCGGATCTCACAGAATGGGATGGGC-3′ (by IDT Technology, USA). The stock oligonucleotide solution was prepared by diluting to a concentration of 100 μM by in 40 μl of TE buffer (Tris–HCl 10 mM, 5 mM EDTA, pH 7.4). Prior to each experiment, the oligonucleotide is desalted by incubating in 25 mM TCEP (Fisher Scientific, USA) for 1 hr followed by a subsequent purification step using a NAP-5 column (GE Healthcare, UK). The desalting step breaks of the

disulfide bond of the oligonucleotide making it reactive to the gold surface. The final oligonucleotide concentration for all experiment was 10 μM.

Cantilever preparation

Silicon cantilevers (CSC38/tipless/no Al-coating, Mikromash) are solvent cleaned with acetone, isopropanol and methanol, the cantilevers before thermal evaporation. A 2 nm thick titanium adhesion layer is evaporated onto the cantilever with a rate of 0.9 Å/s followed by 100 nm of gold at a rate of 1 Å/s at a pressure of $< 3 \times 10^{-6}$ mBar and stored under ambient condition before use. To define the gold area that is exposed to the elec-trochemical set-up, a thin layer of apiezon wax (Apiezon wax W, APWK, USA) that is dissolved in trichloroethylene (TCE) (Fisher Scientific, USA) is applied to the base of the cantilever leaving an area of 1.0 mm^2 exposed.

Electrochemical cleaning

Argon is injected into potassium perchlorate (50 mM $KClO_4$) to remove any oxygen in solution. Subsequently, prior to each experiment the cantilever is electrochemically cleaned in 50 mM $KClO_4$ (Fisher Scientific, USA) by cycling the potential between –0.8 to 1.4 V at 20 mV/sec until a repeatable gold cyclic voltammogram peak is observed (CHI 1000, CH Instruments, USA). The cantilever is set up as the working electrode, a platinum wire (1 mm thick, Alfa Aesar, USA) is used as the counter electrode and a standard Ag/AgCl (sat. KCl) reference electrode (BASi, USA).

Abbreviations

UHV: Ultra high vacuum; DNA: Deoxyribonucleic acid; CV: Cyclic voltammetry; TN buffer: Tris–HCl 10 mM NaCl 50 mM pH 7.4 tris(hydroxymethyl)aminomethane hydrogen chloride natrium chloride; XPS: X-ray photoelectron spectroscopy; SAM: Self-assembled monolayer; MCH: 6-mercapto-1-hexanol.

Competing interests

The authors declare that they have no competing interests.

Authors' contributions

ALH carried out the experiments, data analysis and wrote the manuscript. YN prepared the oligonucleotide sample and buffer solutions. The experiments were supervised and directed by RBL and PG. All authors have read, approved and provided critical revisions for the final manuscript.

Acknowledgments

The authors would like to thank Robert Gagnon and Yoichi Miyahara for numerous helpful suggestions and technical support as well as Robert Sladek for fruitful discussions on oligonucleotide functionalization. YN thanks Genome Canada and Genome Quebec for financial support. This work was supported by Fonds de recherche du Quebec (FQRNT), Canadian Institutes of Health Research (CIHR) and the NSERC-CREATE Integrated Sensor Systems Training Program.

Author details

[1]Department of Physics, McGill University, 3600 Rue University, Montreal, QC H3A 2T8, Canada. [2]Research Institute of the McGill University Health Centre, 2155 Guy Street, Montreal, QC H3H 2R9, Canada. [3]Department of Chemistry and FQRNT Centre for Self Assembled Chemical Structures, McGill University, 801 Sherbrooke Street West, Montreal QC H3A 2K6, Canada.

References

1. Fritz J. Cantilever biosensors. Analyst. 2008;133:855–63.
2. Braun T, Ghatkesar M, Backmann N. Quantitative time-resolved measurement of membrane protein – ligand interactions using microcantilever array sensors. Nat Nanotechnol. 2009;4:179–85.
3. Lang H, Hegner M, Gerber C. Cantilever array sensors. Material Today. 2005;8:30–6.
4. Godin M, Tabard-Cossa V, Grütter P, Williams P. Quantitative surface stress measurements using a microcantilever. Appl Phys Lett. 2001;79:551–3.
5. Hanay MS, Kelber S, Naik AK, Chi D, Hentz S, Bullard EC, et al. Single-protein nanomechanical mass spectrometry in real time. Nat Nanotechnol. 2012;7:602–8.

6.   Svitelskiy O, Sauer V, Vick D, Cheng K-M, Liu N, Freeman MR, et al. Nanoelectromechanical devices in a fluidic environment. Phys Rev E Stat Nonlin Soft Matter Phys. 2012;85:056313.

7.   Ding L, Baker C, Senellart P, Lemaitre A, Ducci S, Leo G, et al. High Frequency GaAs Nano-optomechanical disk resonator. Phys Rev Lett. 2010;105:263903.

8.   Berger R. Surface Stress in the Self-Assembly of Alkanethiols on Gold. Science. 1997;276:2021–4.

9.   Godin M, Tabard-Cossa V, Miyahara Y, Monga T, Williams PJ, Beaulieu LY, et al. Cantilever-based sensing: the origin of surface stress and optimization strategies. Nanotechnology. 2010;21:75501.

10. Yang YT, Callegari C, Feng XL, Ekinci KL, Roukes ML. Zeptogram-scale nanomechanical mass sensing. Nano Lett. 2006;6:583–6.

11. Jensen K, Kim K, Zettl A. An atomic-resolution nanomechanical mass sensor. Nat Nanotechnol. 2008;3:533–7.

12. Boisen A, Thundat T. Design & fabrication of cantilever array biosensors. Material Today. 2009;12:32–8.

13. Tabard-Cossa V, Godin M. Microcantilever-based sensors: effect of morphology, adhesion, and cleanliness of the sensing surface on surface stress. Anal Chem. 2007;79:8136–43.

14. Haiss W, Nichols RJ, Sass JK, Charle KP. Linear correlation between surface stress and surface charge in anion adsorption on Au(111). J Electroanal Chem. 1998;452:199–202.

15. Nagai Y, Carbajal JD, White JH, Sladek R, Grutter P, Lennox RB. An electrochemically controlled microcantilever biosensor. Langmuir. 2013;29:9951–7.

16. Bard A, Faulkner L. Electrochemical methods. New York: John Wiley & Sons, Inc; 2001.

17. Lipkowski J, Shi Z, Chen A, Pettinger B, Bilger C. Ionic adsorption at the Au(111) electrode. Electrochim Acta. 1998;43:2875–88.

18. Stoney GG. The tension of metallic films deposited by electrolysis. Proc R Soc Lond A Math Phys Sci. 1909;82:172–5.

19. Beaulieu LY, Godin M, Laroche O, Tabard-Cossa V, Grütter P. A complete analysis of the laser beam deflection systems used in cantilever-based systems. Ultramicroscopy. 2007;107:422–30.

20. Beaulieu LY, Godin M, Laroche O, Tabard-Cossa V, Grütter P. Calibrating laser beam deflection systems for use in atomic force microscopes and cantilever sensors. Appl Phys Lett. 2006;88:083108.

21. Fischer LM, Tenje M, Heiskanen AR, Masuda N, Castillo J, Bentien A, et al. Gold cleaning methods for electrochemical detection applications. Microelectron Eng. 2009;86:1282–5.

22. Lee LYS, Lennox RB. Electrochemical desorption of n-alkylthiol SAMs on polycrystalline gold: studies using a ferrocenylalkylthiol probe. Langmuir. 2007;23:292–6.

23. Steel AB, Herne TM, Tarlov MJ. Electrochemical quantitation of DNA immobilized on gold. Anal Chem. 1998;70:4670–7.

24. Norman LL, Badia A. Microcantilevers modified with ferrocene-terminated self-assembled monolayers: effect of molecular structure and electrolyte anion on the redox-induced surface stress. J Phys Chem C. 2011;115:1985–95.

25. Gorodetsky A, Buzzeo MC, Barton JK. DNA-mediated electrochemistry. Bioconjug Chem. 2008;19:2285–96.

9

Real-time gas identification on mobile platforms using a nanomechanical membrane-type surface stress sensor

Ricardo Jose S Guerrero[1,2†], Francis Nguyen[1,2†] and Genki Yoshikawa[1*]

*Correspondence:
YOSHIKAWA.Genki@nims.go.jp
†Equal contributors
[1] World Premier International (WPI)
Research Center Initiative,
International Center for Materials
Nanoarchitectonics (MANA),
National Institute for Materials
Science (NIMS), 1-1 Namiki,
305-0044 Tsukuba, Japan
Full list of author information is
available at the end of the article

Abstract

Here we show real-time multiple gas identification on a mobile platform through the use of an array of nanomechanical membrane-type surface stress sensors (MSS). Commercially available hardware is used to integrate the MSS array into a portable unit with wireless capability. This unit transmits data to a consumer mobile tablet where data is displayed and processed in real-time. To achieve real-time processing with the limited computational power of commercial mobile hardware, a machine learning algorithm known as Random Forest is implemented. We demonstrate the real-time identification capability of the device by measuring the vapours of *water*, *ethanol*, *isopropanol*, and *ambient air*.

Keywords: Random forest; MSS; Piezoresistive; Gas identification; Android; Mobile phone

Correspondence

Hardware implementation

The basis of our platform is a membrane-type surface stress sensor (MSS) that involves piezoresistors laid out in a full Wheatstone bridge configuration as described in previous papers [1,2] (Figure 1a). The MSS has a theoretical detection limit of ~0.1 mN/m. By varying the properties of the coated polymer over each individual membrane such as its hydrophobicity and functional groups, differences in polymer-gas affinity can be utilized for identification of a wide range of VOC analytes.

Two piezoelectric micropumps (Bartels Microtechnik mp6) flowed sample gases and ambient air over the MSS chip near their maximum rate of 0.3 mL/s [3]. A commercially available analog to digital converter with a resolution of 632 nV (ADS1258 EVM) measured the differential output voltage of the MSS with a bias voltage of -1.0 V. An Arduino Mega 2560 received this data via a Serial Peripheral Interface (SPI) to the analog-to-digital converter. The Arduino Mega 2560 also controlled the micropump switching. A custom breakout board mounted the MSS chip, which was then encased in a 3D-printed enclosure designed to maximize gas flow over the polymer receptor layers (Figure 1b). The Arduino Mega then sent the data to a consumer tablet (Google Nexus 7) over WiFi, using an Arduino WiFi shield.

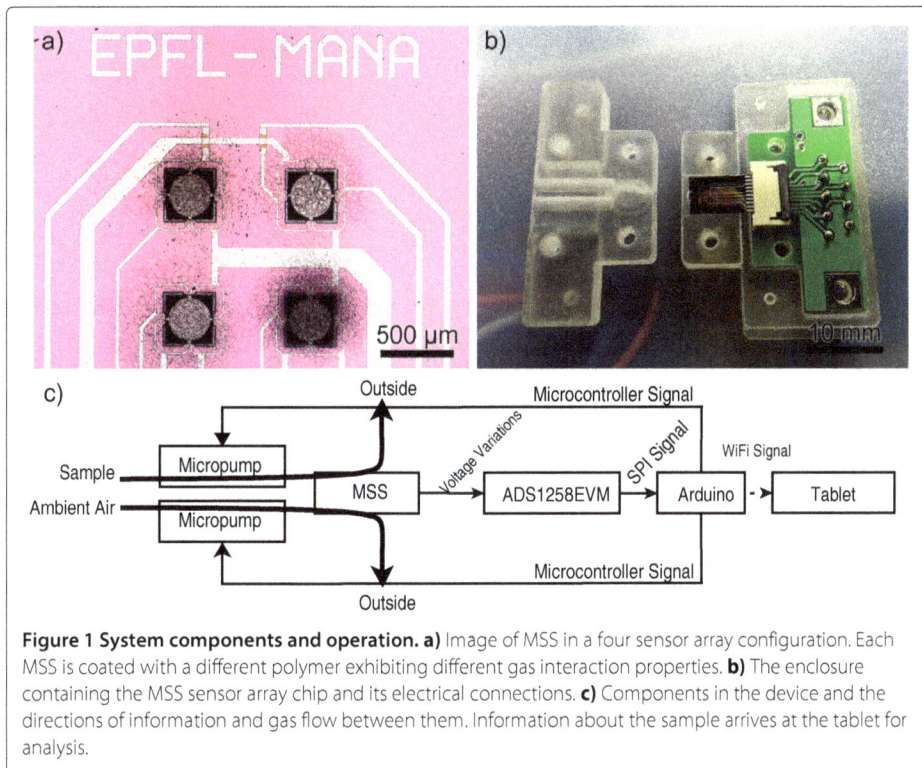

Figure 1 System components and operation. a) Image of MSS in a four sensor array configuration. Each MSS is coated with a different polymer exhibiting different gas interaction properties. **b)** The enclosure containing the MSS sensor array chip and its electrical connections. **c)** Components in the device and the directions of information and gas flow between them. Information about the sample arrives at the tablet for analysis.

Data processing

Random Forests [4] allow short characterization times of arbitrary input; characterization time is tunable through the size of the *Forest*. Each *Forest* can be tailored to complete its task on hardware of arbitrary speeds while maintaining a real-time analysis. Once generated offline, this machine learning algorithm can be moved to a target platform for quick, real-time analysis. Classification of data using a Random Forest simply involves traversal of many decision *trees*, which can be multithreaded easily for fast computation on multicore processors. While this approach is sometimes coupled with Principle Component Analysis (PCA) to determine better candidates for predictors [5], the device is capable of identifying the chosen samples without requiring the full dataset in contrast to PCA. Voltage variations as a result of sample flowing through the device form unique curves when measured over time. These curves have several identifying characteristics, which can be extracted quickly by splitting the input into several windows, obtaining the difference of their averages, and using these as predictors for the Random Forest analysis.

The Random Forest was trained on sample data collected with the device using Scikit-learn [6] with Python 3.2. Converting the *Forest* into a custom, portable file format allows a consumer handheld tablet to predict outcomes with the CPU to be the only limiting factor in prediction speed.

Verification

The device was verified by identifying between ethanol, isopropanol, water, and ambient air. A training set consisting of 196 sample/purge cycles (48 cycles per sample type, and 52 for ambient air) was used to train a 100 tree Random Forest. Though the ADS1258 measured at a rate of 460 samples per second (SPS) per channel, data transmission was

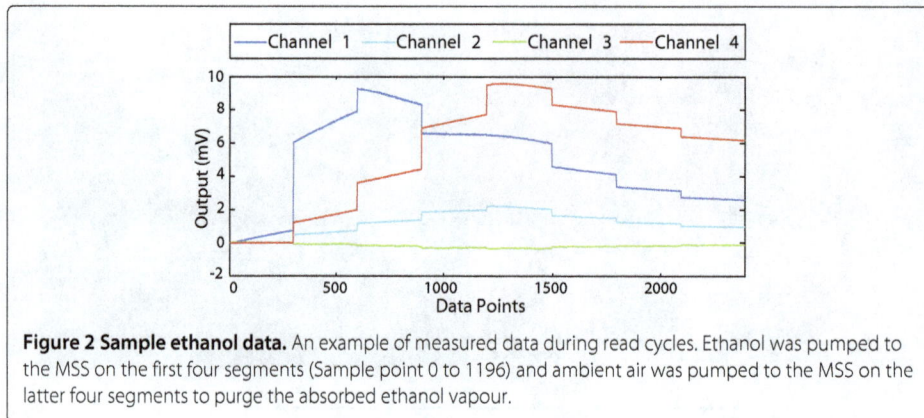

Figure 2 Sample ethanol data. An example of measured data during read cycles. Ethanol was pumped to the MSS on the first four segments (Sample point 0 to 1196) and ambient air was pumped to the MSS on the latter four segments to purge the absorbed ethanol vapour.

limited to 4800 bytes every 2 seconds (approximately 400 samples per second per channel) due to problems inherent within the Arduino WiFi module [7]. This caused the signal to appear discontinuous (Figure 2).

The training data was found to be easy to differentiate as a series of slopes in a voltage vs time series. Since the signal was divided evenly into eight segments per purge/sample cycle, the slopes between the averages of each segment indicated the general trend of the curve. Using these slopes as predictors for the Random Forest allowed the algorithm to identify sample gases with a high degree of accuracy (Additional files 1, 2).

The data classification technique traditionally performs well at avoiding overfitting of training data [4]. Using the out-of-bag estimator built-in to our Random Forest creation library, the *Forest* scored approximately 0.939, representing a 93.9% accuracy in predicting outcomes from inputs for which the *Forest* was not trained. The training set used to grow

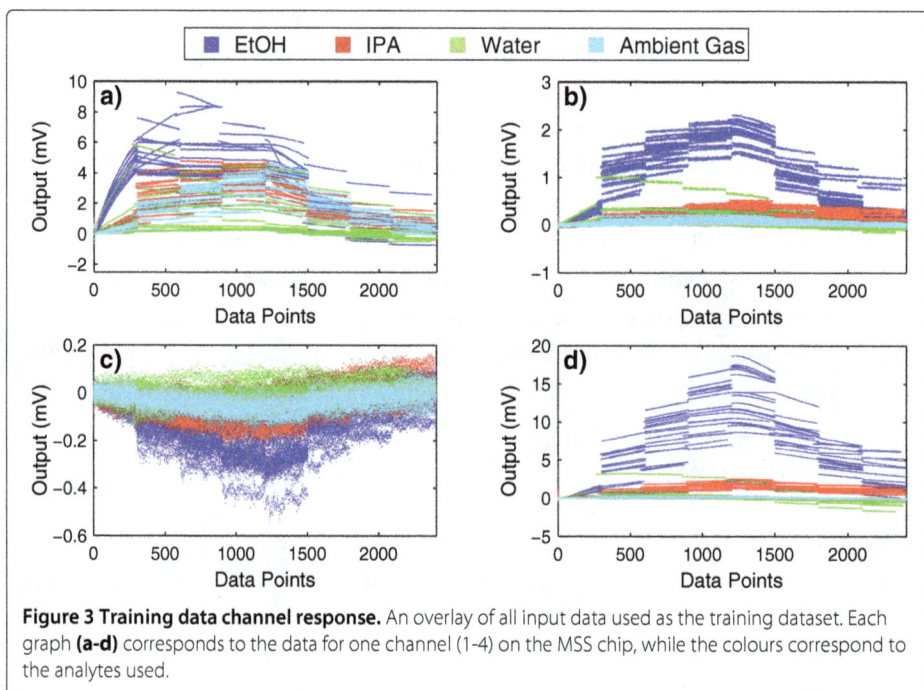

Figure 3 Training data channel response. An overlay of all input data used as the training dataset. Each graph **(a-d)** corresponds to the data for one channel (1-4) on the MSS chip, while the colours correspond to the analytes used.

the Random Forest had a high amount of variability in its input (Figure 3), forcing the *Forest* to identify a corresponding sample from a wide variety of input data.

Conclusion

We have demonstrated that the combination of an advanced algorithm (Random Forest) and the optimized nanomechanical sensor (MSS) can achieve real-time gas identification with commerical off-the-shelf hardware. Since the peripheral electronic components can be miniaturized by the introduction of application specific integrated circuits (ASIC) or field programmable gate arrays (FPGA), the present demonstration indicates the feasibility of integrating a real-time nanomechanical olfactory system into virtually any type of mobile platforms such as smartphones. Future developments towards real world applications will include a larger dataset with proper selection of parameters from output signals, effective receptor layers, and optimization of system components including the chamber and pumps.

Additional files

Additional file 1: Android screenshots. Four screenshots of the Google Nexus 7 analyzing different samples. Each sample shows a distinct pattern in its voltage measurements, and the Random Forest prediction is displayed beneath each graph.

Additional file 2: Receiver operating characteristic (ROC) curve. ROC analysis of the generated *Forest*. Graph was generated using PyLab by using out-of-bag error estimates per predictor to determine the sensitivity/specificity response. The area under the water, ethanol, isopropanol, and ambient air curves are 0.9908, 0.9956, 0.9914, and 0.9900 respectively.

Competing interests
The authors declare that they have no competing interests.

Authors' contributions
RJSG contributed to the construction of the device, polymer coating of the chip, hardware design, and drafted the manuscript. FN contributed to the construction of the device, hardware design, and performed all programming involved. GY managed the study, contributed to the hardware design, and helped to draft the manuscript. All authors read and approved the final manuscript.

Acknowledgements
The authors express gratitude to Dr. Heinrich Rohrer, Dr. Terunobu Akiyama, Dr. Frederic Loizeau, Dr. Sebastian Gautsch, Dr. Peter Vettiger, Dr. Kota Shiba, Mr. Cory J. Y. Lee, Mr. Mayuran Saravanapavanantham, and Mr. Max Palumbo for their indispensable contributions to the development of the MSS platform and related devices, and Prof. Masakazu Aono, Dr. Tomonobu Nakayama, and Prof. Nico F. de Rooij for their help and support. This work was supported by WPI Research Center Initiative for Materials Nanoarchitectonics (MANA); the Grant-in-Aid for Young Scientist (A) 23685017 (2011), MEXT, Japan; Research Foundation for Opto-Science and Technology (REFOST); TEPCO Memorial Foundation; and Japan Science and Technology Agency (JST).

Author details
[1]World Premier International (WPI) Research Center Initiative, International Center for Materials Nanoarchitectonics (MANA), National Institute for Materials Science (NIMS), 1-1 Namiki, 305-0044 Tsukuba, Japan. [2]University of Waterloo, 200 University Ave W, N2L 3G1 Waterloo, Canada.

References
1. Yoshikawa G, Akiyama T, Gautsch S, Vettiger P, Rohrer H: **Nanomechanical membrane-type surface stress sensor.** *Nano Lett* 2011, **11**(3):1044–1048. doi:10.1021/nl103901a.
2. Yoshikawa G, Akiyama T, Loizeau F, Shiba K, Gautsch S, Nakayama T, Vettiger P, Rooij NFd, Aono M: **Two dimensional array of piezoresistive nanomechanical membrane-type surface stress sensor (MSS) with improved sensitivity.** *Sensors* 2012, **12**(11):15873–15887. doi:10.3390/s121115873.
3. **Operating manual micropump Mp6/mp6-pp and controller.** [http://www.micro-components.com/image/pdf/EN_Manual%20mp6_mp6-pp_15052013.pdf]
4. Breiman L: **Random forests.** *Mach Learn* 2001, **45**(1):5–32. doi:10.1023/A:1010933404324.

5. Haindl M, Kittler J, Roli F (eds.): *Multiple Classifier Systems. Lecture Notes in Computer Science*, Vol. 4472. Berlin, Heidelberg: Springer; 2007. doi:10.1007/978-3-540-72523-7.
6. Pedregosa F, Varoquaux G, Gramfort A, Michel V, Thirion B, Grisel O, Blondel M, Prettenhofer P, Weiss R, Dubourg V, Vanderplas J, Passos A, Cournapeau D, Brucher M, Perrot M, Duchesnay E: **Scikit-learn: Machine learning in Python.** *J Mach Learn Res* 2011, **12:**2825–2830.
7. *Notes regarding behaviour of arduino wifi shield as a TCP client.* [http://mssystems.emscom.net/helpdesk/knowledgebase.php?article=51]

10

Direct IR excitation in a fast ion beam: application to NO⁻ photodetachment cross sections

Rico Otto, Amelia W Ray, Jennifer S Daluz and Robert E Continetti[*]

* Correspondence:
rcontinetti@ucsd.edu
Department of Chemistry and
Biochemistry, University of
California, San Diego, 9500 Gilman
Drive, La Jolla, CA, 92093–0340, USA

Abstract

Background: Optical access to a travelling ion packet is required in many ion beam experiments that study ion-photon interactions.

Methods: An approach is described for carrying out direct infrared excitation of a fast ion beam that uses an optical-quality reflective beam blocker to illuminate a counter propagating pulsed ion beam in a collinear configuration. This arrangement provides optical access along the axis of ion beam propagation by placing a mirror in the beam path at a 25 degree angle. The ion packet is bumped over the mirror, which is also used to block fast neutral particles produced during ion beam acceleration that also propagate along the beam path.

Results: The efficiency of this setup is demonstrated in a photodetachment experiment on NO⁻ anions, where a photoinduced depletion of up to 90% of the beam is achieved in a single laser shot. To demonstrate the application of this configuration, the relative photodetachment cross section for NO⁻ has been measured in the range of 2800 – 7200 cm-1. The measured relative cross section shows a set of sharp peaks that are identified as vibrational autodetachment resonances.

Conclusion: The new setup paves the way for future experiments where parent anionic species are vibrationally excited via direct infrared excitation first and undergo photodetachment/photodissociation in a subsequent step.

Keywords: Infrared excitation; Fast ion beam; Nitric oxide anion photodetachment
PACS: 33.15.Ry; 33.80.Eh; 33.80.-b

Background

Optical access to a travelling ion packet is required in many ion beam experiments that study ion-photon interactions. Ion storage rings can provide optical access tangentially to the beam path, thereby enabling lifetime measurements of metastable ions [1]. Multipass cell [2,3] or resonator [4] arrangements are used to increase the overlap between a short laser pulse and a molecular beam. Developments in manufacturing microchannel plate (MCP) ion detectors have enabled photofragmentation experiments where a detector featuring a center hole allows laser access to a fast moving ion packet [5]. Cluster predissociation studies often make use of a spatial and temporal focus of the ion bunch to maximize the overlap with a laser that crosses the beam from the side [6]. However, due to conservation of phase space, such a focus will always create an axial energy spread of the beam that might be undesirable if the kinetic energy release of the fragments is of interest. The work presented here was inspired by the need to illuminate an outstretched ion packet (~30 cm) in a completely collinear ion beam experiment, in which photoelectrons and photofragments created from a fast moving ion beam are collected in coincidence (PPC) [7].

The fast-beam apparatus used in these studies features a plasma discharge ion source where anions are created in a supersonic expansion at a 10 Hz duty cycle. The ions pass through a skimmer together with the gas jet and can be accelerated up to 10 keV, with an energy spread of less than 0.1 eV, before being re-referenced to ground while traveling through a cylindrical electrode. The length of this cylinder (30 cm) determines the spatial extension of the ion packet. After a flight distance of ~1 m the ion packet is injected into an electrostatic ion trap (EIBT), where it is repetitively probed perpendicularly with the pulsed output from a 1 kHz Ti:Sapphire laser system for studies of photodetachment and dissociative photodetachment processes. As part of an effort to expand the present set of experiments towards direct infrared excitation of the parent anionic species prior to the photodetachment step, it has become essential to irradiate the entire outstretched ion packet with the output of a 10 Hz infrared (IR) optical parametric oscillator/optical parametric amplifier (OPO/OPA) laser system before it enters the EIBT. In addition, such an arrangement opens up the possibility to study photodetachment processes in weakly bound anionic species.

Here we demonstrate a simple configuration where a single gold mirror is placed in the ion beam, with the surface normal of the mirror at an angle of 25 degrees relative to the ion beam propagation direction. The pulsed IR laser light enters the vacuum through a viewport located on the side of the vacuum chamber such that it forms an angle of 50 degrees with the beam axis. The laser pulse reflects off the gold mirror and illuminates the entire ion bunch in a single shot. The ions are then electrostatically bumped over the mirror and corrected to the ion beam axis to continue travelling towards the EIBT. As a secondary effect, the gold mirror acts as a beam blocker for fast neutral particles in the beam, preventing this source of background from striking the neutral particle detector used in the PPC experiments. This also ensures ultra-high vacuum conditions in the EIBT and detection regions. The performance of the new setup was demonstrated in a photodetachment experiment, making use of the small electron affinity (EA = 26 ± 5 meV) [8] of the NO molecule to directly deplete ions from a fast ion beam. A measurement of the wavelength dependence of the depletion, which is proportional to the photodetachment cross section, reveals sharp resonance features that are associated with vibrational transitions in the NO^- anion, followed by vibrational autodetachment.

Results and discussion
Design and performance of the reflective beam blocker
Depicted in Figure 1 is the reflective beam blocker design, consisting of a rectangular gold mirror (3 × 1 cm) that is attached to an L-bracket aluminum holder. The assembly is located in an electrode arrangement that bumps the ion packet over the mirror and returns it to the incident beam axis. To ensure good electrical conductivity with the mirror's surface, the edges of the mirror were connected to the aluminum substrate using a colloidal silver paint. For practical reasons the mirror is placed in the beam centerline at an angle of 25 degrees, allowing for coupling in an IR laser beam through

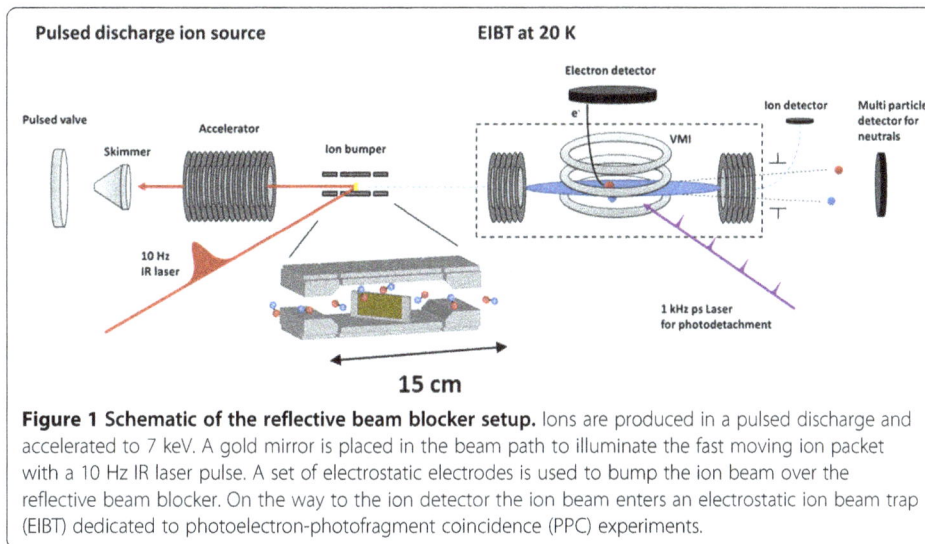

Figure 1 Schematic of the reflective beam blocker setup. Ions are produced in a pulsed discharge and accelerated to 7 keV. A gold mirror is placed in the beam path to illuminate the fast moving ion packet with a 10 Hz IR laser pulse. A set of electrostatic electrodes is used to bump the ion beam over the reflective beam blocker. On the way to the ion detector the ion beam enters an electrostatic ion beam trap (EIBT) dedicated to photoelectron-photofragment coincidence (PPC) experiments.

a CaF vacuum viewport located on the side of the ion-beam transport chamber. The ion deflection unit consists of three sets of electrostatic electrodes that are used to transfer the ion packet over the mirror before returning it to the original trajectory. The first and last deflector electrodes are held at identical potentials (V1, V3) while the electric field is reversed for the center deflector (V2). The voltages for all deflectors are in the range of 700 – 850 V, with typical values for V1 and V3 of about $0.9 \cdot V2$. The voltages that need to be employed in order to effectively transfer the ion packet over the mirror were optimized using SIMION [9]. In addition to its vertical deflection properties, the simulations revealed a small focusing effect on the vertical axis of the ion deflector unit. In the experiment this effect is accounted for with an additional set of ion optics after the unit. Further details of the experimental approach are provided in the Methods section below.

Figure 2 shows the depletion N_{IR}/N_0 as a function of the temporal delay of the incoming IR pulse with respect to the ion source, measured at an IR wavelength of

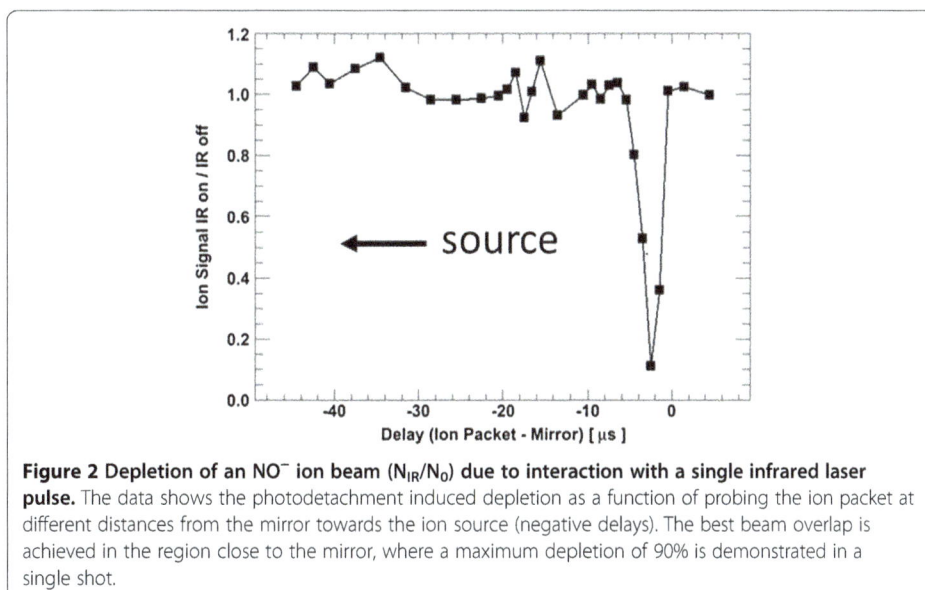

Figure 2 Depletion of an NO− ion beam (N_{IR}/N_0) due to interaction with a single infrared laser pulse. The data shows the photodetachment induced depletion as a function of probing the ion packet at different distances from the mirror towards the ion source (negative delays). The best beam overlap is achieved in the region close to the mirror, where a maximum depletion of 90% is demonstrated in a single shot.

5300 cm^{-1}. Changing this delay probes the ion packet at different positions along the beam axis, where a delay of 0 μs corresponds to a laser pulse entering the chamber as the ion packet travels over the mirror, while at earlier timings the packet is still closer to the ion source. A maximum in the depletion is achieved shortly before the ion packet is transferred over the mirror. This can be attributed to the shape and trajectory of the travelling ion packet, which is collimated using an Einzel lens before approaching the mirror. The measured depletion gives an upper limit to the fraction of ions detached in a single laser shot of 90%. Note that this result also provides a lower limit to the fraction of ions that are illuminated, regardless of the laser power.

The depletion measurements outlined above can be used to measure the relative photodetachment cross section of the NO$^-$ molecule. In general the ion-laser interaction leads to an exponential decay of the signal N_0 so that the signal after a short laser pulse of temporal length t can be written as

$$N(t) = N_0 \cdot \exp(-kt) \tag{1}$$

with a photodetachment-induced decay rate k that can be expressed as

$$k = F_L \cdot \sigma \cdot \rho. \tag{2}$$

This decay rate depends only on the total photon flux F_L (cm^{-2} s^{-1}), the photodetachment cross section σ (Mbarn $= 10^{-18}$ cm^2) and the geometrical overlap ρ between the ion packet and the laser beam. While the geometrical overlap and the exact beam profile of the laser in the chamber are unknown, it is assumed that they remain constant within $\pm 20\%$ over the course of the experiment, as inferred by projecting the IR beam in the far field. An upper limit for the laser beam diameter is given by the apertures of about 1 cm in the ion time-of-flight region. In order to determine the total photon flux F_L the power P_{IR} at each wavelength was measured at the output port of the IR laser using a power meter (Ophir Nova). The IR wavelength λ was determined via the OPO signal and idler wavelength using a spectrometer (Ocean Optics HR2000+) and independently calibrated using a photo acoustic spectroscopy setup. Furthermore, the fast moving ion packet gives rise to a Doppler shift of the counter propagating IR pulse, which accounts for $2 - 5$ cm^{-1} over the range of the experiment and has been corrected for. The Doppler spread due to the axial energy uncertainty of the ion beam is in the range of a few MHz and is therefore much smaller than the bandwidth of the pulsed IR laser system. The laser beam had to undergo reflections from a total of six gold mirrors and cross a CaF vacuum window before counter propagating into the ion beam. The reflectivity for the mirrors ($98 \pm 0.5\%$) as well as the transmittance of the CaF window ($94 \pm 0.5\%$) can be considered constant over the measured wavelength range, so that the measured laser power multiplied by the photon wavelength is proportional to the photon flux in the ion interaction region. A relative photodetachment cross section is then given by

$$\sigma_{PD} \propto -\log(N_{IR}/N_0)/(P_{IR} \cdot \lambda). \tag{3}$$

Shown in Figure 3 is the relative photodetachment cross section measured as a function of the photon energy in the range $2800 - 7200$ cm^{-1}. These data represent the average of a large number of datasets that have been concatenated such that

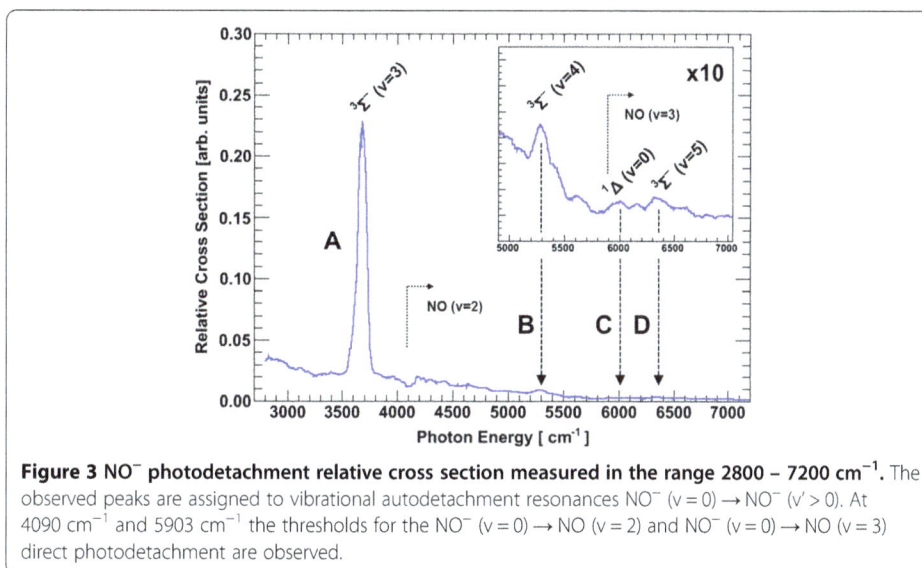

Figure 3 NO⁻ **photodetachment relative cross section measured in the range 2800 – 7200 cm⁻¹.** The observed peaks are assigned to vibrational autodetachment resonances NO⁻ (v = 0) → NO⁻ (v′ > 0). At 4090 cm⁻¹ and 5903 cm⁻¹ the thresholds for the NO⁻ (v = 0) → NO (v = 2) and NO⁻ (v = 0) → NO (v = 3) direct photodetachment are observed.

overlapping frequency ranges match up. For these measurements the optimum temporal delay, derived from Figure 2, has been used.

The shape of the cross section makes it obvious that two different processes are observed in the experiment. The first process is a bound-free transition between the molecular anion and a neutral NO molecule plus an electron in the continuum. This process is responsible for the continuous part of the spectrum that slowly decreases with increasing photon energy. On top of that resonant peaks are observed that are identified as vibrational transitions induced in the NO⁻ molecule by IR absorption, followed by autodetachment.

Direct photodetachment of NO⁻

Molecular photodetachment has been shown to provide a sensitive tool to probe both the initial anion [10] and final neutral state [11]. According to Wigner's law the energy dependence of the cross section close to threshold scales as

$$\sigma(E) \propto (E-E_0)^{2l+1} \tag{4}$$

with the threshold energy E_0 and the angular momentum l of the outgoing electron. The more complex zero core contribution (ZCC) model [12] has been used to describe the shape of the cross section above threshold for atomic systems. While the Wigner law is well suited to describe the rising cross section behavior close to the threshold, the ZCC model also reproduces a decaying cross section at higher energies. Al-Za'al *et al.* applied the model to NO⁻ photodetachment and predicted a sharp rise in the cross section at 0.507 eV (4090 cm⁻¹), which is associated with the channel to produce NO (v = 2) opening up [13]. Surprisingly they could not verify this threshold experimentally. Instead a continuous spectrum without sharp increases was observed that slowly decreased over the experimental range. The authors based their analysis on the electron affinity for the NO⁻ (v = 0) → NO (v = 2) transition measured by Siegel *et al.* [14], and added a rotational correction of 12.5 meV. Based on those values another rise in the cross section is expected at 0.732 eV (5903 cm⁻¹) where the NO (v = 3) state becomes accessible. More recent photoelectron spectroscopy experiments suggest values of 0.488 eV and 0.714 eV for the NO (v = 2) and NO (v = 3) electron affinity relative to NO⁻ (v = 0) [8].

In the results reported here a slowly decaying cross section attributed to direct photodetachment that is accompanied by several sharp resonance features is observed. The continuous part of the cross section decays by a factor of 10 over the measured range. Upon closer inspection, two regions in the spectrum at 4100 cm⁻¹ and 5900 cm⁻¹ can be identified where a sudden increase in the cross section is observed. Both of these features are assigned to the opening of new product channels, leading to NO (v = 2) and NO (v = 3) products respectively, as indicated in the figure. The cross section at the (v = 2) threshold rises by almost a factor of two, in accordance with the predictions from the ZCC model [13]. It is interesting to note that above this threshold the cross section reaches a maximum after only 100 cm⁻¹ before starting to decrease again. The (v = 3) threshold is not analyzed here, since it is located close to one of the resonance features that will be discussed in the next section.

Photoinduced Vibrational Autodetachment of NO⁻

Two previous studies have examined photodetachment of NO⁻ in the IR [13,15]. Maricq *et al.* used diode lasers to study the photodetachment cross section in the range 1100 – 1500 cm⁻¹. To cover the frequency range of the experiment three different diode lasers were used that had line widths between 1 and 10 cm⁻¹. They found a vibrational autodetachment resonance centered at 1284 cm⁻¹ assigned to the NO⁻ (v = 0) → NO⁻ (v = 1) transition and reported a width of the observed resonance of 95 cm⁻¹, attributed to lifetime broadening. Al-Za'al *et al.* studied the NO⁻ cross section in the range 3000 – 4150 cm⁻¹ using an F-center laser. They reported a continuous spectrum without any evidence for vibrational resonances. The observed monotonic decrease of 50% over the measured range was interpreted as the high energy tail of the NO⁻ (v = 0) → NO⁻ (v = 2) vibrational resonance. This scenario, however, requires the resonance to have a width of 600 cm⁻¹ that seems unlikely. Prior data from photodetachment studies is scarce but vibrational autodetachment resonances have been studied in numerous electron scattering experiments [14,16-19] arising from electron attachment to NO(v = 0) forming an intermediate NO⁻(v' > 0) excited state. The energy of the electron beam is thereby given relative to the ground state of NO. Therefore, in order to compare these values to a photodetachment experiment it is essential to take into account the electron affinity of the NO molecule, measured to be 26 meV in previous photodetachment experiments [8]. The variation among these earlier experiments is summarized in Figure 4 for the NO⁻ resonances associated with the vibrational levels (v = 1 – 5) of the NO⁻ ³Σ⁻ state.

The data presented in this work shows a series of distinct resonant peaks in the photodetachment cross section (see Figure 3). The positions of the peaks labeled A – D are listed in Table 1. Based on the comparison with the electron scattering data, the peaks A and B are assigned to the vibrational transitions NO⁻ (v = 0) → NO⁻ (v = 3, 4) within the ³Σ⁻ anion ground state. In previous experimental studies by Tronc *et al.* [19] the width of the NO⁻ (v = 5) resonance was observed to be much broader than the NO⁻ (v = 1 – 4) series, which was attributed to a superposition of the NO⁻ (³Σ⁻, v = 5) and NO⁻ (¹Δ, v = 0) state. Ziesel *et al.* [17] resolved the splitting caused by the two states more clearly, but observed a line shape that required a superposition of three peaks to

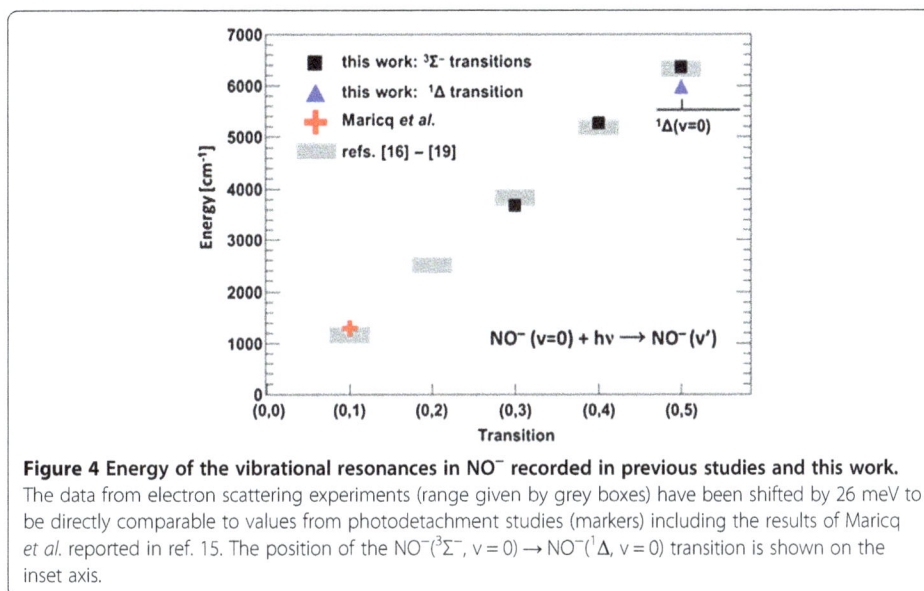

Figure 4 Energy of the vibrational resonances in NO⁻ recorded in previous studies and this work. The data from electron scattering experiments (range given by grey boxes) have been shifted by 26 meV to be directly comparable to values from photodetachment studies (markers) including the results of Maricq *et al.* reported in ref. 15. The position of the NO−($^3\Sigma^-$, v = 0) → NO−($^1\Delta$, v = 0) transition is shown on the inset axis.

achieve the best fit to their data. Based on these previous studies, peaks C and D are assigned to the NO⁻ ($^1\Delta$, v = 0) and NO⁻ ($^3\Sigma^-$, v = 5) resonances, respectively. A comparison of this new photodetachment data with the range of previous measurements is shown in Figure 4, where it is seen that the positions of the peaks observed are well within the range of previously measured resonance features.

The observed intensity of the resonance peaks is largest for the (v = 3) resonance and decreases for higher vibrational excitation, which can be understood in terms of a decreasing Franck-Condon overlap [20]. The fact that the NO⁻ (v = 3) level is nearly degenerate with NO (v = 2) also contributes to the high intensity of the (v = 3) resonance. Finally, it is interesting to note that the spacing found between the (v = 3) and (v = 4) resonances is significantly larger ($\Delta E = 1600$ cm⁻¹) than that between the (v = 4) and (v = 5) feature ($\Delta E = 1100$ cm⁻¹). The resonance features detected in electron scattering show a consistent spacing of about 0.16 eV (1300 cm⁻¹) at least up to (v = 9), pointing to only weak anharmonicity in the NO⁻ potential. However, it has to be pointed out that the underlying mechanism in those experiments is different in nature than the direct bound-bound transition probed in vibrational autodetachment. Also, it cannot be ruled out that rotational effects might play an important role in the autodetachment process [21]. An analysis of such effect is beyond the scope of this work.

Conclusions

A new experimental setup has been presented here that allows direct IR excitation in a completely collinear pulsed ion beam experiment that does not feature any other

Table 1 Peak positions observed in the photodetachment cross section of NO⁻

Peak	Energy (cm⁻¹)	Width (cm⁻¹)	Assignment
Ref. ([15])	1284 ± 10	95 ± 15	$^3\Sigma^-$ (v = 1)
A	3687 ± 2	28 ± 3	$^3\Sigma^-$ (v = 3)
B	5290 ± 3	41 ± 3	$^3\Sigma^-$ (v = 4)
C	5976 ± 13	108 ± 24	$^1\Delta$ (v = 0)
D	6355 ± 10	115 ± 16	$^3\Sigma^-$ (v = 5)

optical access along the ion beam path. This has been achieved by placing an optical-quality mirror in the beam path at an angle that allows coupling in an IR laser from the side of the setup. Photodetachment-induced depletion of an NO^- beam was used to demonstrate the overlap that can be achieved between the laser pulses and a fast moving ion packet. Furthermore, measurements of the relative photodetachment cross section of the NO^- molecule in the range of $2800 - 7200$ cm^{-1} were made. It was found that the cross section in this range is a combination of direct photodetachment and vibrational autodetachment. The observed autodetachment resonances are within the range of previous experimental results from electron scattering experiments.

This new setup paves the way for future PPC studies with vibrationally excited molecules. In these experiments molecular anions will be prepared in specific vibrational states before entering the ion beam trap, and the fragmentation dynamics in dissociative photodetachment processes will be studied. The systems amenable to study by this approach will in general be strongly bound anions where the lifetimes of the excited vibrational modes are long enough to allow for transferring the ions over the mirror and carrying out PPC experiments on a millisecond timescale.

Methods

To demonstrate the capabilities of the reflective beam blocker design for coupling a light source with a travelling ion packet, a photodetachment experiment using a beam of NO^- molecules was carried out. These experiments exploit the low electron affinity of NO^- that allows for efficient photodetachment at wavelengths between $2 - 5$ μm. The NO^- anions were generated from a 10 Hz pulsed discharge (20% N_2O seeded in a 1:2 He/Ne mixture, 20 psi stagnation pressure). Typical rotational temperatures for this ion source have been measured to be $50 - 100$ K using near threshold photodetachment of OH^- in a different set of experiments. The ions were accelerated to 7 keV before approaching the reflective beam blocker. The ion signal was monitored 2.5 m behind the beam blocker using an off-axis MCP ion detector. Tunable IR laser pulses from a 10 Hz Nd:YAG (Surelite III EX) pumped OPO/OPA system (LaserVision, 5 ns FWHM, 3 cm^{-1} bandwidth) were coupled into the approaching ion packet at a time delay synchronized with the pulsed ion source. A typical output power between 100 and 300 mW can be achieved over the wavelength range covered in this work. The depletion of the beam caused by photodetachment of the NO^- was derived from consecutively measuring the IR on and IR off ion signals (denoted as N_{IR} and N_0 respectively) at a given wavelength. To acquire each signal an average over 32 source cycles was recorded before the status of the IR laser was switched (IR on/off).

Abbreviations
MCP: Microchannel plate; PPC: Photoelectron photofragment coincidence; EIBT: Electrostatic ion beam trap; IR: Infrared; ZCC: Zero core contribution.

Competing interests
The authors declare that they have no competing interests.

Authors' contributions
RO designed the apparatus, carried out the experiments and drafted the manuscript. AR participated in the experiments. JD participated in the experiments and the wavelength calibration measurements. AR and JD both participated in manuscript preparation. RC conceived of the study, participated in design and coordination and helped draft the manuscript. All authors read and approved the final manuscript.

Acknowledgements

This work was supported by the U.S. Department of Energy under grant number DE-FG03-98ER14879. RO thanks the German Academic Exchange Service (DAAD) for a postdoctoral research fellowship. AWR and JSD acknowledge support from GAANN grant P200A120223 from the U.S. Dept. of Education.

References

1. Andersen JU, Hvelplund P, Nielsen SB, Tomita S, Wahlgreen H, Møller SP, Pedersen UV, Forster JS, Jørgensen TJD: The combination of an electrospray ion source and an electrostatic storage ring for lifetime and spectroscopy experiments on biomolecules. *Rev Sci Instrum* 2002, **73**:1284–1287.
2. Gough TE, Gravel D, Miller RE: Multiple crossing devices for laser-molecular beam spectroscopy. *Rev Sci Instrum* 1981, **52**:802.
3. Lethbridge PG, Stace AJ: Design considerations for the construction of a reflecting symmetric multipass cell for use in laser molecular-beam experiments. *Rev Sci Instrum* 1987, **58**:2238–2243.
4. Hlavenka P, Plašil R, Bánó G, Korolov I, Gerlich D, Ramanlal J, Tennyson J, Glosík J: Near infrared second overtone cw-cavity ringdown spectroscopy of D_2H^+ ions. *Int J Mass Spectrom* 2006, **255**:170–176.
5. Toker Y, Aviv O, Eritt M, Rappaport ML, Heber O, Schwalm D, Zajfman D: Radiative cooling of Al_4^- clusters. *Phys Rev A* 2007, **76**:053201.
6. Ayotte P, Johnson MA: Electronic absorption spectra of size-selected hydrated electron clusters: $(H_2O)_n^-$, n = 6–50. *J Chem Phys* 1997, **106**:811.
7. Johnson CJ, Shen BB, Poad BLJ, Continetti RE: Photoelectron-photofragment coincidence spectroscopy in a cryogenically cooled linear electrostatic ion beam trap. *Rev Sci Instrum* 2011, **82**:105105.
8. Travers MJ, Cowles DC, Ellison GB: Reinvestigation of the electron affinities of O_2 and NO. *Chem Phys Lett* 1989, **164**:449–455.
9. Dahl DA: **SIMION 3D, Version 7.0 User's Manual INEEL-95/0403.** In *Idaho National Engineering and Environmental Laboratory, Bechtel BWXT Idaho, LLC.* 2000.
10. Otto R, von Zastrow A, Best T, Wester R: Internal state thermometry of cold trapped molecular anions. *Phys Chem Chem Phys* 2013, **15**:612–618.
11. Schulz PA, Mead RD, Jones PL, Lineberger WC: OH and OD threshold photodetachment. *J Chem Phys* 1982, **77**:1153.
12. Stehman RM, Woo SB: Zero-core-contribution model and its application to photodetachment of atomic negative ions. *Phys Rev A* 1979, **20**:281.
13. Al-Za'al M, Miller HC, Farley JW: Measurement of the infrared photodetachment cross section of NO⁻. *Phys Rev A* 1986, **33**:977.
14. Siegel MW, Celotta RJ, Hall JL, Levine J, Bennett RA: Molecular Photodetachment Spectrometry. I. The Electron Affinity of Nitric Oxide and the Molecular Constants of NO⁻. *Phys Rev A* 1972, **6**:607.
15. Maricq MM, Tanguay NA, O'Brien JC, Rodday SM, Rinden E: Vibrational autodetachment of NO⁻. *J Chem Phys* 1989, **90**:3136.
16. Spence D, Schulz GJ: Vibrational Excitation and Compound States in NO. *Phys Rev A* 1968, **1971**:3.
17. Randell J, Lunt SL, Mrotzek G, Field D, Ziesel JP: Low-energy negative-ion states of NO at high electron energy resolution. *Chem Phys Lett* 1996, **252**:253–257.
18. Alle DT, Brennan MJ, Buckman SJ: Low-energy total electron scattering cross section and electron affinity for NO. *J Phys B Atomic Mol Phys* 1996, **29**:L277.
19. Tronc M, Huetz A, Landau M, Pichou F, Reinhardt J: Resonant vibrational excitation of the NO ground state by electron impact in the 0.1–3 eV energy range. *J Phy B* 1975, **8**:1160.
20. Teillet-Billy D, Fiquet-Fayard F: The NO⁻ $^3\Sigma^-$ and $^1\Delta$ resonances: theoretical analysis of electron scattering data. *J Phy B* 1977, **10**:L111.
21. Neumark DM, Lykke KR, Andersen T, Lineberger WC: Infrared spectrum and autodetachment dynamics of NH⁻. *J Chem Phys* 1985, **83**:4364.

11

Applications of Raman micro-spectroscopy to stem cell technology: label-free molecular discrimination and monitoring cell differentiation

Adrian Ghita[1], Flavius C Pascut[1], Virginie Sottile[2], Chris Denning[2] and Ioan Notingher[1*]

* Correspondence:
ioan.notingher@nottingham.ac.uk
[1]School of Physics and Astronomy,
University of Nottingham,
Nottingham NG7 2RD, UK
Full list of author information is
available at the end of the article

Abstract

Stem cell therapy is widely acknowledged as a key medical technology of the 21st century which may provide treatments for many currently incurable diseases. These cells have an enormous potential for cell replacement therapies to cure diseases such as Parkinson's disease, diabetes and cardiovascular disorders, as well as in tissue engineering as a reliable cell source for providing grafts to replace and repair diseased tissues. Nevertheless, the progress in this field has been difficult in part because of lack of techniques that can measure non-invasively the molecular properties of cells. Such repeated measurements can be used to evaluate the culture conditions during differentiation, cell quality and phenotype heterogeneity of stem cell progeny. Raman spectroscopy is an optical technique based on inelastic scattering of laser photons by molecular vibrations of cellular molecules and can be used to provide chemical fingerprints of cells or organelles without fixation, lysis or use of labels and other contrast enhancing chemicals. Because differentiated cells are specialized to perform specific functions, these cells produce specific biochemicals that can be detected by Raman micro-spectroscopy. This mini-review paper describes applications of Raman micro-scpectroscopy to measure moleculare properties of stem cells during differentiation in-vitro. The paper focuses on time- and spatially-resolved Raman spectral measurements that allow repeated investigation of live stem cells in-vitro.

Introduction

Tissue engineering and stem cells technology

The capacity of the human body to recover from injuries or diseases is limited and in some situations almost non-existent. The standard medical procedures, which involve surgery and drug based-therapies, have shown limitations in treating many complex conditions. New alternatives to existing medical treatments are required as the incidence of the medical conditions with clinical impact on society in terms of mortality, morbidity and quality of life is continuously increasing. Tissue engineering is an emerging field in modern medicine that aims to grow tissue in laboratories that can be used for replacement of diseased tissue and organs, repair and regeneration. Tissue replacement involves the growth of tissue and organs in-vitro that can be implanted in the human body [1]. The repair involves medical treatment at the biological and molecular level. The main advantages of these approaches are related to an increased availability of grafts and reducing the risks associated with viral infections (e.g. tissue rejection) or negative immunological response [2].

A number of different sources of cells are considered for tissue engineering and regenerative medicine. Mature fully differentiated somatic cells, adult and embryonic stem cells have been used in different applications. The somatic cells extracted from the patients have no risks associated with the integration and immune response, but these cells can be obtained in limited numbers and have low proliferation rates. Adult stem cells are immature undifferentiated cells extracted from the human body, usually bone marrow, and are capable of generating daughter cells. The self-renewal process occurs over the entire lifetime of the biological host. However, the limited number of available adult stem cells in the human body gives rise to impediments in the self-healing capabilities of the human body. Embryonic stem cells (ESCs) are derived from the inner cell mass of mammalian blastocyst and can differentiate spontaneously in-vitro giving rise to pluripotent cells [3]. ESCs can serve as a potential research model for tissue development, cancer formation and metastasis, phenotype commitment, stem cells based therapy, gene therapy strategies and drug design [4]. Despite their clinical potential, the progress of stem cell therapy has been slow, mainly due to issues related to cell quality, phenotype heterogeneity, delivery, integration, proliferation and differentiation inside the host [5].

Review

Raman Micro-Spectroscopy (RMS)

During the last century, a plethora of chemical and imaging techniques have been developed to understand the molecular dynamics underpinning the fundamental cellular processes such as proliferation, differentiation or apoptosis [6,7]. Each of these techniques can address certain applications or particular biological problems or processes. Optical microscopy techniques are particularly attractive due to their ability to provide information with high-spatial resolution while using compact and cost-effective instrumentation. Nevertheless, conventional microscopy relies on refractive-index contrast in a sample and does not provide molecular information. Fluorescence microscopy relies on selective imaging of cellular molecules labelled with specific fluorescent dyes [8,9]. Thus, its capabilities are affected by limited stability and photo bleaching of the fluorophores, while the free radical released inside the cells during photo-excitation can induce toxicity. Certain fluorescence labels may influence biochemical processes in cells or cannot enter live cells to access cytoplasmic molecules (often fluorescence imaging requires fixation of the cells).

The Raman effect is based on inelastic scattering of photons when electromagnetic waves interact with molecules. As a consequence of the energy transfer between the vibrating molecules in the sample and the electromagnetic waves, the Raman scattered photons have different frequencies compared to the incident photons, frequencies that are related to the vibrational levels of the molecules. Thus, Raman spectroscopy is a label-free analytical technique and molecular information can be obtained in a non-destructive way with no sample preparation: light is used for excitation of the molecules and the inelastically scattered light is detected for molecular analysis. Raman spectroscopy has been extensively used for molecular analysis of biological tissues and disease diagnostic [10-13].

An important advantage of RMS is that the information regarding molecular vibrations can be obtained using microscopy instrumentation operating in the visible and near-infrared spectral regions. RMS has benefited enormously from the recent advances in high-power laser technology, optical microscopy, fibre optics and detectors with high quantum efficiency in this spectral range. Another important advantage of Raman micro-spectroscopy for live cell imaging is that the background Raman signals from water are significantly weaker compared to signals obtained from cell's molecules. Therefore, RMS can be used to measure repeatedly the molecular properties of live cells maintained in culture media.

Label-free Raman spectral imaging of cells

The first Raman micro-spectrometer was demonstrated by Dhamelincourt and Delhaye in 1973 [14] but it was not until 1990 when Puppels et al. reported the first Raman spectra of individual cells [15]. Since then, Raman spectroscopy has been extensively used for studies of cells, including detection of protein conformation in living cells [16], chemical differences at different stages of the cell cycle [17,18], differences between cells of different phenotypes [19,20] or apoptotic cells [21]. Raman spectroscopy can probe molecular information from living cells revealing distinct chemical features between nucleus and cytoplasm [15,22-24], lipid structures inside cytoplasm [25], and cell response to diverse stress stimuli [26,27]. High resolution Raman imaging provided new insights into the spatial distribution of chemical species and organelles inside cells [28-34]. RMS was used to follow the cellular uptake of the drugs [35-38] and cellular dynamics [39]. Apoptosis, also known as programmed cell death, is an important biological phenomenon that involves morphological and chemical changes of the cell organelles. Raman spectral mapping of apoptotic cells showed changes in the distribution of the nucleic acids [40,41] and detected the accumulation of lipids when cells were treated with different apoptotic drugs [35,42-44].

Nevertheless, one of the most important features of the Raman spectroscopy is the ability to measure time-course molecular changes in live cells, revealing biochemical processes not attainable with other imaging methods [43-47]. Such experiments are of particular relevance to regenerative medicine and tissue engineering applications as repeated Raman measurements can provide time- and spatially-resolved information regarding the molecular properties of stem cells during differentiation or the growth and development of the engineered grafts. Such information may be used for better understanding the biological processes as well as for improving the bioprocesses.

Raman spectroscopy was also used to study molecular changes at the early stages of differentiation of stem cells [48] and for phenotypic identification of stem cell progeny. The differentiation process describes a series of biochemical changes that are meant to transform a stem cell into a fully functional somatic cell. These changes can be correlated to changes in the Raman spectra of cells measured without using labels or affecting their viability [49]. Recent studies have shown that Raman spectral markers can be used to discriminate fully committed mature cells from undifferentiated cells [23,24,50] and to assess the differentiation status of murine stem cells [51]. In addition, RMS may be used for quality control of cell and tissue grafts prior to clinical use. Recent reports discussed the potential of RMS as a non-invasive method for enrichment of cardiomyocyte populations obtained by differentiation of human stem cells [52]. In this mini-review paper we present three applications of RMS for time- and spatially-resolved molecular characterisation of stem cells during in-vitro differentiation.

Instrumentation

Raman micro-spectroscopy is based on coupling a Raman spectrometer to an optical microscope (Figure 1). Inverted optical microscopes are in particular suitable for time-course cell imaging as the cells can be cultured in dedicated enclosed cell-chambers that allow efficient collection of the Raman scattered light while maintaining the cells in culture media and avoiding bacterial contamination. The instrument used for the applications described in this paper was based on an inverted microscope (IX 71, Olympus, Essex, UK) equipped with an environmental enclosure (Solent, Segensworth, UK) that maintains the live cells at 37°C and 5% CO_2 atmosphere. A water immersion microscope objective (Olympus) 60×/1.2 N.A. was used to focus the 785 nm laser beam (Toptica Xtra) and collect the Raman scattered light, which was then analysed by an optical spectrometer (830nm diffraction grating, IDus 401 CCD, Andor Technologies, Belfast, UK). The spectrometer was connected to the microscope through a 50 μm diameter optical fibre, providing a spectral resolution of ~4 cm^{-1}. A step motorize stage (Prior, Cambridge, UK) was attached to the microscope to raster scan the sample area. The microscope also had an integrated wide-field fluorescence imaging system that allowed imaging of the cells at the end of the Raman spectral measurements, after staining with specific labelles. For typical experiments for cell imaging, the laser power at the sample was in the 100–150 mW range and the exposure time per spot 0.5–2 s (total time required for the aquisiot of a Raman image of a cell ~20–40 minutes at 1 μm step-size).

Applications

Assessment of differentiation status of neuro-progenitor stem cells

Neural stem cells are responsible for the generation of neurons and glial cells and offer great promise to develop treatments for Parkinson's, Alzheimer's diseases, chronic inflammatory disorders of the central nervous system, spinal cord injuries or strokes

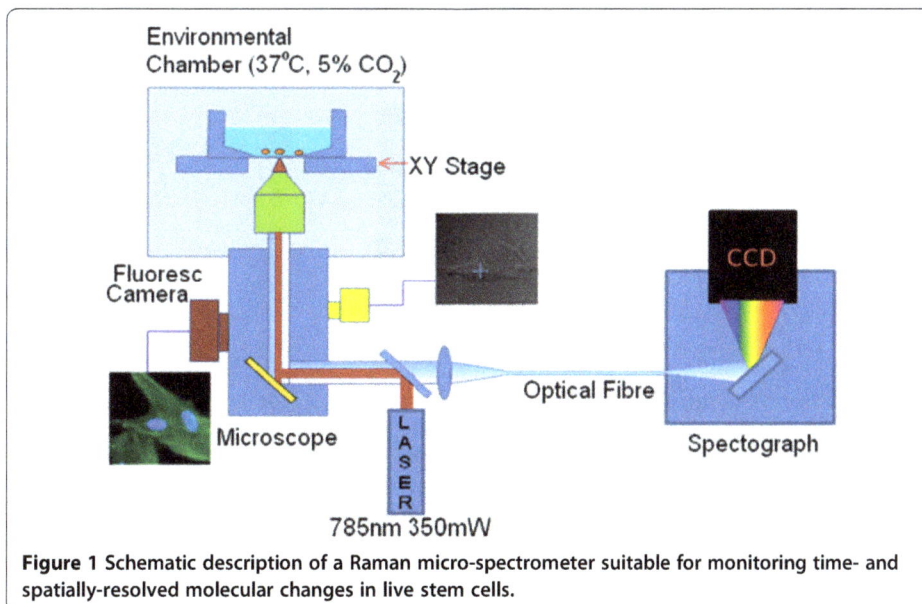

Figure 1 Schematic description of a Raman micro-spectrometer suitable for monitoring time- and spatially-resolved molecular changes in live stem cells.

[53]. The capacity of endogenous neural stem cells to proliferate and replace neural cells in-vivo may be affected by chronic inflammatory processes [53]. Hence the repair potential of endogenous stem cells may be limited and any medical procedures designed to mobilize neural stem towards inflammatory sites may be limited. Thus, therapies based on transplantation of neural stem cells from exogenous sources have been developed recently [54].

Raman micro-spectroscopy was proposed as a label-free method for identification of neural stem cells and discrimination of glial cells. Raman spectra of neuro-progenitor cells and glial cells showed significant differences in the fingerprint spectral region (600–1800 cm^{-1}) (Figure 2A,B). A multivariate statistical model was developed that allowed discrimination between neural stem cells and glial cells with 89.4% sensitivity and 96.4% specificity [55]. The computed difference between the average Raman spectra of neuro-progenitor and glial cells shows a close similarity with the Raman spectrum of purified RNA (Figure 2B).

Focusing on the 700–830 cm^{-1} region, distinct spectral signatures from nucleic acid base pairs can be observed: adenine (729 cm^{-1}), uracil and cytosine (782 cm^{-1} and 785 cm^{-1}). Raman bands corresponding to the stretching vibrations of phosphate O-P-O backbone of nucleic acids can also be identified: 788 cm^{-1} for B-DNA, and 788 cm^{-1} and 813 cm^{-1} for RNA [56]. Raman spectra measured at different locations inside live neuro-progenitor stem cells indicate that regions rich in RNA are found mostly in the cytoplasms of these cells (Figure 2E, F).

Raman micro-spectroscopy can measure detailed molecular maps of cells to obtain spatially-resolved chemical information inside the cells with spatial resolution down to

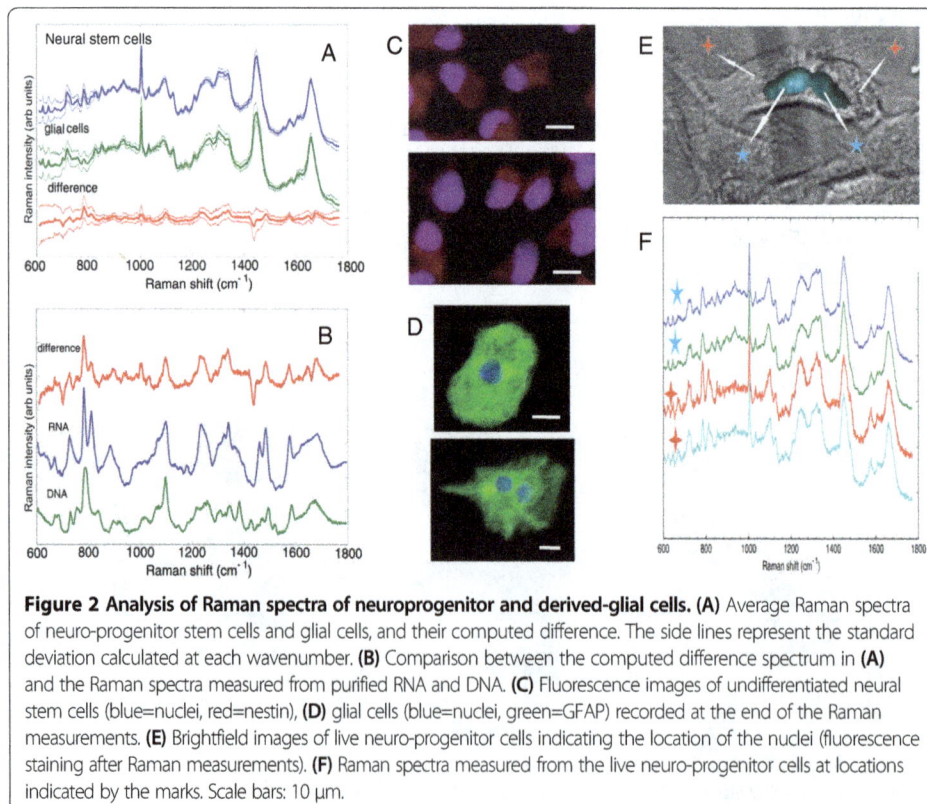

Figure 2 Analysis of Raman spectra of neuroprogenitor and derived-glial cells. (A) Average Raman spectra of neuro-progenitor stem cells and glial cells, and their computed difference. The side lines represent the standard deviation calculated at each wavenumber. **(B)** Comparison between the computed difference spectrum in **(A)** and the Raman spectra measured from purified RNA and DNA. **(C)** Fluorescence images of undifferentiated neural stem cells (blue=nuclei, red=nestin), **(D)** glial cells (blue=nuclei, green=GFAP) recorded at the end of the Raman measurements. **(E)** Brightfield images of live neuro-progenitor cells indicating the location of the nuclei (fluorescence staining after Raman measurements). **(F)** Raman spectra measured from the live neuro-progenitor cells at locations indicated by the marks. Scale bars: 10 μm.

the diffraction limit ($\sim$750 nm in this case). Figure 3 presents Raman spectral maps along with fluorescence and phase contrast images of neuro-progenitor and glial cells. The Raman maps presented in Figure 3 were constructed by computing the peak area of the 788 cm^{-1} and 813 cm^{-1} bands for each Raman spectrum after subtracting the local baseline. The cells were raster scanned with a step size of 500 nm.

For neuro-progenitor cells (Figure 3), the Raman maps corresponding to the 788 cm^{-1} band is slightly larger than the DAPI staining of the cell nucleus as it contains spectral overlap signature from DNA and RNA. On the other hand, the band at 813 cm^{-1}, specifically assigned to RNA, reveals high amount of RNA in the cytoplasm. However, Raman maps recorded for glial cells (Figure 3H) fail to detect bands corresponding to RNA in the cytoplasm. By developing a solution-based calibration model for RNA, the maximum concentration of RNA in the cytoplasm of neuro-progenitor stem cells ranged from 3–5 mg/ml (accuracy in this range is $\pm$0.4 mg/ml) while for glial cells the concentration becomes lower than the detection limit of RNA for our instrument, which is $\sim$1 mg/ml

This finding was somehow surprising considering that the ribosomal RNA represents the dominant type of RNA in cells and most somatic cells have abundant ribosomes. However, the estimates of the cell volumes indicate that the increase in the cytoplasm volume for glial cells by a factor of 4.5 may account for a decrease in the cytoplasm RNA concentration during the differentiation of neuroprogenitor cells to the glial phenotype. The higher intensities of Raman bands corresponding to RNA in neural stem cells may also be related to a higher amount of RNA in the cytoplasm of undifferentiated cells as suggested by earlier histological analysis of embryonic brain explants. These studies showed that neuroepithelial progenitor populations in the ependymal layer have a higher total RNA content than their mature differentiated progeny [57]. Increased concentration of non-translated mRNAs corresponding to the post-transcriptional control of genes has been related to neurogenesis [58], neuronal function [59], as well as stem cell proliferation and embryogenesis [60,61]. For example, high abundance of proteins which repress the translation of mRNAs and maintained the undifferentiated state of neuroprogenitor stem cells have been found in the

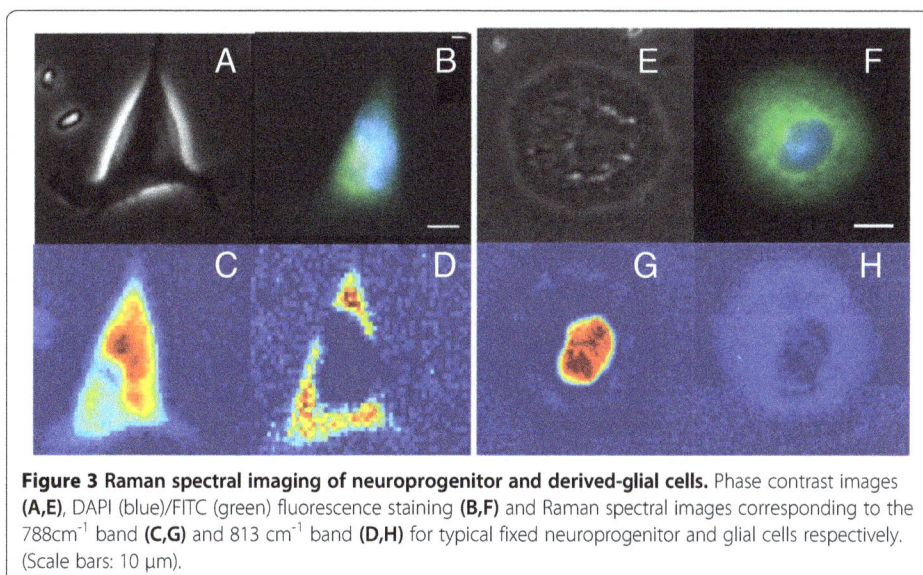

Figure 3 Raman spectral imaging of neuroprogenitor and derived-glial cells. Phase contrast images **(A,E)**, DAPI (blue)/FITC (green) fluorescence staining **(B,F)** and Raman spectral images corresponding to the 788cm^{-1} band **(C,G)** and 813 cm^{-1} band **(D,H)** for typical fixed neuroprogenitor and glial cells respectively. (Scale bars: 10 μm).

cytoplasms of these cells [58]. Nevertheless, RMS has the ability to detect and quantify the concentration of RNA in neuroprogenitor cells, with high spatial (less than 1 micron) and temporal resolution (minutes).

Time- and spatially-resolved monitoring of mineralisation of bone nodules in-vitro

Bone is one of the largest organs in the human body with numerous mechanical and haematological functions. In addition to age-related diseases, there are congenital bone deformation, bone cancer and bone trauma that may require a bone transplant. Current bone reconstruction and replacement surgical procedures are based on allogeneic tissue grafts. Because of a limited supply of allogeneic bone, tissue engineering and regeneration of the bone based on mesenchymal stem cells has raised a major research interest. The bone grafts obtained in-vitro can be used for clinical applications to restore the functionality of the skeletal system of the patients [1]. The growth of bone in-vitro requires a culture of mesenchymal stem cells (MSCs) in the osteogenic culture medium. Following proliferation and differentiation, MSCs gives rise to the osteoblasts, which are specialized cells responsible for the formation of the bone nodules. Changes during formation and mineralisation of bone nodules can be observed in Raman bands associated to vibrations of hydroxyapatite, mostly in the 950–960 cm^{-1} range [29,30]. RMS was used to record time- and spatially-resolved molecular information during differecntiation of MSCs and formation of bone nodules over 28 days time period.

Figure 4 presents phase contrast images of MSCs grown in osteogenic and non-osteogenic media at different time-points. Mineralisation of the cultures grown in osteogenic media was confirmed by alizarin red staining performed at the end of the experiments. The difference between Raman spectra of MSCs grown in osteogenic and non-osteogenic media are presented in Figure 5A. Mineralisation of bone nodules is indicated by the strong Raman band at ~959 cm^{-1} corresponding to the PO_3^- symmetric stretching vibrations.

Principal component analysis (PCA) was used to extract temporal and spatial chemical information during differentiation of MSCs. The first four principal components (PCs) containing 99% from the total variance are presented in the Figure 5B. The loading of PC2 resembles the Raman spectrum of hydroxyapatite (HA) [62], which containes a sharp peak at 958 cm^{-1} assigned to PO_3^- symmetric stretching and represents a strong indication of mineralization. Images obtained by plotting the PC2 scores provides information regarding the phase (disordered and crystalline) and spatial distribution of HA. Crystalline HA elicits a sharp Raman band at 958 cm^{-1} while the disordered HA phase is indicated by a broader band in the 940-945 cm^{-1} region. The intensity peak ratio I_{958}/I_{945} represents a strong indicator of spatial and temporal evolution of the HA phase (Figure 5C). This ratio was found to increase over time, indicating a higher concentration of disordered phase at the beginning of bone nodule formation leading to a phase transition towards crystalline phase at day 19.

Monitoring cardiac differentiation of human embryonic stem cells

The human heart is considered to be a non-regenerative tissue and the permanent loss of cardiomyocytes can lead to cardiac muscle failure. For this reason stem cell therapy is considered a desirable alternative to classical heart transplants for replacing the damaged cardiac tissue. Although these technologies offer great promise for patient with cardiac failure, in-vitro differentiation protocols are not optimised and currently produce cell

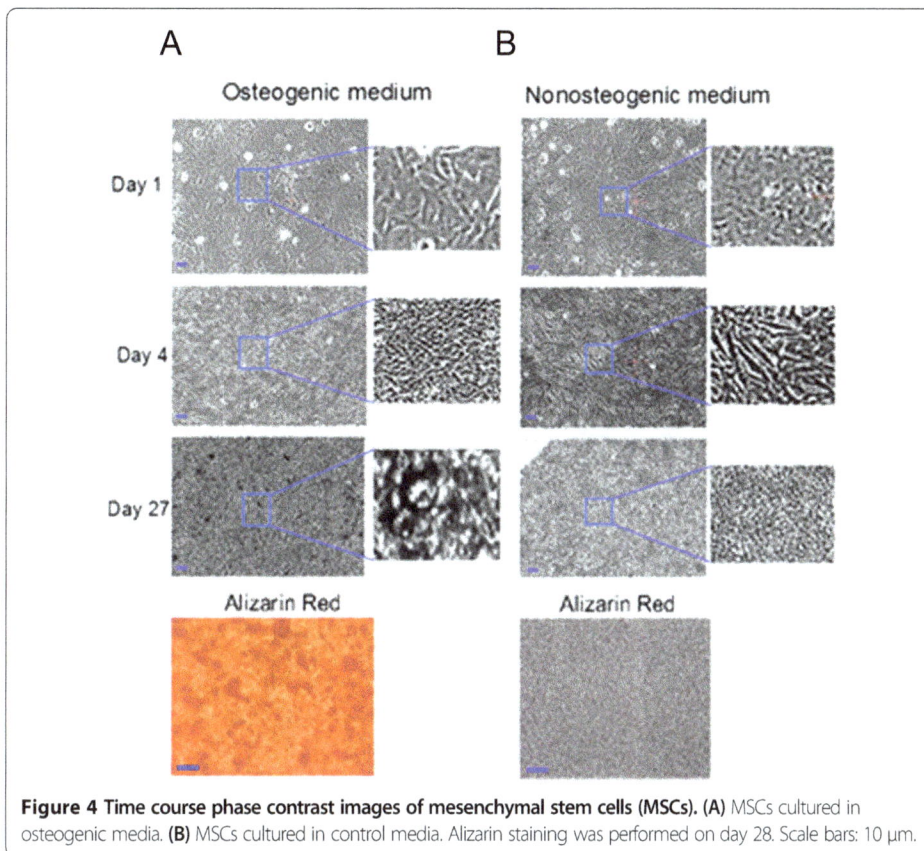

Figure 4 Time course phase contrast images of mesenchymal stem cells (MSCs). (A) MSCs cultured in osteogenic media. **(B)** MSCs cultured in control media. Alizarin staining was performed on day 28. Scale bars: 10 μm.

populations with high phenotypic heterogeneity. Such cell populations require further enrichment and purification prior to clinical use [63].

Pascut *et. al.* showed that RMS can be used for label-free discrimination of individual live cardiomyocytes (CM), derived in-vitro from human embryonic stem cells [64] and investigated the potential for developing Raman-activated cell-sorting of individual cells [52]. A statistical multivariate model using principal component analysis of Raman spectra from stem cell-derived CMs and non-CMs achieved 97% specificity and 96% sensitivity [64]. It was found that the main spectral features that provided the discrimination were related to Raman bands associated to glycogen and proteins [28]. Furthermore, online analysis of the beat frequency of individual cardiomyocytes analysed by RMS showed no significant differences when exposed to the Raman laser, when compared to control cells. In a recent study, RMS was used to monitor the molecular changes in live stem cells during cardiac differentiation, and then to correlate these changes to gold-standard fluorescence staining for the cardiac phenotype [65].

Figure 6A presents schematically the design of the experiment in which embryoid bodies (EB) formed by aggregation of human embryonic were grown in conditioned medium on the Raman microscope. The live EBs were grown in conditioned medium to induce differentiation towards cardiac phenotype. Raman spectral maps were acquired by raster-scanning at 24 hour intervals between days 5 and 9 of differentiation, a window wherein cardiac markers were expected to be expressed (immuno-fluorescence on fixed EBs). Flow cytometry and immuno-fluorescence analysis using the CM markers α-actinin and/or cardiac troponin I carried out on individual cells dispersed

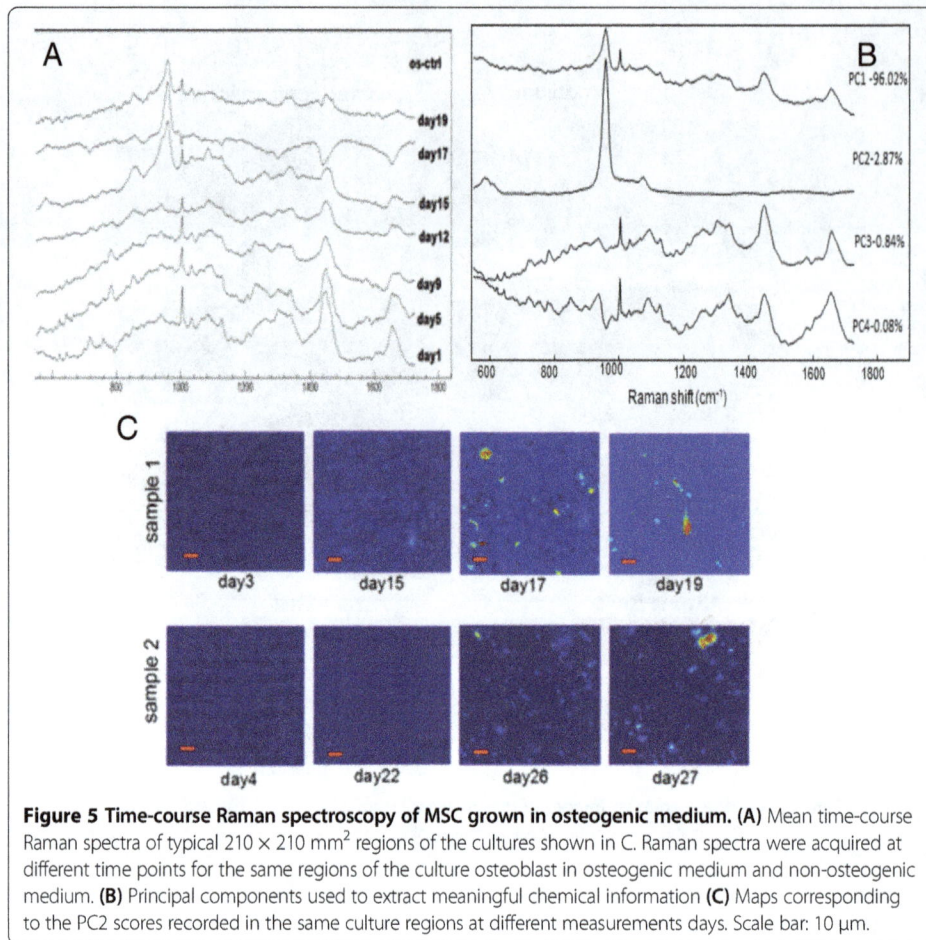

Figure 5 Time-course Raman spectroscopy of MSC grown in osteogenic medium. (A) Mean time-course Raman spectra of typical 210 × 210 mm² regions of the cultures shown in C. Raman spectra were acquired at different time points for the same regions of the culture osteoblast in osteogenic medium and non-osteogenic medium. **(B)** Principal components used to extract meaningful chemical information **(C)** Maps corresponding to the PC2 scores recorded in the same culture regions at different measurements days. Scale bar: 10 μm.

from EBs at day 12 of differentiation indicated that approximately 85% of the cells present in beating EBs were CMs, whereas less than 1% were identified as CM in non-beating EBs.

Figure 7 compares the time-course average Raman spectra of typical EBs, beating (successful cardiac differentiation) and non-beating (unsuccessful cardiac differentiation) with the Raman spectra of individual hESC-derived CMs and a non-CM. Significant differences between the Raman spectra of EBs can be observed starting on day 7, when the spectra of the beating EBs showed an increase in the intensity of the bands at 482 cm^{-1}, 577 cm^{-1}, 858 cm^{-1}, 937 cm^{-1}, 1083 cm^{-1} and 1340 cm^{-1}. The increase in the intensity of these bands was directly correlated with the increase in the number of CMs, and intense bands at the same frequencies can be observed in the Raman spectra of isolated beating CMs derived from hESCs (Figure 7A). These spectral changes were attributed to the formation of myofibrils and accumulation of glycogen in the CMs. These molecular changes are hallmarks for the formation of cardiac tissue and reflect the development of the contractile machinery of the CMs [66,67]. A high accumulation of glycogen in hESC-derived CMs was observed by transmission electron microscopy for CMs derived from several hESCs lines [67] and was related to the increase in fuel demand following the switch from the glycolytic metabolism to oxidative phosphory-laion [68,69].

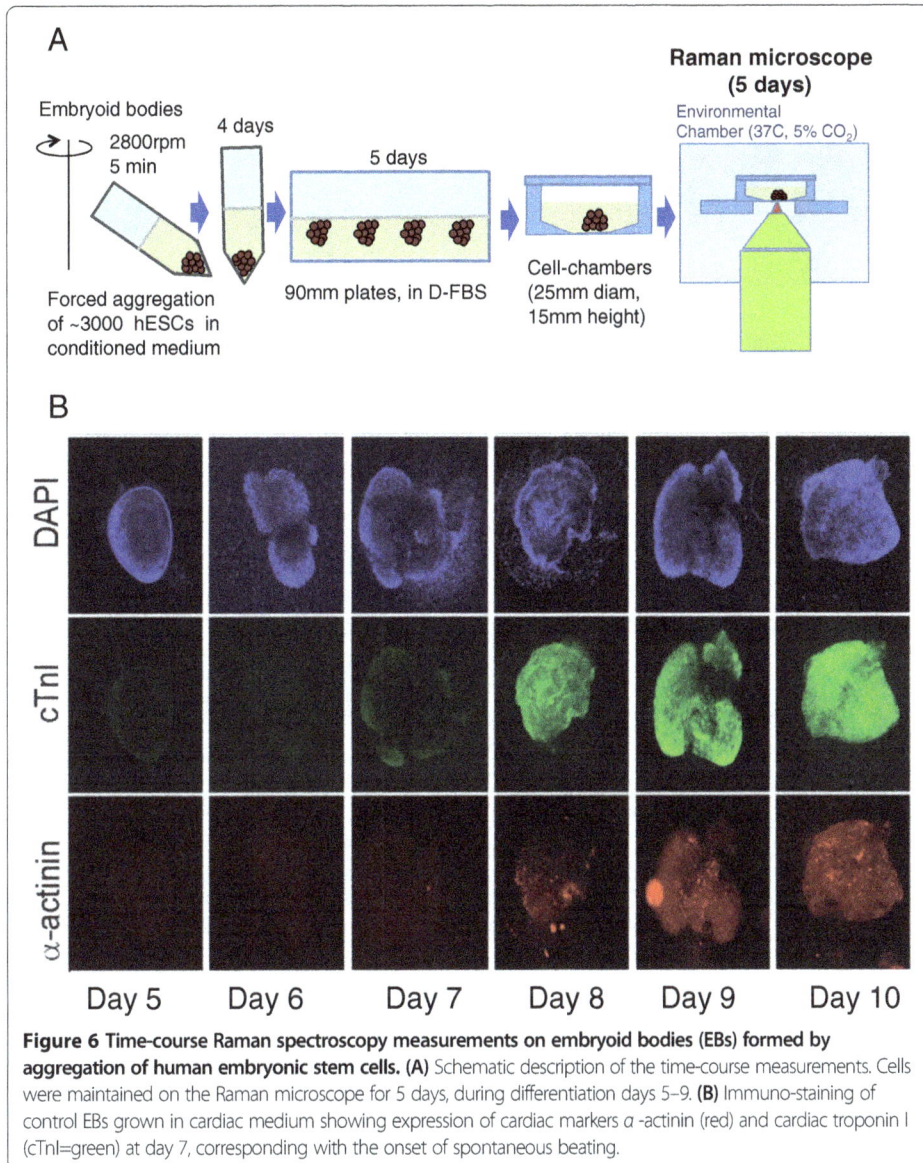

Figure 6 Time-course Raman spectroscopy measurements on embryoid bodies (EBs) formed by aggregation of human embryonic stem cells. (A) Schematic description of the time-course measurements. Cells were maintained on the Raman microscope for 5 days, during differentiation days 5–9. **(B)** Immuno-staining of control EBs grown in cardiac medium showing expression of cardiac markers a -actinin (red) and cardiac troponin I (cTnI=green) at day 7, corresponding with the onset of spontaneous beating.

RMS can also provide time-resolved 2D spectral maps of the EBs and is able to detect the increase in the number of CMs during differentiation. Starting with day 7, the spectral maps show the appearance of CMs in the beating EBs, while no relative increase in the PC1 scores was detected in the non-beating EBs. It was also found that the areas of high PC1 scores also matched the regions of the EBs where the beating was most pronounced, as well as the expression of α-actinin obtained by immuno-fluorescence imaging at the end of the time-course experiments [65]. These recent studies show the potential of RMS for non-invasive monitoring of stem cell differentiation, which may enable a more efficient optimization of the relevant bioprocesses.

Conclusion

This paper reviews recent applications of Raman micro-spectroscopy for time- and spatially-resolved molecular imaging of stem cells during differentiation in-vitro. By

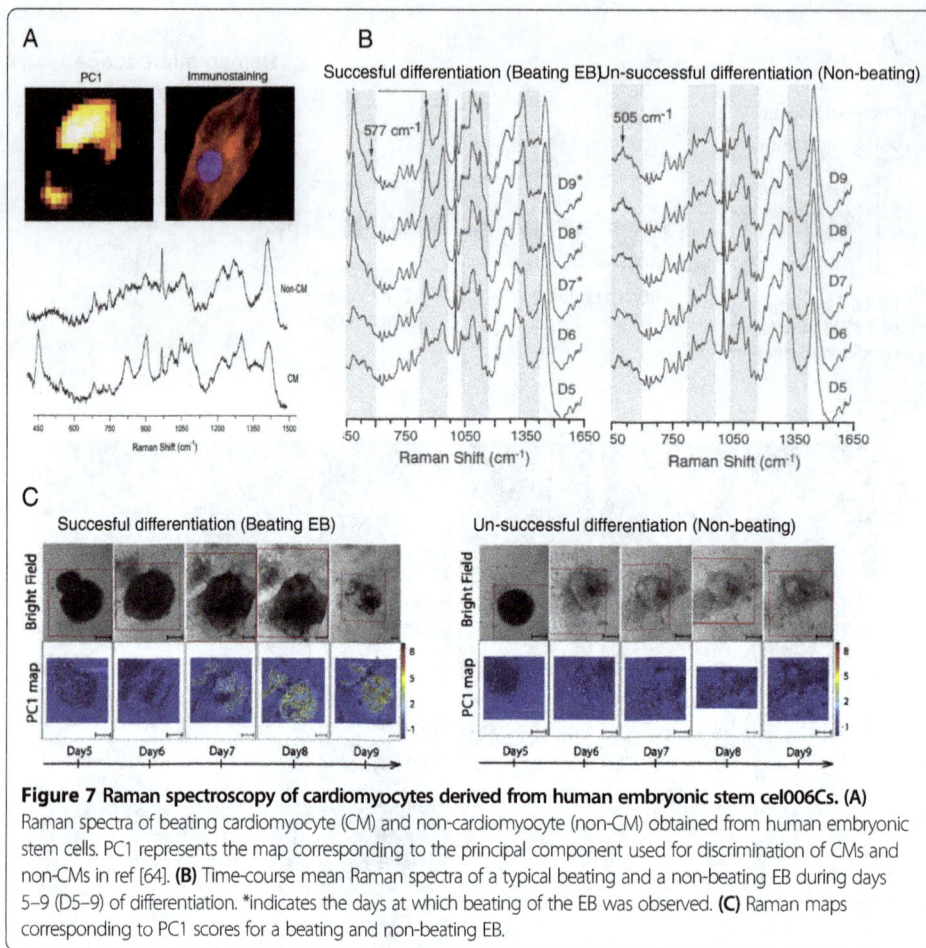

Figure 7 Raman spectroscopy of cardiomyocytes derived from human embryonic stem cel006Cs. (A) Raman spectra of beating cardiomyocyte (CM) and non-cardiomyocyte (non-CM) obtained from human embryonic stem cells. PC1 represents the map corresponding to the principal component used for discrimination of CMs and non-CMs in ref [64]. **(B)** Time-course mean Raman spectra of a typical beating and a non-beating EB during days 5–9 (D5–9) of differentiation. *indicates the days at which beating of the EB was observed. **(C)** Raman maps corresponding to PC1 scores for a beating and non-beating EB.

integrating the Raman micro-spectrometer with an environmental enclosure, RMS can be used for non-invasive monitoring time-dependent molecular changes in live cells and can provide on-line information regarding the cells and their phenotypic characteristics. RMS may be a useful technique for monitoring bioprocesses and help the refinement and standardisation of differentiation protocols to induce the efficient differentiation of pluripotent stem cells. Such non-invasive techniques are needed to help overcoming the current bottlenecks in the manufacturing and quality assessment of stem cell populations, which are key factors for the future advancement and widespread clinical use of regenerative medicine therapies. In addition, information regarding molecular changes during differentiation can advance the understanding of stem cell differentiation and the development of in vitro models for embryo development. RMS can also provide an invaluable platform for further fundamental studies on stem cells and the effect of various stimuli on their differentiation (eg. mechanical stimulation for osteoblasts), as well as in-vitro testing of new pharmaceuticals on cell models.

Competing interests
The authors declare that they have no competing interests.

Authors' contributions
AG carried out the Raman spectroscopy experiments on neuroprogenitor and mesenchymal stem cells, and drafted the paper. FCP carried out the Raman spectroscopy on hESCs ad cardiomyocytes. VS and CD conceived the study and supervised the cell-cultures. IN conceived the study, supervised the Raman spectroscopy measurements and drafted the paper. All authors read and approved the paper.

Acknowledgements

This work was supported by the Biotechnology and Biological Sciences Research Council, UK (BB/G010285/1).

Author details

[1]School of Physics and Astronomy, University of Nottingham, Nottingham NG7 2RD, UK. [2]School of Medicine, University of Nottingham, Nottingham NG7 2RD, UK.

References

1. Frohlich M, Grayson WL, Wan LQ, Marolt D, Drobnic M, Vunjak-Novakovic G. Tissue engineered bone grafts: biological requirements, tissue culture and clinical relevance. Curr Stem Cell Res Ther. 2008;3:254–64.
2. Williams D. Benefit and risk in tissue engineering. Mater Today. 2004;7:24–9.
3. Evans MJ, Kaufman MH. Establishment in culture of pluripotential cells from mouse embryos. Nat. 1981;292:154–6.
4. Dean M. Cancer stem cells: redefining the paradigm of cancer treatment strategies. Mol Interv. 2006;6:140–8.
5. Choumerianou DM, Dimitriou H, Kalmanti M. Stem cells: promises versus limitations. Tissue Eng Part B Rev. 2008;14:53–60.
6. Trouillon R, Passarelli MK, Wang J, Kurczy ME, Ewing AG. Chemical analysis of single cells. Anal Chem. 2013;85:522–42.
7. Lin Y, Trouillon R, Safina G, Ewing AG. Chemical analysis of single cells. Anal Chem. 2011;83:4369–92.
8. Bour-Dill C, Gramain MP, Merlin JL, Marchal S, Guillemin F. Determination of intracellular organelles implicated in daunorubicin cytoplasmic sequestration in multidrug-resistant MCF-7 cells using fluorescence microscopy image analysis. Cytometry. 2000;39:16–25.
9. Rusan NM, Fagerstrom CJ, Yvon AMC, Wadsworth P. Cell cycle-dependent changes in microtubule dynamics in living cells expressing green fluorescent protein-alpha tubulin. Mol Biol Cell. 2001;12:971–80.
10. Caspers PJ, Lucassen GW, Carter EA, Bruining HA, Puppels GJ. In vivo confocal Raman microspectroscopy of the skin: noninvasive determination of molecular concentration profiles. J Invest Dermatol. 2001;116:434–42.
11. Meyer T, Bergner N, Bielecki C, Krafft C, Akimov D, Romeike BF, et al. Nonlinear microscopy, infrared, and Raman microspectroscopy for brain tumor analysis. J Biomed Opt. 2011;16:021113.
12. Kendall C, Stone N, Shepherd N, Geboes K, Warren B, Bennett R, et al. Raman spectroscopy, a potential tool for the objective identification and classification of neoplasia in Barrett's oesophagus. J Pathol. 2003;200:602–9.
13. Kong K, Rowlands CJ, Varma S, Perkins W, Leach IH, Koloydenko AA, et al. Diagnosis of tumors during tissue-conserving surgery with integrated autofluorescence and Raman scattering microscopy. Proc Natl Acad Sci U S A. 2013;110:15189–94.
14. Delhaye M, Dhamelincourt P. Raman microprobe and microscope with laser excitation. J Raman Spectrosc. 1975;3:33–43.
15. Puppels GJ, Demul FFM, Otto C, Greve J, Robertnicoud M, Arndtjovin DJ, et al. Studying single living cells and chromosomes by confocal Raman microspectroscopy. Nat. 1990;347:301–3.
16. Ye J, Fox SA, Cudic M, Rezler EM, Lauer JL, Fields GB, et al. Determination of penetratin secondary structure in live cells with Raman microscopy. J Am Chem Soc. 2010;132:980–8.
17. Matthews Q, Jirasek A, Lum J, Duan XB, Brolo AG. Variability in Raman spectra of single human tumor cells cultured in vitro: correlation with cell cycle and culture confluency. Appl Spectrosc. 2010;64:871–87.
18. Swain RJ, Jell G, Stevens MA. Non-invasive analysis of cell cycle dynamics in single living cells with Raman micro-spectroscopy. J Cell Biochem. 2008;104:1427–38.
19. Notingher I, Jell G, Lohbauer U, Salih V, Hench LL. In situ non-invasive spectral discrimination between bone cell phenotypes used in tissue engineering. J Cell Biochem. 2004;92:1180–92.
20. Brown KL, Palyvoda OY, Thakur JS, Nehlsen-Cannarella SL, Fagoaga OR, Gruber SA, et al. Raman spectroscopic differentiation of activated versus non-activated T lymphocytes: an in vitro study of an acute allograft rejection model. J Immunol Methods. 2009;340:48–54.
21. Notingher I, Verrier S, Haque S, Polak JM, Hench LL. Spectroscopic study of human lung epithelial cells (A549) in culture: living cells versus dead cells. Biopolymers. 2003;72:230–40.
22. Draux F, Gobinet C, Sule-Suso J, Manfait M, Jeannesson P, Sockalingum GD. Raman imaging of single living cells: probing effects of non-cytotoxic doses of an anti-cancer drug. Analyst (Cambridge, U K). 2011;136:2718–25.
23. Hedegaard M, Krafft C, Ditzel HJ, Johansen LE, Hassing S, Popp J. discriminating isogenic cancer cells and identifying altered unsaturated fatty acid content as associated with metastasis status, using k-means clustering and partial least squares-discriminant analysis of Raman maps. Anal Chem. 2010;82:2797–802.
24. Nawaz H, Bonnier F, Meade AD, Lyng FM, Byrne HJ. Comparison of subcellular responses for the evaluation and prediction of the chemotherapeutic response to cisplatin in lung adenocarcinoma using Raman spectroscopy. Analyst (Cambridge, U K). 2011;136:2450–63.
25. Pully VV, Lenferink A, Otto C. Hybrid Rayleigh, Raman and two-photon excited fluorescence spectral confocal microscopy of living cells. J Raman Spectrosc. 2010;41:599–608.
26. Konorov SO, Jardon MA, Piret JM, Blades MW, Turner RFB. Raman microspectroscopy of live cells under autophagy-inducing conditions. Analyst (Cambridge, U K). 2012;137:4662–8.
27. Hosoda A, Maruyama A, Oikawa D, Oshima Y, Komachi Y, Kanai G, et al. Detection of ER stress in vivo by Raman spectroscopy. Biochem Biophys Res Commun. 2011;405:37–41.
28. Konorov SO, Schulze HG, Piret JM, Turner RFB, Blades MW. Evidence of marked glycogen variations in the characteristic Raman signatures of human embryonic stem cells. J Raman Spectrosc. 2011;42:1135–41.
29. McManus LL, Burke GA, McCafferty MM, O'Hare P, Modreanu M, Boyd AR, et al. Raman spectroscopic monitoring of the osteogenic differentiation of human mesenchymal stem cells. Analyst (Cambridge, U K). 2011;136:2471–81.
30. McManus LL, Bonnier F, Burke GA, Meenan BJ, Boyd AR, Byrne HJ. Assessment of an osteoblast-like cell line as a model for human primary osteoblasts using Raman spectroscopy. Analyst (Cambridge, U K). 2012;137:1559–69.

31. Gentleman E, Swain RJ, Evans ND, Boonrungsiman S, Jell G, Ball MD, et al. Comparative materials differences revealed in engineered bone as a function of cell-specific differentiation. Nat Mater. 2009;8:763–70.
32. van Manen HJ, Kraan YM, Roos D, Otto C. Single-cell Raman and fluorescence microscopy reveal the association of lipid bodies with phagosomes in leukocytes. Proc Natl Acad Sci U S A. 2005;102:10159–64.
33. Krafft C, Knetschke T, Funk RHW, Salzer R. Identification of organelles and vesicles in single cells by Raman microspectroscopic mapping. Vib Spectrosc. 2005;38:85–93.
34. Uzunbajakava N, Lenferink A, Kraan Y, Willekens B, Vrensen G, Greve J, et al. Nonresonant Raman imaging of protein distribution in single human cells. Biopolymers. 2003;72:1–9.
35. Uzunbajakava N, Lenferink A, Kraan Y, Volokhina E, Vrensen G, Greve J, et al. Nonresonant confocal Raman imaging of DNA and protein distribution in apoptotic cells. Biophys J. 2003;84:3968–81.
36. Matthaus C, Chernenko T, Newmark JA, Warner CM, Diem M. Label-free detection of mitochondrial distribution in cells by nonresonant Raman microspectroscopy. Biophys J. 2007;93:668–73.
37. Draux F, Jeannesson P, Beljebbar A, Tfayli A, Fourre N, Manfait M, et al. Raman spectral imaging of single living cancer cells: a preliminary study. Analyst (Cambridge, U K). 2009;134:542–8.
38. Miljkovic M, Chernenko T, Romeo MJ, Bird B, Matthaus C, Diem M. Label-free imaging of human cells: algorithms for image reconstruction of Raman hyperspectral datasets. Analyst (Cambridge, U K). 2010;135:2002–13.
39. Chernenko T, Matthaus C, Milane L, Quintero L, Amiji M, Diem M. Label-free Raman spectral imaging of intracellular delivery and degradation of polymeric nanoparticle systems. ACS Nano. 2009;3:3552–9.
40. Ling J, Weitman SD, Miller MA, Moore RV, Bovik AC. Direct Raman imaging techniques for study of the subcellular distribution of a drug. Appl Optics. 2002;41:6006–17.
41. Meister K, Niesel J, Schatzschneider U, Metzler-Nolte N, Schmidt DA, Havenith M. Label-free imaging of metal-carbonyl complexes in live cells by Raman microspectroscopy. Angew Chem Int Ed. 2010;49:3310–2.
42. Fukunaga H, Yoshimura H, Nishina Y, Nagashima Y, Tachibana M. Label-free biomedical imaging of hydrodynamics in single human cells. Biomed Res-Tokyo. 2010;31:177–81.
43. Huang YS, Karashima T, Yamamoto M, Hamaguchi H. Molecular-level investigation of the structure, transformation, and bioactivity of single living fission yeast cells by time- and space-resolved Raman spectroscopy. Biochem. 2005;44:10009–19.
44. Zoladek A, Pascut FC, Patel P, Notingher I. Non-invasive time-course imaging of apoptotic cells by confocal Raman micro-spectroscopy. J Raman Spectrosc. 2011;42:251–8.
45. Eder SH, Gigler AM, Hanzlik M, Winklhofer M. Sub-micrometer-scale mapping of magnetite crystals and sulfur globules in magnetotactic bacteria using confocal Raman micro-spectrometry. PLoS One. 2014;9:e107356.
46. Okada M, Smith NI, Palonpon AF, Endo H, Kawata S, Sodeoka M, et al. Label-free Raman observation of cytochrome c dynamics during apoptosis. Proc Natl Acad Sci U S A. 2012;109:28–32.
47. Venkata HNN, Shigeto S. Stable isotope-labeled Raman imaging reveals dynamic proteome localization to lipid droplets in single fission yeast cells. Chem Biol. 2012;19:1373–80.
48. Swain RJ, Kemp SJ, Goldstraw P, Tetley TD, Steyens MM. Assessment of cell line models of primary human cells by Raman spectral phenotyping. Biophys J. 2010;98:1703–11.
49. Chan JW, Taylor DS, Zwerdling T, Lane SM, Ihara K, Huser T. Micro-Raman spectroscopy detects individual neoplastic and normal hematopoietic cells. Biophys J. 2006;90:648–56.
50. Zuser E, Chernenko T, Newmark J, Miljkovic M, Diem M. Confocal Raman microspectral imaging (CRMI) of murine stem cell colonies. Analyst (Cambridge, U K). 2010;135:3030–3.
51. Notingher L, Bisson I, Polak JM, Hench LL. In situ spectroscopic study of nucleic acids in differentiating embryonic stem cells. Vib Spectrosc. 2004;35:199–203.
52. Pascut FC, Goh HT, George V, Denning C, Notingher I. Toward label-free Raman-activated cell sorting of cardiomyocytes derived from human embryonic stem cells. J Biomed Opt. 2011;16:045002.
53. Martino G, Pluchino S. The therapeutic potential of neural stem cells. Nat Rev Neurosci. 2006;7:395–406.
54. Giusto E, Donega M, Cossetti C, Pluchino S. Neuro-immune interactions of neural stem cell transplants: From animal disease models to human trials. Exp Neurol. 2014;260:19–32.
55. Ghita A, Pascut FC, Mather M, Sottile V, Notingher I. Cytoplasmic RNA in undifferentiated neural stem cells: a potential label-free raman spectral marker for assessing the undifferentiated status. Anal Chem. 2012;84:3155–62.
56. Tu AT. Raman spectroscopy in biology: principles and applications. New York: Wiley; 1982.
57. Birge WJ. A histochemical study of ribonucleic acid in differentiating ependymal cells of the chick embryo. Anat Rec. 1962;143:147–55.
58. Okano H, Kawahara H, Toriya M, Nakao K, Shibata S, Imai T. Function of RNA-binding protein Musashi-1 in stem cells. Exp Cell Res. 2005;306:349–56.
59. Finkenstadt PM, Kang WS, Jeon M, Taira E, Tang W, Baraban JM. Somatodendritic localization of Translin, a component of the Translin/Trax RNA binding complex. J Neurochem. 2000;75:1754–62.
60. Kuersten S, Goodwin EB. The power of the 3' UTR: translational control and development. Nat Rev Genet. 2003;4:626–37.
61. Macnicol MC, Macnicol AM. Developmental timing of mRNA translation-integration of distinct regulatory elements. Mol Reprod Dev. 2010;77:662–9.
62. Ghita A, Pascut FC, Sottile V, Notingher I. Monitoring the mineralisation of bone nodules in vitro by space- and time-resolved Raman micro-spectroscopy. Analyst (Cambridge, U K). 2014;139:55–8.
63. Musunuru K, Domian IJ, Chien KR. Stem cell models of cardiac development and disease. Annu Rev Cell Dev Biol, Vol 26. 2010;26:667–87.
64. Pascut FC, Goh HT, Welch N, Buttery LD, Denning C, Notingher I. Noninvasive detection and imaging of molecular markers in live cardiomyocytes derived from human embryonic stem cells. Biophys J. 2011;100:251–9.
65. Pascut FC, Kalra S, George V, Welch N, Denning C, Notingher I. Non-invasive label-free monitoring the cardiac differentiation of human embryonic stem cells in-vitro by Raman spectroscopy. Biochim Et Biophys Acta-Gen Subj. 1830;2013:3517–24.
66. Snir M, Kehat I, Gepstein A, Coleman R, Itskovitz-Eldor J, Livne E, et al. Assessment of the ultrastructural and proliferative properties of human embry- onic stem cell-derived cardiomyocytes. Am J Physiol Heart Circ Physiol. 2003;285:H2355–63.

67. Gherghiceanu M, Barad L, Novak A, Reiter I, Itskovitz-Eldor J, Binah O, et al. Cardiomyocytes derived from human embryonic and induced pluripotent stem cells: comparative ultrastructure. J Cell Mol Med. 2011;15:2539–51.

68. St John JC, Ramalho-Santos J, Gray HL, Petrosko P, Rawe VY, Navara CS, et al. The expression of mitochondrial DNA transcription factors during early cardiomyocyte in vitro differentiation from human embryonic stem cells. Cloning Stem Cells. 2005;7:141–53.

69. Chung S, Dzeja PP, Faustino RS, Perez-Terzic C, Behfar A, Terzic A. Mitochondrial oxidative metabolism is required for the cardiac differentiation of stem cells. Nat Clin Pract Cardiovasc Med. 2007;4:S60–7.

12

Energy influx measurements with an active thermal probe in plasma-technological processes

Ruben Wiese[1*], Holger Kersten[1], Georg Wiese[2] and René Bartsch[2]

* Correspondence: wiese@physik.
uni-kiel.de
[1]Institut für Experimentelle und
Angewandte Physik, Leibnizstr. 19,
Kiel D-24098, Germany
Full list of author information is
available at the end of the article

Abstract

Many plasma-technological applications are based on plasma wall interaction, which can be characterised by calorimetric probes to measure the energy influx from the plasma to the substrate surface. Passive probes are based on the principle of recording the temperature course during heating and cooling of the probe for calculating the energy influx. The disadvantages of these probes are that the energy influx has to be interrupted by switching off the energy source or by using suitable apertures and by the necessity of knowing the exact heat capacity of the probe.

A continuously operating active probe is, therefore, developed which does not need to be calibrated and which compensates the environmental effects as well as the heat conduction by the probe holder. By means of controlled electrical heating the probe is set to a given working temperature and then the energy supply supporting the fixed operating temperature is measured. The energy influx by the plasma is compensated by decreasing the heating power and is directly displayed in J/cm^2s. Some practical measurements are presented. Even, if the probe is designed as double probe the directionality of the energy influx can be determined.

Keywords: Sensors; Thermal probe; Energy influx; Process control

Introduction

A large variety of plasma-technological applications are based on plasma wall interaction which occurs via the generated plasma sheath. A good possibility for its characterisation is offered by calorimetric thermal probes for energy influx measurements from the plasma to the substrate, e.g. for the measurement of the deposited power.

Passive thermal probes are often used by several groups, based on the principle of recording the temperature temporal course at the heating and the cooling phase to calculate the energy influx [1,2]. A disadvantage of such passive probes is that the energy influx has to be interrupted by switching off the energy source or by using suitable apertures.

Active probes are also established, constructed as a planar substrate or as a rotationally-symmetric membrane, whereby the energy influx is determined by the measurement of a temperature gradient [3-5]. However, the temperature of these probes is not freely adjustable and its distribution along the surface is not constant. A calibration is necessary before the measurement which may constitute an additional error source.

A novel continuously operating active probe is presented which does not need to be calibrated and which compensates the environmental effects as well as the heat conduction by the probe holder. The change in the heat capacity of the probe by thin film deposition also does not influence the measured energy influx. With a double probe configuration it is even possible to measure the directionality of the energy influx.

Different methods are established for determination the energy flux at plasma-technological processes: for example by measuring the temperature difference generated on a substrate [1,6], by measuring the increasing temperature in the centre of a membrane being exposed to an energy flux and clamped and cooled by a guard ring [4,7], or by recording the temporal heating and cooling temperature course of a dummy substrate [5,8,9].

These measuring methods often produce problems by measurement errors, caused by undesirable heat transport processes that are difficult to control. This concerns for example the heat conduction of the probe and/or the change of the heat capacity by coating the probe. Furthermore, the probe has to be calibrated by an energy source of known intensity, which is associated with uncertainty again. For example, the frequently unknown environmental conditions during calibration usually differ and changing reflexion and emission properties of the thermal probe occur.

Ellmer et al. carried out measurements with a Gardon sensor and tried to minimize the errors by blackening the probe [4,7]. However, the calibration results deviated strongly from the expected value because of the heat conduction of the thermal probe. [4] Even through calibrating the passive thermal probe, errors of at least 20% were estimated [5,10,11].

Stahl et al. have calibrated passive probes by an electron beam to minimize the measurement error [12]. Furthermore, Bornholdt et al. used a transient method for measurements with a passive thermal probe [13]. In addition, Ellmer et al. investigated a method for calibrating their calorimetric sensor by charged particles emitted from the plasma [14].

Theoretical considerations and principle of measurement

The problems of calibration were the reason for investigating other methods to measure at a constant thermal balance where heat conduction should not influence the results and calibration is not required.

The following principle of the active thermal probe is applied: likewise as a "dummy substrate" the probe is used to determine the energy flux to a given area. The probe is placed in the region of the plasma where the energy influx needs to be measured.

At the beginning, the probe is heated until the temperature balance reaches the equilibrium temperature T_{equ} without any plasma exposure.

The heating energy $J_h|_{cool}$ is supplied as needed, e.g. by an electrical heater. The incoming heat flux received at the probe (Q_{in}) and the outgoing heat flux (Q_{out}) in this "cold" phase (without plasma) are

$$Q_{in}|_{cool} = \int_A \left(J_h|_{cool} + J_{env}(T_{env}) \right) dA \tag{1}$$

$$Q_{out}|_{cool} = \int_A \left(J_{rad}\left(T_{equ}\right) + J_{con}\left(T_{equ} - T_{car}\right) \right) dA \tag{2}$$

Here J_{env} is the heat conduction from the environment, T_{env} the environmental temperature, J_{rad} the heat loss at the probe by radiation, J_{con} the heat flux from and to the probe generated by heat conduction, T_{car} the temperature of the probe holder or the connection cable and A is the probe area. Since thermal balance is assumed, the following equation must be valid

$$Q_{in}|_{cool} = Q_{out}|_{cool} \tag{3}$$

and, thus,

$$J_h|_{cool} = J_{rad}(T_{equ}) + J_{con}(T_{equ} - T_{car}) - J_{env}(T_{env}) \tag{4}$$

The probe is conditioned by adjusting the temperature balance and determining the required energy input $J_h|_{cool}$.

In the case of an additional energy flux J_{in} from the plasma, the temperature rises and the thermal balance is disturbed. This effect is counteracted by lowering the input heat ΔJ_h to a certain amount until the temperature balance T_{equ} is achieved again. To maintain this state, the energy $J_h|_{cool}$ becomes redundant and is replaced by a lower energy.

$$J_h|_{heat} = J_h|_{cool} - \Delta J_h, \tag{5}$$

since some of the heat loss is compensated by the energy flux. It can easily be demonstrated that now the heating power is

$$J_h|_{heat} = -J_{gain} - J_Q - J_{env}(T_{env}) + J_{rad}(T_{equ}) + J_{con}(T_{equ} - T_{car}) \tag{6}$$

Here, J_Q is the energy influx at the probe, which is produced by the heat radiation of the source and J_{gain} is the energy influx at the probe produced by the plasma process (without the heat radiation of the source).

If this relation together with equation (4) is combined with equation (5), the actual energy J_{in} at the probe can be obtained at the equilibrium temperature T_{equ}:

$$J_{in} = \Delta J_h = J_{gain} + J_Q \quad |_{Tequ} \tag{7}$$

There are no terms in equation (7) which depend on probe temperature, environmental temperature or the temperature of the probe holder, respectively.

Furthermore, the measured energy is independent of the heat capacity of the probe. That is an important fact because now the determination of the energy by calibrating a known radiation source is not necessary. Another advantage is the independence of the measurement on the heat conduction J_{con} along the probe holder.

Because the measured energy influx is, in principle, the disturbing variable of the temperature balance, the probe becomes less sensitive due to higher heat loss through the holder and connection cable. The compensation of this disturbance is directly proportional to the efficiency of the disturbance, i.e. all other heat flows in equation (6) which influence the heat flux $J_h|_{heat}$. If the thermal radiation J_{rad} of the probe or the heat conduction J_{con} through the connecting cables is high, all other terms remain ineffective.

These conclusions can be found in the given balance of equation 6: The heating power $J_h|_{heat}$ cannot be negative. Therefore, considering equation (5) the heating power must always be set higher than the energy flux to be measured. The sensitivity and the maximum of the measurable energy flux can be influenced by the geometry of the probe holder.

For accurate measurements of the energy influx all incoming and outgoing heating fluxes to the probe which are not involved in measurement value have to be assumed constant in time.

This requires specifically constructed probe holders. They contain one or more heated zones, which are set to a given constant operation temperature close to the probe temperature. Thus, the heat fluxes between holder and probe are zero or constant.

A similar configuration is used for the construction of the double probe. Two identical probes are bonded together with their back sides. If both probes are at the same temperature, the heating flux between them is zero. Hence, each probe detects only the incoming energy flux from one half-space. By rotation of the probe this can be used to obtain a "heating radar image" of the environment.

In particular, the ability for continuous measurement is a great advantage of the active probe. After reaching the temperature balance, any change of energy influx from plasma is registered and the temperature balance is restored again. However, setting the temperature balance level as well as the energy influx measurement occurs with a time delay. Note that only the change of the energy influx can be measured and be used as a control variable at the same time when used in process controlling.

Configuration of the probe and measurement procedure

A specially constructed PT100-cell heated by an electrical current, whose temperature can be measured and controlled by a supplied external power is exposed to the energy flux of the plasma. The supplied power and the temperature can be calculated from the potential difference and the current. The probe is connected through a shunt to the output of an amplifier which is controlled by a digital-analog converter. Customised software is installed to filter the readings, control the temperature and calculate the energy influx. For thermal isolation, a second heating is located between the sensor area and the holder, which works in the same way and is set to the same temperature as the sensor. This is necessary to ensure that during the measurement the energy fluxes between holder and probe are zero or constant in time. This is important for the validity of equation 7.

Figure 1 shows a photograph of a sensor with a size of about 7×10 mm.

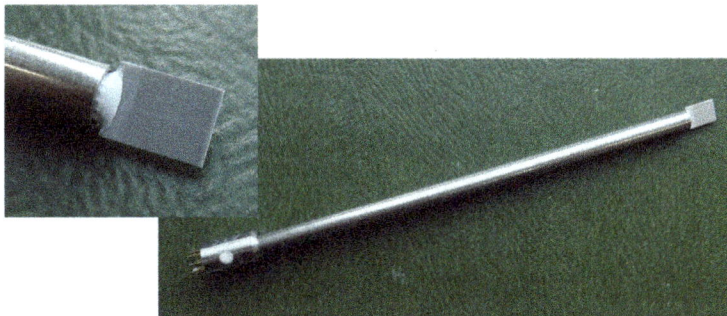

Figure 1 Active Probe, e.g. Pt100-cell on the manipulator arm.

The procedure stores the value of power necessary to maintain the temperature balance level at set-point temperature without energy influx and controls the temperature balance level by analysing the value and the changing rate of the probe temperature. Also it calculates the difference between the power currently needed and the power needed without energy influx from plasma. It displays directly if measurement starts the incoming energy influx at the probe in mW.

Experimental results

Test measurements were obtained by a prototype of the active probe in a plasma chamber at a pressure of 0.04 Pa. The first test should show if the active thermal probe can provide reliable and repeatable data of the energy influx in a plasma environment. For that reason an ion beam is quite suitable because it is a very constant energy source without coating the probe.

An ion beam source at beam voltage of 500 V and with beam diameter of 160 mm was used at a distance of 225 mm in front of the probe. Typical measurements are shown in Figure 2, e.g. the temporal course of the input heating power at the probe and the resulting temperature versus time. After switching the energy flux (ion beam) at t = 300 s the probe temperature increases shortly from 343°C to 355°C and the control unit responds by decreasing the heating power from about 580 mW to about 40 mW. After a certain time the temperature balance level (343°C) is reached again. At this point the input power is approximately 540 mW lower then at the "plasma off" point (at the time from 100 to 300 s) which is equivalent to the energy influx from the plasma ion source to the probe. After switching-off the source (at t = 950 s), the probe temperature decreases shortly but reaches its set point again after about 10 sec.

The energy influx at the probe can be measured with an accuracy of approximately 1 mW/cm^2, which has been confirmed by the fluctuations of the used prototype. Smaller changes in the energy influx are not measurable because of the stochastic fluctuations of the data.

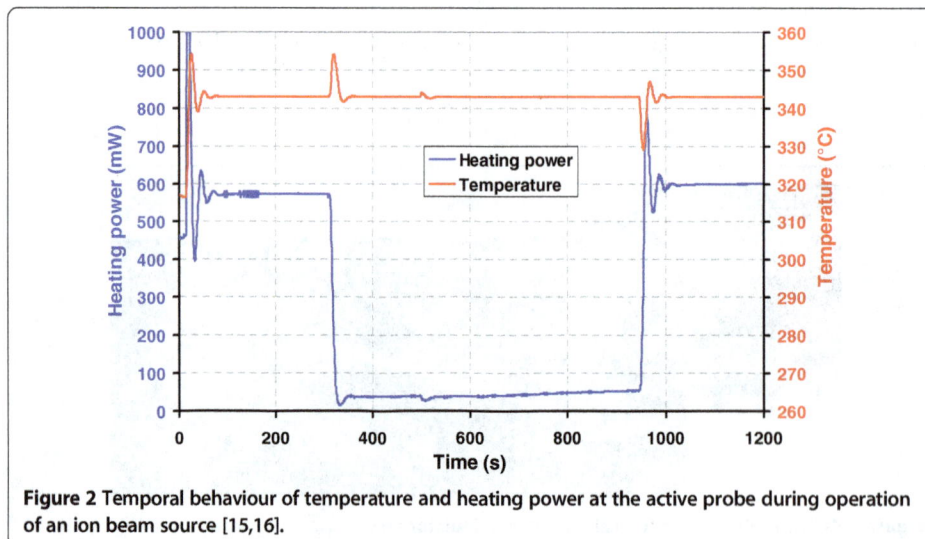

Figure 2 Temporal behaviour of temperature and heating power at the active probe during operation of an ion beam source [15,16].

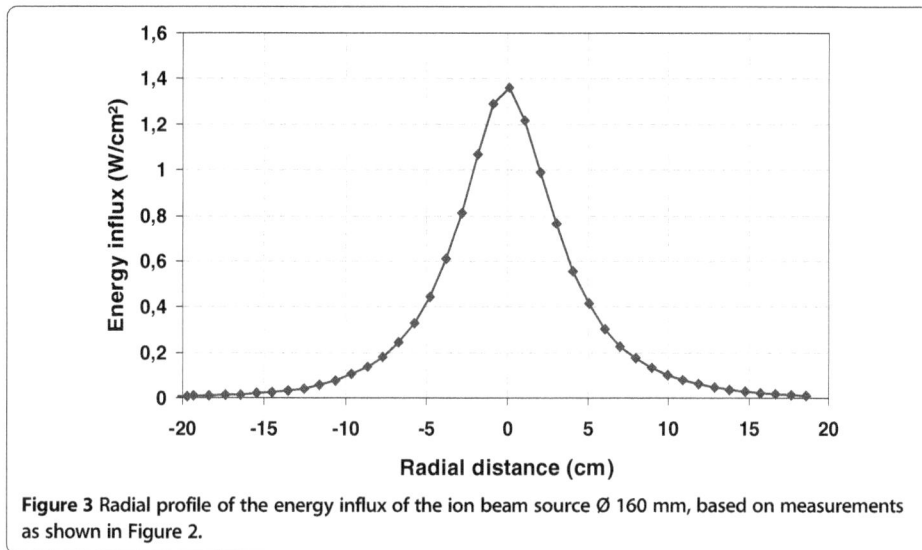

Figure 3 Radial profile of the energy influx of the ion beam source Ø 160 mm, based on measurements as shown in Figure 2.

The time needed to reach the temperature balance level of the probe was in the range of 30 s which may be for many applications, especially for controlling and regulating long-time plasma processes, an acceptable value. The time for reaching the equilibrium depends on the heat capacity of the sensor and the quality of the temperature control technique.

However, by the process control the reaction time of the probe can be much shorter. A dysfunction of the process can already be detected by small variations of the probe heating power. That means, the temporal gradient of the heating power can be an indicator for the process regulation. Typical dimensions of the probe (from 2 × 2 mm to 7 × 7 mm) allow for measurements with sufficient spatial resolutions.

Finally, the result of the first test was that the active thermal probe is suited to provide reliable data of the energy influx. As a useful example the radial profile of the energy influx in the ion beam of the described source was measured and is shown in Figure 3. Although the beam diameter is 160 mm in a distance of 225 mm there is a broader profile due to divergence effect. By means of this measured profile the symmetry of the ion source can be evaluated.

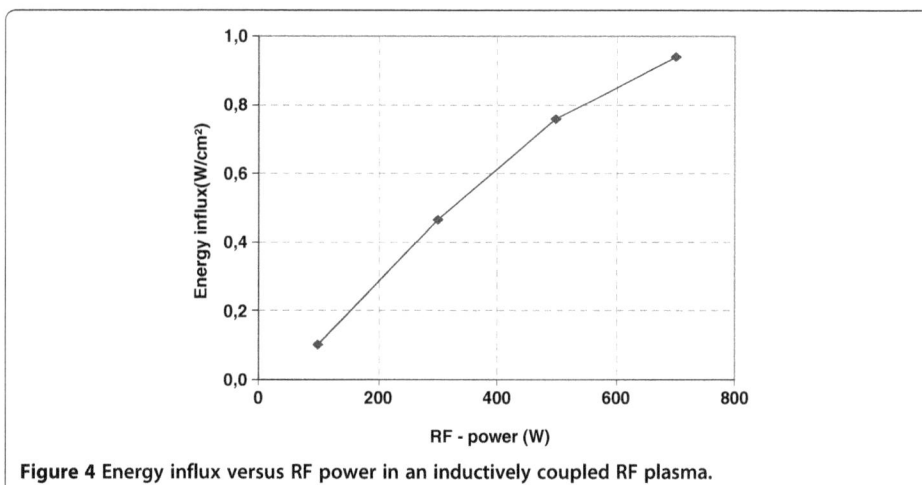

Figure 4 Energy influx versus RF power in an inductively coupled RF plasma.

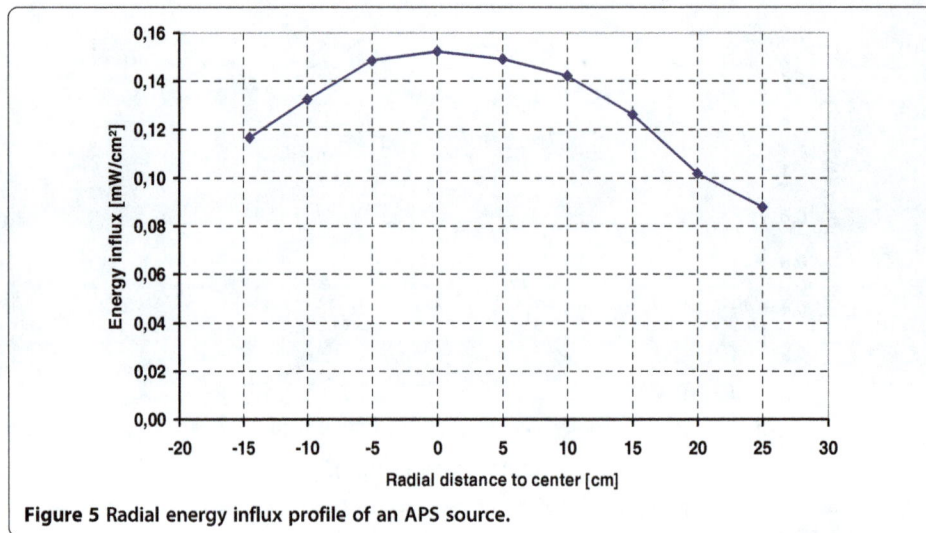

Figure 5 Radial energy influx profile of an APS source.

Additional test measurements have been performed in an inductively coupled rf-plasma and a dc-plasma source (APS, so-called advanced plasma source [17]). These experiments should prove if the electromagnetic field of the plasma disturbs the measurements. For the inductively coupled rf-plasma in argon (pressure 0.5 Pa, rf-power 300 W) used for plasma etching the dependence of the energy influx on the rf-power is shown in Figure 4. For the APS source (bias voltage 133 V, power 7.3 kW, pressure 0.02 Pa) a radial profile has been measured, too, in order to illustrate the applicability of the sensor, see Figure 5. Both plasma sources are commonly used in coating devices for the production of optical films. Also a measurement in such as plasma environment is possible.

In thin film deposition processes the probe may be coated itself – and its behaviour under such conditions is of special interest. For this purpose, the probe was alternately coated by copper and titanium. The results are shown in Figure 6. The coating experiments were carried out in a chamber with two magnetrons at a pressure of 0,5 Pa, an argon flux of 100 sccm, a magnetron power of 1000 W and a probe temperature of about 320°C. Before the measurements the probe was coated with a titanium layer. At 250 s (see Figure 6) the copper coating process starts. It could be observed that the heating power of the probe is decreased by the incoming energy influx. Since the reflection coefficient of copper is lower then in the case of titanium the incoming energy increases as the copper film is growing. This effect causes the decrease of the heating

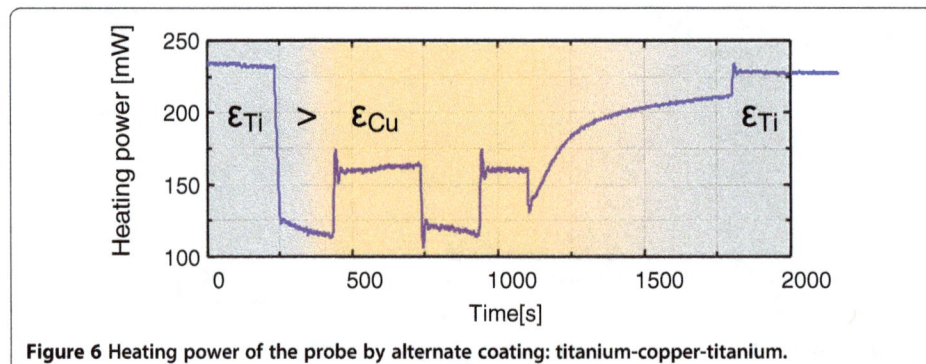

Figure 6 Heating power of the probe by alternate coating: titanium-copper-titanium.

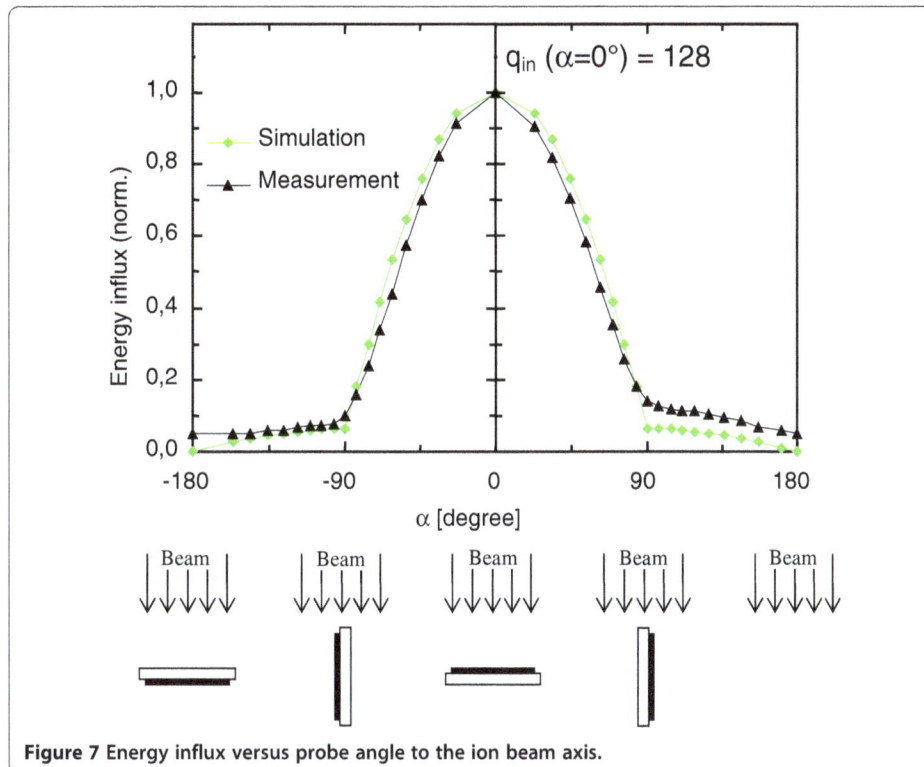

Figure 7 Energy influx versus probe angle to the ion beam axis.

power in the time interval t = 250 s to 450 s (see Figure 6). After closing of the aperture, the energy flux is interrupted. But the base level at 230 mW is not reached again. Assuming that the probe temperature is constant. This observation means, that the radiation loss of the copper layer has significantly decreased. The initial level (source switched off) of about 160 mW is also conserved after additional coatings with copper (t = 500 s to 1200 s). At t = 1200 s the titanium coating starts again. The cumulative coating with titanium changes the radiation coefficient of the probe surface again. The radiation losses increase and the heating power of the probe increases too (t = 1200 s until 1600 s). If a welded layer is produced, then the heating power will be constant. After closing of the aperture (1900 s) the coating process is finished and the base level of the experiment is almost reached.

Hence, the application of the thermal probe at different coating processes is possible. However, the coating material should be not changed during the measurement.

A quite interesting benefit of the probe is the application as a double probe. By means of such a probe setup the direction of the incoming energy influx can be detected. Two identical probes are bonded by their back sides. The practical experience has shown that all control cycles on the probe chips operate accurately. Obviously, the thermal resistance between both chips is sufficiently high for the thermal decoupling.

With this configuration angle-resolved measurements with an ion source (type Vecco ALS340) has been performed. The process parameters were as follows: pressure at 0,4 pa argon with 10% oxygen, a beam voltage of 2 kV and a beam current at 415 mA. The probe is located in front of the ion source in a distance of 30 cm. During the

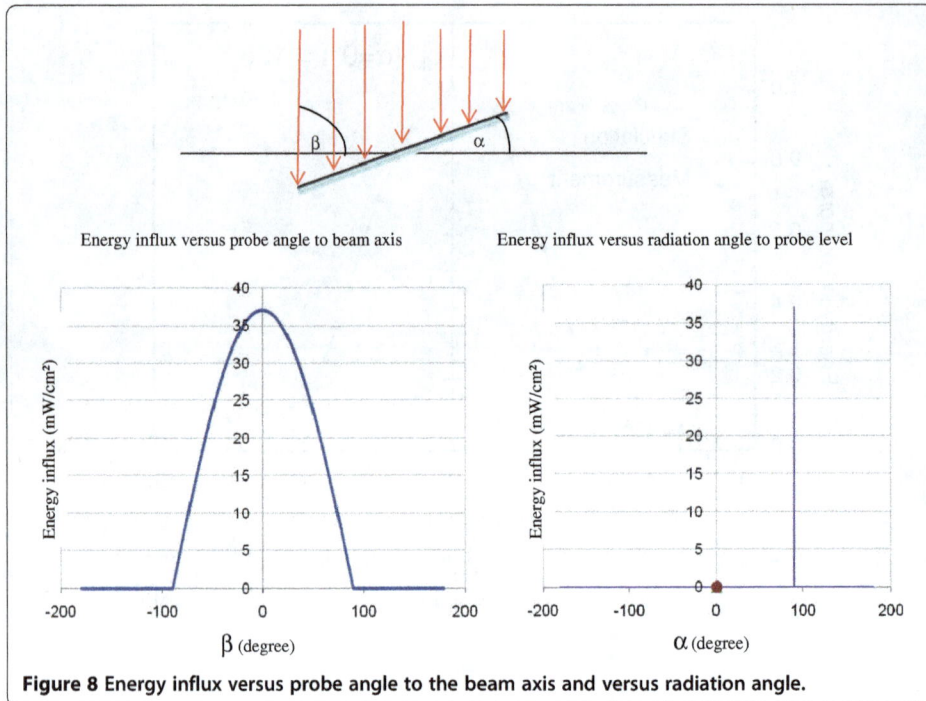

Figure 8 Energy influx versus probe angle to the beam axis and versus radiation angle.

measurement the probe was rotated around the roll-axis. The data in Figure 7 shows the expected behaviour which depends on the cosine of the probe angle to the beam axis. Also a calculation of the angular values is included at angles from 90° to 180° the probe edges are exposed by the ion beam.

The observed behavior of the double probe indicates clearly that there is no heat flux from the back side to the measurement area on the front side of the probe.

By the use of the Fourier transform the dependence of the energy influx on the probe angle to the beam axis is converted to the dependence on the angle between probe level and the direction of the incoming energy influx. The result is shown in Figure 8.

Thus, the double probe can be used to obtain a "heating radar image" of the probe environment.

Conclusion

By means of the described prototype it could be verified that the principle of an active heated thermal probe is applicable for the determination of the energy influx in several plasma-technological processes. The principle is based on the decrease of the external input heating power at the probe, which is needed to compensate the incoming energy flux from the plasma. The attained sensitivity almost reaches the level of passive probes. Further optimisation in terms of sensitivity and time resolution should be made for broader applications. An increase in sensitivity while measuring lower energy influxes could be achieved by minimising the heat capacity of the probe and improving the temperature regulation. For special applications, a miniaturised version of the probe in the form of a microchip is conceivable, which would lead to a considerable decrease in the probe's response time.

Competing interests
The authors declare that they have no competing interests.

Authors' contributions
All authors read and approved the final manuscript.

Acknowledgment
We like to thank the AIF for supporting the investigations in the frame of the ZIM-project VP2345701DF9 (E-impact).

Author details
[1]Institut für Experimentelle und Angewandte Physik, Leibnizstr. 19, Kiel D-24098, Germany. [2]Formerly Institut für Plasmaforschung und Technologie, Felix-Hausdorff-Str. 2, Greifswald D-17489, Germany.

References
1. Bornholdt S, Peter T, Strunskus T, Zaporojtchenko V, Faupel F, Kersten H. IL-6 release after intestinal ischemia/reperfusion in rats is under partial control of TNF. Surf Coat Techn. 2011;205:388–92.
2. Steffen H, Kersten H, Wulff H. Investigation of the energy transfer to the substrate during titanium deposition in a hollow cathode arc. J Vac Sci Technol A. 1994;12:2780.
3. Kersten H, Rohde D, Berndt J, Deutsch H, Hippler R. Investigations on the energy influx at plasma processes by means of a simple thermal probe. Thin Solid Films. 2000;377–378:585–91.
4. Cormier PA, Stahl M, Thomann AL, Dussart R, Wolter M, Semmar N, et al. On the measurement of energy fluxes in plasmas using a calorimetric probe and a thermopile sensor. JPhysD: Appl Phys. 2010;43:465201.
5. Lundin D, Stahl M, Kersten H, Helmersson U. Energy flux measurements in high power impulse magnetron sputtering. J Phys D Appl Phys. 2009;42:185202.
6. Ellmer K, Mientus R. Calorimetric measurements with a heat flux transducer of the total power influx onto a substrate during magnetron sputtering. Surf Coat Techn. 1999;116–119:1102–6.
7. Gardon R. An Instrument for the direct measurement of intens thermal radiation. Rev Sci Instrum. 1953;24:366.
8. Thornton JA. Substrate heating in cylindrical magnetron sputtering sources. Thin Solid Films. 1978;54:23.
9. Kersten H, Deutsch H, Steffen H, Kroesen GMW, Hippler R. The energy balance at substrate surfaces during plasma processing. Vacuum. 2001;63:385–431.
10. R.Wiese. Neue Methoden der Diagnostik von Plasmaquellen, Dissertation 2007, Library Ernst-Moritz-Arndt Universität Greifswald
11. Wiese R, Kersten H. Einsatz einer aktiven Thermosonde zur Diagnostik von Prozessplasmen. Fachzeitschrift Galvanotechjnik. 2008;6:1502–7.
12. Stahl M, Trottenberg T, Kersten H. A calorimetric probe for plasma diagnostics. Rev Sci Instrum. 2010;81:023504.
13. Bornholdt S, Kersten H. Transient calorimetric diagnostics for plasma processing. Eur Phys J D. 2013;67:176.
14. Wendt R, Ellmer K, Wiesemann K. Thermal power at a substrate during ZnO:Al thin film deposition in a planar magnetron sputtering system. J Appl Phys. 1997;85:2115–22.
15. Zeuner M, Scholze F, Neumann H, Chassé T, Otto G, Roth D, et al. A unique ECR broad beam source for thin film processing. Surf Coat Technol. 2001;142–144:11–20.
16. Wiese R, Kersten H, Wiese G, Häckel M. Aktive Thermosonde zur Messung des Energieeinstromes. Vakuum in Forschung und Praxis. 2011;23:20–3.
17. Schroeder B, Peter R, Harhausen J, Ohl A. Modelling and simulation of the advanced plasma source. J Appl Phys. 2011;110:043305.

13

A microfluidic AFM cantilever based dispensing and aspiration platform

Ralph van Oorschot[1*], Hector Hugo Perez Garza[2], Roy J S Derks[1], Urs Staufer[2] and Murali Krishna Ghatkesar[2*]

* Correspondence:
r.vanoorschot@ma3solutions.com;
m.k.ghatkesar@tudelft.nl
[1]MA3 Solutions, Eindhoven, The Netherlands
[2]Department of Precision and Microsystems Engineering, Delft University of Technology, Delft, The Netherlands

Abstract

We present the development of a microfluidic AFM (atomic force microscope) cantilever-based platform to enable the local dispensing and aspiration of liquid with volumes in the pico-to-femtoliter range. The platform consists of a basic AFM measurement system, microfluidic AFM chip, fluidic interface, automated substrate alignment, external pressure control system and controlled climate near the dispensing area. The microfluidic AFM chip has a hollow silicon dioxide (SiO_2) cantilever connected to an on-chip fluid reservoir at one end and a silicon nitride (Si_3N_4) tip with an aperture on the other end. A 3D printed plastic fluidic interface glued over the on-chip reservoir was used to connect microfluidics and macrofluidics. The fluidics is connected to an external pressure control system ranging from −0.8 bar to 5 bar with 0.1 bar resolution. This pressure range allows dispensing and aspiration of liquids through the cantilever tip aperture. The controlled climate with a temperature control range between 25°C – 40°C and humidity up to 95% near the dispensing area keeps the droplets for sufficiently long time before they evaporate. An array of droplets can be programmed to be dispensed automatically and access them again with a position accuracy of 1 micron. Experiments were performed with two types of cantilevers with different geometrical configurations. A minimum flow rate control of 50 fL/s was obtained and also frequency shift was monitored as the cantilever was filled with liquid. This platform will be used for various chemical and biological applications.

Keywords: *Microfluidic*; *Femtoliter*; Hollow cantilever; Dispensing; Aspiration; Pressure control system; AFM

Introduction

The ability to controllably infuse or withdraw different types of reagents with nanometer precision unlocks the door for new processes that can lead to novel opportunities like cell surgery. Lately, more applications, particularly in chemistry and biology are continuously demanding the development of certain tools that would allow the precise dispensing and aspiration of extremely small amounts of liquids at predefined positions. In on-going efforts to achieve this, different types of fluid dispensing devices, such as micropipettes and ink-jet printers, have been explored. Nevertheless, these devices do not offer the degree of precision and control that is required to explore the intended applications. Furthermore, they are limited to handle volumes in the order of micro-to-picoliter [1,2], which constitutes at least three orders of magnitude bigger than the suggested (sub)femtoliter regime for most of the upcoming applications. In attempts to manipulate liquid in the smallest possible regime, scientists have started to develop cantilever-based techniques (i.e. DPN [3], NADIS [4], NanoFountain Probe [5] and FluidFM [6]) which have been proved to surpass the picoliter barrier and handle volumes as small as few zeptoliters [7]. These devices, compatible with Atomic Force Microscopy (AFM), combine the versatility of microfluidics with the high precision provided by the cantilever [6]. The challenge, nevertheless, remains on achieving controlled dispensing of the desired volume despite the complications given by external factors such as evaporation [8], humidity [9], hydrophobicity of the tip [10], surface energy of the substrate [10] and viscosity of the liquid [9]. Additionally, the incapability to aspirate controlled amounts of liquid has slowed down the potential break-through applications. In order to overcome these challenges and get closer towards the envisioned goals, we have developed a microcantilever AFM-based platform to enable the dispensing and aspiration of liquids with volumes in the femtoliter range under controlled environment. The various parts in this system constitute a basic AFM measurement system, a microfluidic chip, an automated substrate alignment, a humidity chamber and a microfluidic interface connected to an external pressure control system. The microfluidic chip consisted of a hollow silicon nitride (Si_3N_4) tip and channelled silicon dioxide (SiO_2) cantilever which were connected to a fluidic reservoir located in the handling part of the chip. After filling the reservoir with liquid, a positive or negative pressure relative to the ambient is applied in the reservoir to infuse or withdraw liquids through an aperture located at the apex of the tip. Constant volume systems such as a syringe pump used for applying pressure differences suffer from the effect of shock waves produced by the stepper motors and have a slow response at low flow rates. Moreover, it can take up to few minutes for the pressure to stabilize. Therefore, we have chosen the constant pressure system which is pulse free, has a fast response and can also produce high flow rates. The results shown here represent an improvement over our earlier work on thermal pumping [8], evaporation based pumping and aspiration [11] and syringe-pump based pipetting [12].

Results and discussion

Microfluidic chip

The fabrication of the microfluidic chips resulted in a chip of 1.5 mm × 3 mm, connecting a fluidic reservoir located on the front side of the chip, with storage capacity of 19.5 nL, to the hollow cantilever located on the backside. The cantilever had a total length of 155 µm. Two cantilever types used are shown in Figure 1. Type-A had an outer width of 42 µm (Figure 1a), while type-B had two legs in parallel with an outer width of 6.4 µm (Figure 1b). The hollow channels had a width of 3.7 µm and a height of 2.2 µm. The SiO_2 walls were measured to be 1.5 µm. Similarly, the hollow Si_3N_4 tip had a thickness of 200 nm and a height of 5.65 µm, over which an aperture of 750 nm was milled. Overall, these dimensions resulted in a spring constant of 2.4 N/m for type-A and 9.4 N/m for type-B.

Fluid filling and substrate stage

We found that when mounting and gluing the microfluidic chip to the plastic interface, which are both mountable in standard AFMs, the maximum pressure that can withstand before detaching was >5 bar. The system allowed to regulate the pressure

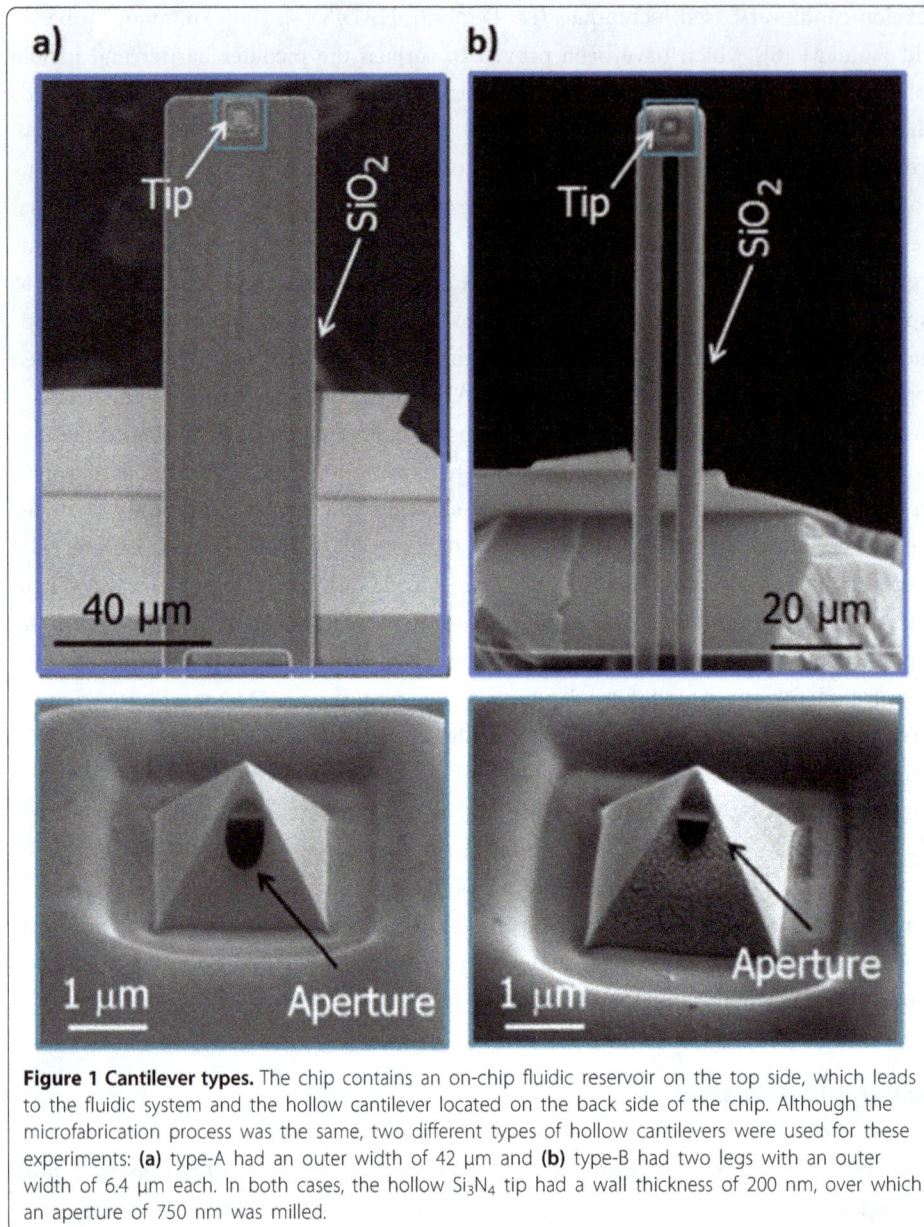

Figure 1 Cantilever types. The chip contains an on-chip fluidic reservoir on the top side, which leads to the fluidic system and the hollow cantilever located on the back side of the chip. Although the microfabrication process was the same, two different types of hollow cantilevers were used for these experiments: **(a)** type-A had an outer width of 42 μm and **(b)** type-B had two legs with an outer width of 6.4 μm each. In both cases, the hollow Si_3N_4 tip had a wall thickness of 200 nm, over which an aperture of 750 nm was milled.

between −0.8 bars and 5.0 bars with better than 0.1 bar resolution, while a valve accurately timed can switch between the regulator set point and ambient pressure within 100 ms.

The temperature and humidity control resulted in precise regulation of temperature between 25-40°C and humidity up to saturation point. Some condensation was noticed around saturation point at high humidity and temperature.

As a first experiment, a chip containing an array of type-B hollow cantilevers independently connected to the fluid reservoir was chosen. A 750 nm aperture was made on the center cantilever. A 3 cm^3 syringe, used as an external liquid reservoir that was filled with deionized water and connected to the pressure system. As a pressure of 1.5bars was applied for 2 minutes, the cantilever was filled and the liquid started to flow out of the aperture as shown in the Figure 2a. The flow speed of the liquid

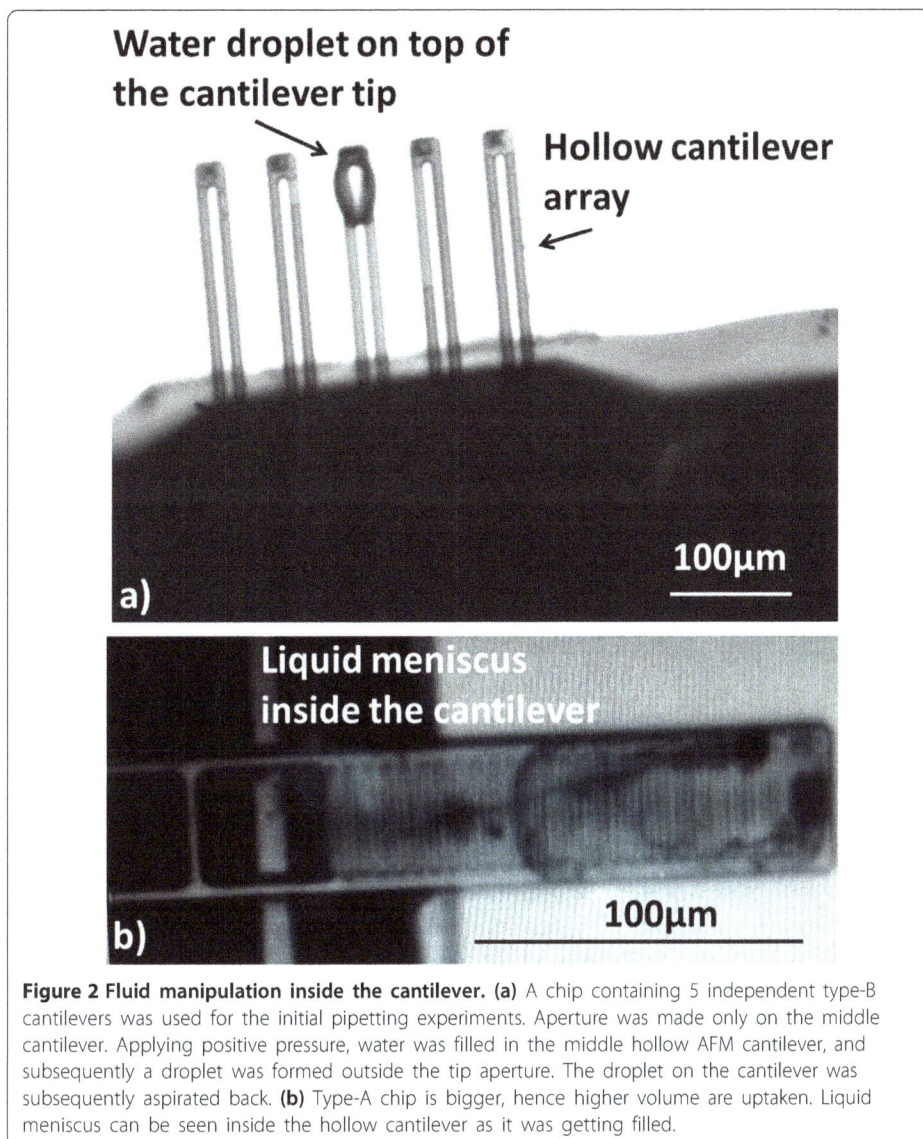

Figure 2 Fluid manipulation inside the cantilever. (a) A chip containing 5 independent type-B cantilevers was used for the initial pipetting experiments. Aperture was made only on the middle cantilever. Applying positive pressure, water was filled in the middle hollow AFM cantilever, and subsequently a droplet was formed outside the tip aperture. The droplet on the cantilever was subsequently aspirated back. **(b)** Type-A chip is bigger, hence higher volume are uptaken. Liquid meniscus can be seen inside the hollow cantilever as it was getting filled.

was determined by following an air bubble inside the fluidic channels and the minimum rate was found to be 50 fL/s. The liquid was eventually aspirated back into the cantilever by switching to vacuum from pressure.

For type-A cantilevers the resonance frequency shift as the fluid was filled (Figure 2b) inside the hollow cantilever was noted. For each pressure increase step, the liquid meniscus advancement was noted, At 1.2 bar of applied external pressure, the entire cantilever was filled. At appropriate applied pressure, equilibrium was established between evaporation of the droplet at the tip and the supplied liquid.

Silicon oxide chips with a 100 μm grid were positioned on the platform and their edges were measured and used as reference by the AFM. We found a re-position accuracy with a standard deviation <2 μm when the substrate was removed and placed back. Similarly, we obtained an accuracy of <1 μm when moving between positions on a substrate. All the characteristics of the platform are given in Table 1.

Table 1 Characteristics of the hollow cantilever and microfluidic cantilever-based platform

Parameter	Units
Cantilever	
Cantilever dimensions (each leg)	Type-A: L:155 μm, W:42 μm, T:4.9 μm Type-B: L:155 μm, W:6.4 μm, T:4.9 μm
Channel dimensions (each leg)	Type-A: L:153.5 μm, W:40 μm, T: 2.2 μm Type-B: L:153.5 μm, W:3.7 μm, T: 2.2 μm
Spring constant	Type-A: 2.4 N/m (when empty); Type-B: 9.4 N/m (when empty)
Resonance frequency	Type-A: 153.94 kHz (when empty); Type-B: 110 kHz (when empty)
Aspiration and dispensing flow rate	50 fL/s @1.5 bars
External dispenser	
Pressure range	0.85 bars under pressure to 5.0 bar overpressure; 0.01 bar resolution
Settling time at dispenser out	<100 ms
Climate control	
Temperature range	Ambient to 40°C ±0.5
Humidity range	30% – 90% RH ±5% non condensing 90% – 100% RH ±5% condensation
Substrate alignment	
Alignment accuracy	Between chip positions on the same substrate: <1 μm Between different chips: <2 μm

Conclusion

To conclude, we have developed a hollow AFM cantilever-based platform able to dispense and aspirate liquids in the picoliter to femtoliter volumes of liquid. Besides regular AFM measurements, the AFM chips and the system are designed to handle these liquid volumes in a controlled climate suitable for numerous chemical and biological applications.

Methods and materials

Microfluidic chip and fluidic interface

The microfabrication of the microfluidic chip and the characterization of the device have been reported elsewhere [12,13]. Briefly, one wafer was KOH-etched to create the fluidic reservoir (Figures 3a-1), while another was DRIE-etched (Figures 3a-2) in order to pattern the cantilever. Then both wafers were bonded together (Figure 3a) and treated in a wet oxidation furnace. Consequently, the O_2 gas entered through the fluidic reservoir and started growing oxide in the patterned cantilever (Figures 3a-4). This resulted in a hollow SiO_2 structure that was buried inside the silicon, which was further removed by wet-etching it in KOH (Figures 3a-5). Later, Focused Ion Beam (FIB) was used to mill the aperture on the tip. For experiments, two different cantilever-designs were used in this paper denoted as rectangular type A and U-shaped type B. Each cantilever differed in the final dimensions, but the fabrication steps were the same.

In order to connect the hollow cantilever to the pressure control system and avoid leakage during liquid transport, a microfluidic plastic interface was manufactured. SolidWorks Software was used to create a suitable model (Figure 3b), which was later printed out of HTM140 polymer using Objet 3D-printer. The polymer has good temperature resistance and high tensile strength.

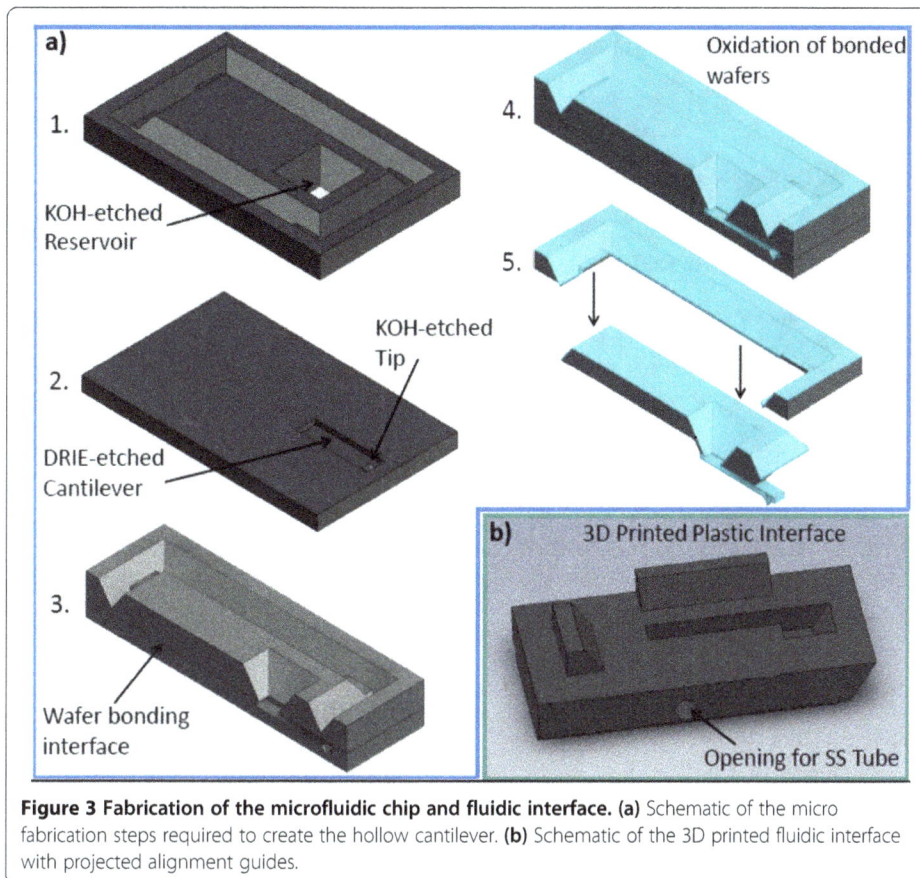

Figure 3 Fabrication of the microfluidic chip and fluidic interface. (a) Schematic of the micro fabrication steps required to create the hollow cantilever. **(b)** Schematic of the 3D printed fluidic interface with projected alignment guides.

Bonding the microfluidic chip to the interface

The fluidic interface is designed with alignment guides to self align on the AFM cantilever chip. Before gluing the interface to the chip, a stainless steel (SS) tube was inserted from the side into the interface and glued with Norland Optical Adhesive 86H. The tube had a length of 12.5 mm, an outer diameter of 240 μm and inner diameter of 100 μm. The tube was further connected to external pressure control system using Tygon tubing. The cantilever chip was then glued to the resulting plastic interface using the same glue. The glue could be cured either by UV light or temperature or both. The chip could be easily debonded from the microfluidic interface by dipping in acetone for few seconds without leaving residues. Thus, making the chip and interface reusable. To avoid excessive amount of glue and prevent clogging of the fluidic reservoir during the bonding, the glue was dispensed in controlled amounts using a foot pedal operated pressure–time based dispensing system, FISNAR Inc., Model SL101N at a dispense time of 50 ms at 0.5 bar. The same glue was dispensed using 30 gauge EFD Nordson dispensing tip with an outer diameter of 310 μm (see Figure 4a). After mounting and aligning, the glue was pre-cured by UV light for 10 seconds, using EFOS ACTICURE A4000 with a 100 W mercury vapor short arc light bulb which produces light in the spectral range of 250-450 nm. To prevent uncured glue which could potentially block the fluidic channels, the parts are cured in a convection oven at 125°C for 15 minutes (Figure 4b).

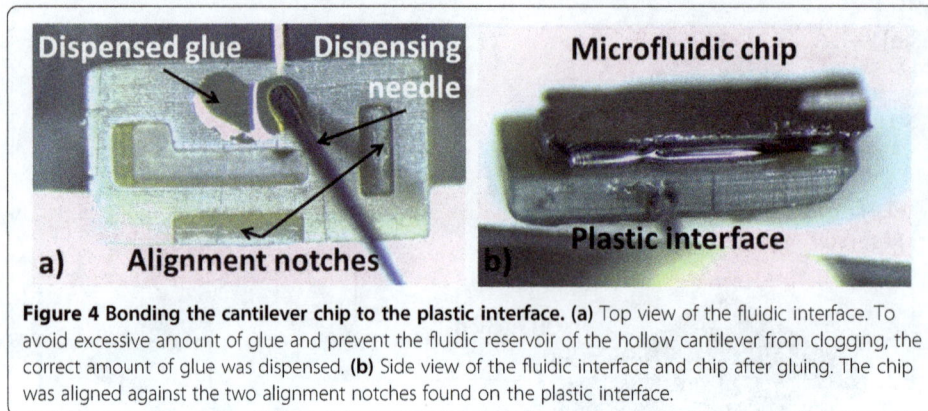

Figure 4 Bonding the cantilever chip to the plastic interface. (a) Top view of the fluidic interface. To avoid excessive amount of glue and prevent the fluidic reservoir of the hollow cantilever from clogging, the correct amount of glue was dispensed. **(b)** Side view of the fluidic interface and chip after gluing. The chip was aligned against the two alignment notches found on the plastic interface.

Microfluidic AFM dispensing and aspiration platform

The system is based on the TT-AFM from AFM workshop company with external peripherals like pressure control, climate control and automated x-y stage for substrate alignment attached to the system. A schematic representation and the corresponding photograph of the entire platform is shown in Figure 5.

External pressure control system

The schematic of the pressure control system is shown in Figure 6. The pressure regulators and the valves control were USB-interfaced via National Instrument NI-6008 interface to the computer. The system includes two MAC valves, type 35A-ACA-DDAA-1BA, which have a switch time of 2-6 ms. Valve-1 switches between pressure and vacuum, while valve-2 switches between regulated and ambient pressure. Both valves have a fast response. A 3 cc syringe is used as a reservoir, which is located at the left side of the platform, see Figure 5b. The pressure and duration of a dispense/aspirate action can be defined in the software application.

Temperature and humidity control

Typically, a droplet in the range of $10\,\mu m^3$ (10 fL) or smaller takes less than 1 s to evaporate [14]. To minimize the evaporation of the droplets and increase the droplet evaporation

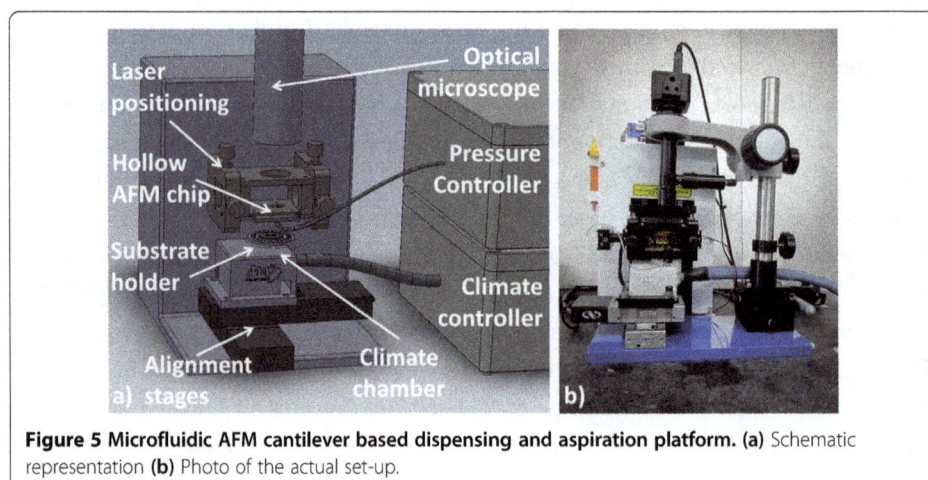

Figure 5 Microfluidic AFM cantilever based dispensing and aspiration platform. (a) Schematic representation **(b)** Photo of the actual set-up.

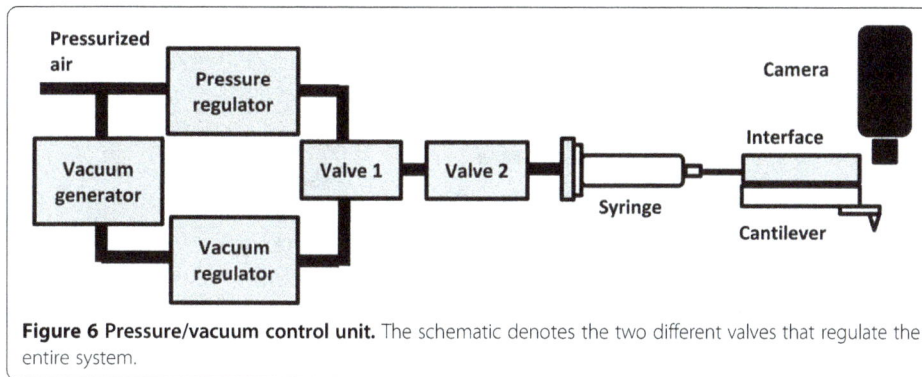

Figure 6 Pressure/vacuum control unit. The schematic denotes the two different valves that regulate the entire system.

time a controllable climate is required. A semi-open climate chamber is designed for convenient access of the sample substrate, while maintaining the climate around the droplet on the sample. A ring of nozzles ensure to produce a curtain of air with the desired temperature and humidity. The airflow is kept sufficiently high such that diffusion of moisture is slower than the supply of the prepared air. This model is validated with a Finite Element Method (FEM) analysis shown in Figure 7.

To supply air with a controlled temperature and humidity, compressed air is pushed through a bubbler (water bath) and a heated tube. The function of the bubbler is to humidify the air to its saturation point. The air temperature is controlled by heating or cooling the water, which is done by two peltiers-elements that can add or extract energy. This saturated air, at a controlled temperature, flows through a heated tube to increase the temperature and therefore decrease the relative humidity (see Figure 8). The

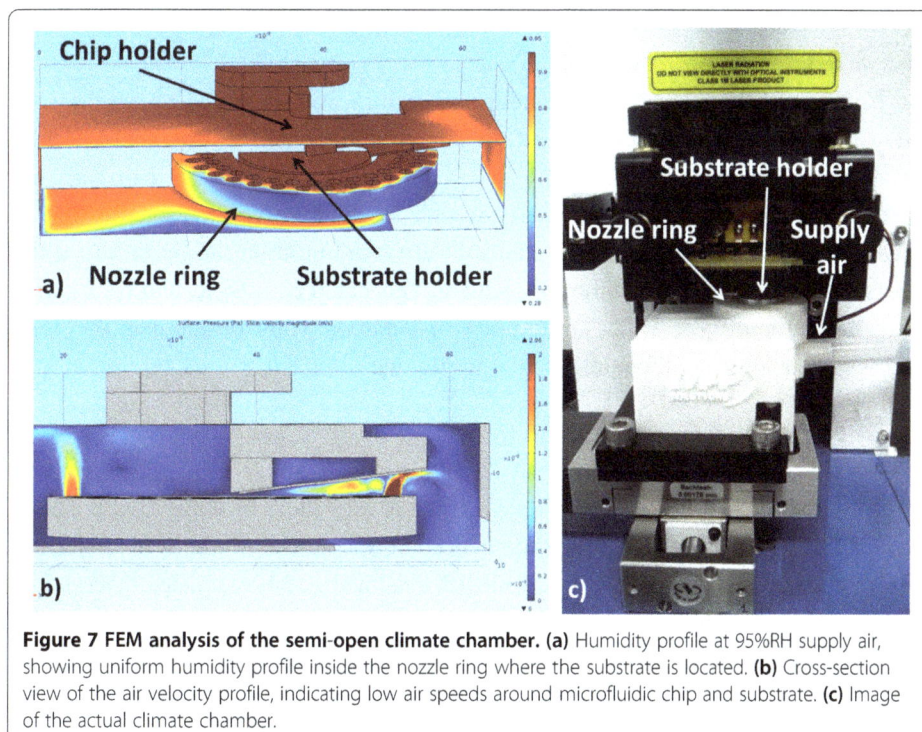

Figure 7 FEM analysis of the semi-open climate chamber. (a) Humidity profile at 95%RH supply air, showing uniform humidity profile inside the nozzle ring where the substrate is located. **(b)** Cross-section view of the air velocity profile, indicating low air speeds around microfluidic chip and substrate. **(c)** Image of the actual climate chamber.

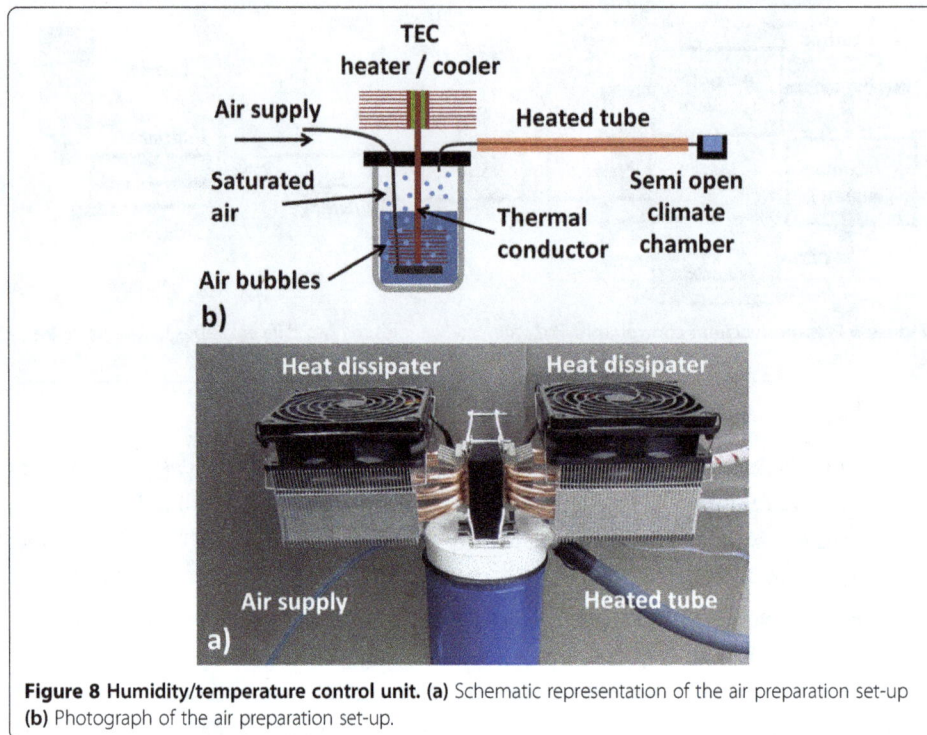

Figure 8 Humidity/temperature control unit. (a) Schematic representation of the air preparation set-up **(b)** Photograph of the air preparation set-up.

software has two PID control loops: 1) Climate chamber temperature is used as an input for the control setpoint of the tube heating, 2) Climate chamber humidity is used as an input for the water temperature setpoint. Note that this setup only increases the droplet lifetime significantly based on the humidity value chosen (upto 95% non-condensing) but doesn't prevent completely.

Substrate to tip alignment

Two Newport SMC100 stages with a range of 25 mm were mounted under the substrate (Figure 5a). These stages are used for XY positioning the substrate and the chip. Silicon oxide chips from Bioforce with a 100 μm square grid were used to align on. The edges of these gridlines were detected with machine vision and software determined the tip and substrate position. The software could position the tip in one of the squares and select a position within this square. This could be used to find back a previously dispensed droplet within an accuracy of 1 micron. After positioning, the AFM piezo stages could be used for sub-micron positioning.

Dispense automation software

A Labview based software program was developed which is able to handle automatic dispensing tasks. Each dispensing action consisted of the following steps: the tip deflection check and a small Z-step. When the deflection was measured, the software stored the corresponding substrate height and moved a defined number of steps towards the substrate. Then the dispense action was performed and the tip was moved back away from the substrate. Figure 9 shows the software controllable parameters. In automatic mode, the software positions the tip to a new XY position and the process is repeated. An array of points can be given to the software via a CSV file.

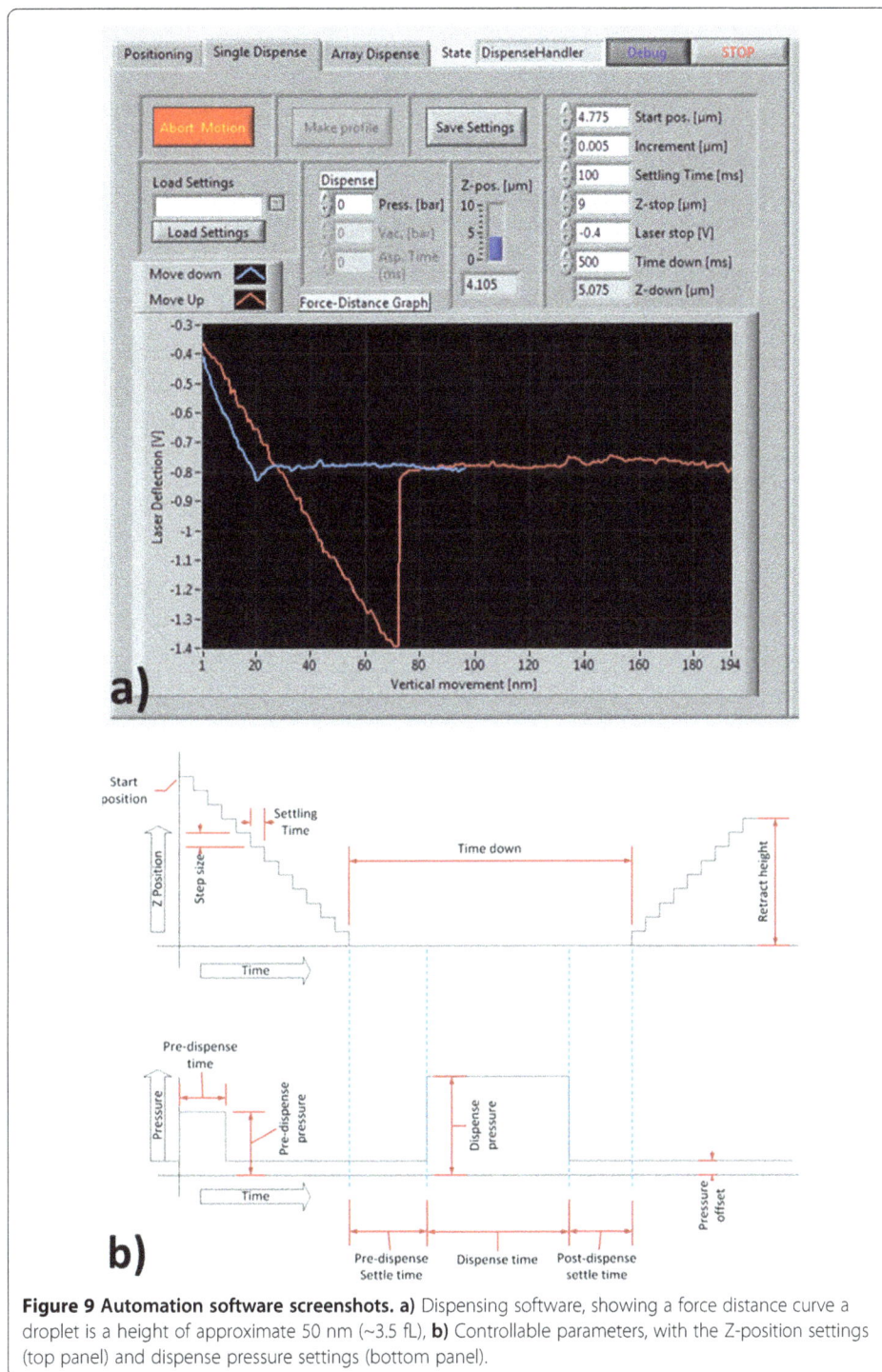

Figure 9 Automation software screenshots. a) Dispensing software, showing a force distance curve a droplet is a height of approximate 50 nm (~3.5 fL), **b)** Controllable parameters, with the Z-position settings (top panel) and dispense pressure settings (bottom panel).

Competing interest

The authors declare that they have no competing interest.

Authors' contribution

RvO and RJSD carried out the implementation of the microfluidic interface and the pheripherals involved in the dispensing and aspiration platform development. HHPG fabricated the microfluidic AFM device. US and MKG were involved in the design of the project. RvO, HHPG, MKG were involved in the experiments and interpretation of the results. All the authors contributed to the writing of the manuscript. All authors read and approved the final manuscript.

Acknowledgements

This work is supported by NanoNextNL, a micro and nanotechnology consortium of the government of the Netherlands and 130 partners

References

1.   Elkins KM. Academic Press Publications; 2013.
2.   Hutchings IMaM. Hoboken, New Jersey: G.D., John Wiley & Sons; 2013.
3.   Piner RD, Zhu J, Xu F, Hong SH, Mirkin CA. Science. 1999; 283:661–3.
4.   Meister A, Jeney S, Liley M, Akiyama T, Staufer U, de Rooij NF, et al. Microelectron Eng. 2003;67–8:644–50.
5.   Kim KH, Moldovan N, Ke C, Espinosa HD. Micro- and Nanosystems. 2004; 782:267–72.
6.   Meister A, Gabi M, Behr P, Studer P, Voros J, Niedermann P, et al. Nano Lett. 2009; 9:2501–7.
7.   Kaisei K, Satoh N, Kobayashi K, Matsushige K, Yamada H. Nanotechnology. 2011; 22:175301.
8.   Perez Garza HH, Ghatkesar M, Staufer U. Journal of Micro-Bio. Robotics. 2013;8:33–40.
9.   Fang AP, Dujardin E, Ondarcuhu T. Nano Lett. 2006; 6:2368–74.
10. Fang AP, Dujardin E, Ondarcuhu T. J Phys Conf Ser. 2007;61:298–301.
11. Perez Garza HH, Ghatkesar M, Staufer U. Micro Nano Lett. 2013; 8:758–61.
12. Ghatkesar MK, Perez Garza HH, Staufer U. Microelectron Eng. 2014; 124:22–5.
13. HugTS, Biss T, de Rooij NF, Staufer Q. Transducers '05, Digest of Technical Papers. 2005; 1 and 2:1191–4.
14. Bonaccurso E, Golovko DS, Bonanno P, Raiteri R, Haschke T, Wiechert W, et al. Atomic Force Microscope Cantilevers Used as Sensors for Monitoring Microdrop Evaporation, Applied Scanning Probe Methods XI. Berlin Heidelberg, Germany: Springer; 2009. p. 17–38.

14

PPMS-based set-up for Raman and luminescence spectroscopy at high magnetic field, high pressure and low temperature

Matthias Hudl[1,2*†], Peter Lazor[3†], Roland Mathieu[4†], Alexander G Gavriliuk[5,6†] and Viktor V Struzhkin[7†]

*Correspondence: hudl@kth.se
†Equal contributors
[1] KTH Royal Institute of Technology, ICT Materials Physics, Electrum 229, SE-164 40 Kista, Sweden
[2] Department of Materials, ETH Zürich, Vladimir-Prelog-Weg 4, CH-8093 Zürich, Switzerland
Full list of author information is available at the end of the article

Abstract

We present an experimental set-up permitting Raman and luminescence spectroscopy studies in a commercial Physical Properties Measurement System (PPMS) from Quantum Design. Using this experimental set-up, gaseous, liquid and solid materials, in bulk or thin film form, may be investigated. The set-up is particularly suitable for the study of the spin-lattice coupling in strongly correlated oxide materials utilizing several different stimuli, e.g. magnetic and electric fields, high pressure and low temperatures. Details for the Raman extension, sample holder assembly and optical design, as well as data acquisition and measurement routine are described. Finally, we present exemplary results collected using the set-up, measured on reference materials, as well as on a correlated transition metal oxide.

Keywords: Raman spectroscopy; Quantum Design PPMS; High pressure; Ruby; Hydrogen conversion; Strongly correlated materials; Langasite

Introduction

Large efforts in the field of material science are spent on the development of novel functional materials. In this context, strongly correlated electron systems are a class of materials that promise a multitude of applications. Several transition metal oxides are being considered for the designing of new devices, owing e.g. to their magnetic, electric and dielectric [1-3] or optical and photovoltaic properties [4,5]. In such materials, the functionalities or physical properties are determined by the localization and delocalization of charge carriers [6]. As a result, magnetic, (di)electric, thermal, and optical properties are correlated with each other [6]. Inelastic light scattering from electronic charge density and phonons in correlated materials is a promising approach to evaluate such correlation properties [7]. This approach is deemed to be particularly fruitful if it can be done under the simultaneous influence of several stimuli that probe these properties such as, e.g. magnetic and/or electric fields, pressure and/or temperature.

Raman spectroscopy for instance is a versatile technique for studying structural properties, lattice distortions, and e.g. the coupling of spin and lattice degrees of freedom. This is due to different types of excitations that can be studied under the influence of external stimuli leading to observation of shifts in vibrational frequencies, changes in peak intensities and polarization states. This was successfully employed to study structure-property relationships in strongly correlated transition metal oxides [8,9]. A particular strength of electronic Raman scattering is the possibility to detect information on the electron dynamics from different regions of the Brillouin zone.

A Raman set-up for in-situ measurements of such properties at low temperature, magnetic field, electrical field and high pressure was constructed as an extension to the Quantum Design, Inc. Physical Properties Measurement System (PPMS) [10] - a standard cryogenic device in many laboratories world wide. An optical probe and an automatized data acquisition system were designed and linked with the already automatized temperature and magnetic field control provided by the PPMS. Several material properties can thus be investigated in the same PPMS set-up, using the standard set of options such as resistivity, magnetization or heat capacity, as well as using extensions permitting e.g. magneto(di)electric measurements [11] or the here proposed Raman spectroscopy set-up. Dedicated custom-made devices for Raman spectroscopy with multiple excitations not using a Quantum Design PPMS were reported in literature previously [12,13].

We present results collected on different reference systems, such as ruby and H_2, as well as a transition metal oxide, $Ba_3NbFe_3Si_2O_{14}$, a langasite material with magnetically induced electronic polarization.

Experimental apparatus and methods
PPMS host system and Raman extension
The Raman set-up is based on a commercial Quantum Design, Inc. PPMS [10] which incorporates (i) a cryostat providing a controllable variable temperature in the 2 - 400 K interval and (ii) a superconducting magnet up to 9 T (superconducting magnets up to 16 Tesla are available). The native hardware control of temperature and magnetic field provided by the PPMS was used and a new extension, permitting the collection of Raman spectroscopy data, was designed and fabricated. The Raman extension comprises the following parts: (a) The custom-made Raman top part (including several imaging lenses, beamsplitter, notch filter, white light illumination and CCD camera) and a modified PPMS Raman insert (using multifunction probe temperature sensor) serving as a sample/DAC holder. (b) A miniature nonmagnetic diamond anvil cell (DAC). (c) A 488 nm argon-ion CW laser system. (d) A Shamrock spectrometer with thermoelectrically cooled CCD detector.

The data acquisition, including Raman spectroscopic module (Solis, Andor Technology Ltd.), as well as software control of PPMS temperature and magnetic fields, was implemented using the QD PPMS third party option. A schematic view of the set-up is shown in Figure 1. It comprises a probe, akin to the commercially available multifunction probe (MFP), albeit mechanically more resistant, providing optical and electrical contacts to the sample, and an optical illumination/data collection system. A single sapphire optical window creates an interface between the evacuated interior of the magnet cryostat and the Raman top part.

Pictures of the upper optical part of the PPMS Raman extension, sample holder and diamond anvil pressure cell are shown in Figure 2(a)-(c), respectively. Magnetic field may be applied up the (+/-) maximum accessible value of the PPMS system. Electric field can also be applied, as indicated by the arrow in Figure 1, using an external voltage source. The high pressure conditions are created using miniature nonmagnetic diamond anvil cell (DAC), specially designed to hold pressure stable during temperature scans [14]. The

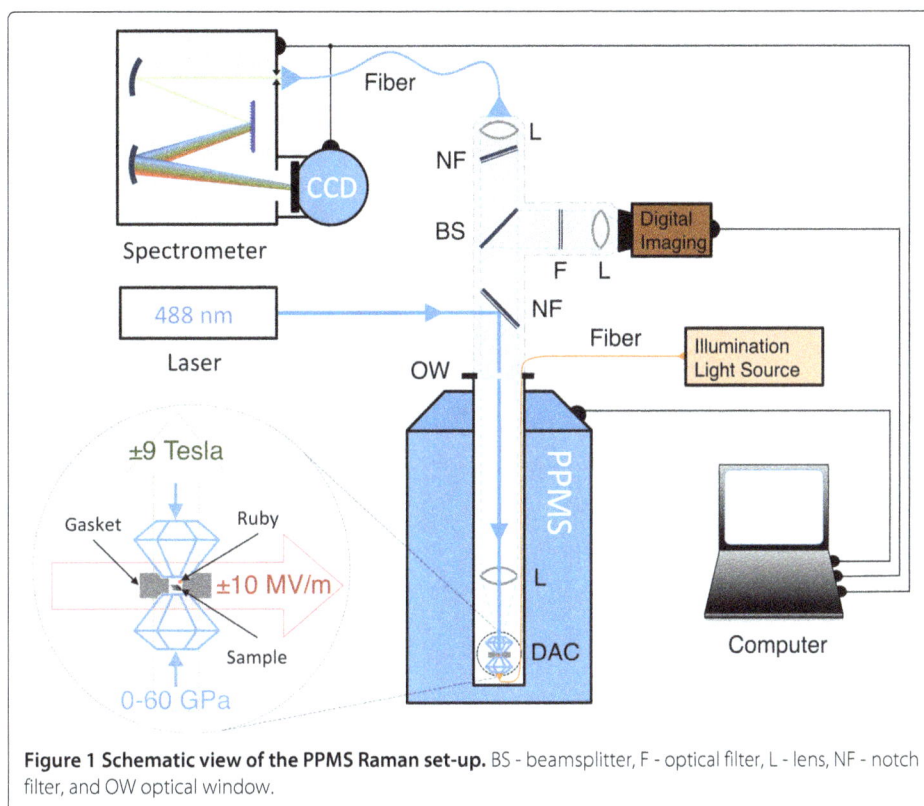

Figure 1 Schematic view of the PPMS Raman set-up. BS - beamsplitter, F - optical filter, L - lens, NF - notch filter, and OW optical window.

DAC is made in a piston-cylinder configuration with a vertical optical axis and a maximum diameter of 10-13 mm. The maximal pressure for diamonds with 0.3 mm culets is ~60 GPa. Our Raman extension is suitable for long-term experiments lasting several days, exploiting the typical operational parameter space of the PPMS.

Optical design

Raman spectra are recorded with a SR-303I-A Shamrock 303i spectrometer (Andor Technology) coupled by a single mode optical fiber to the PPMS Raman extension. The spectrometer is equipped with 1200l/mm grating and thermoelectrically cooled multichannel CCD detector (iDus, Andor Technology, 1024×256 pixels, -50°C). A linearly polarized argon ion laser (488 nm line) is used for the excitation at powers up to 15 mW. The spectral axis is calibrated by the fluorescence lines of a neon lamp. Inside the PPMS, Raman spectra are collected in the back scattering geometry, at a resolution of about 4 cm^{-1}. Elastically backscattered fundamental light (488 nm) is filtered out using a narrow band notch filter (Semrock Inc.). The accuracy of the spectral measurements, resulting from the wavelength calibration procedure and experimental conditions, is estimated to be about 2 cm^{-1}. Typical acquisition times vary between 20 - 120 sec. Experimentally accessible ranges of temperature, pressure, magnetic and electric fields available for spectroscopic acquisitions correspond to 3 - 350 K (higher temperatures with additional heating possible), 0 - 60 GPa, 0 - ±9 Tesla and 0 - ±10 MV/m, respectively. So far, spectral acquisitions involving strong electric fields have been carried out outside the PPMS only. Typically, the experimental arrangement involves application of hundreds of volts across a sample which is few tens of micrometers thick.

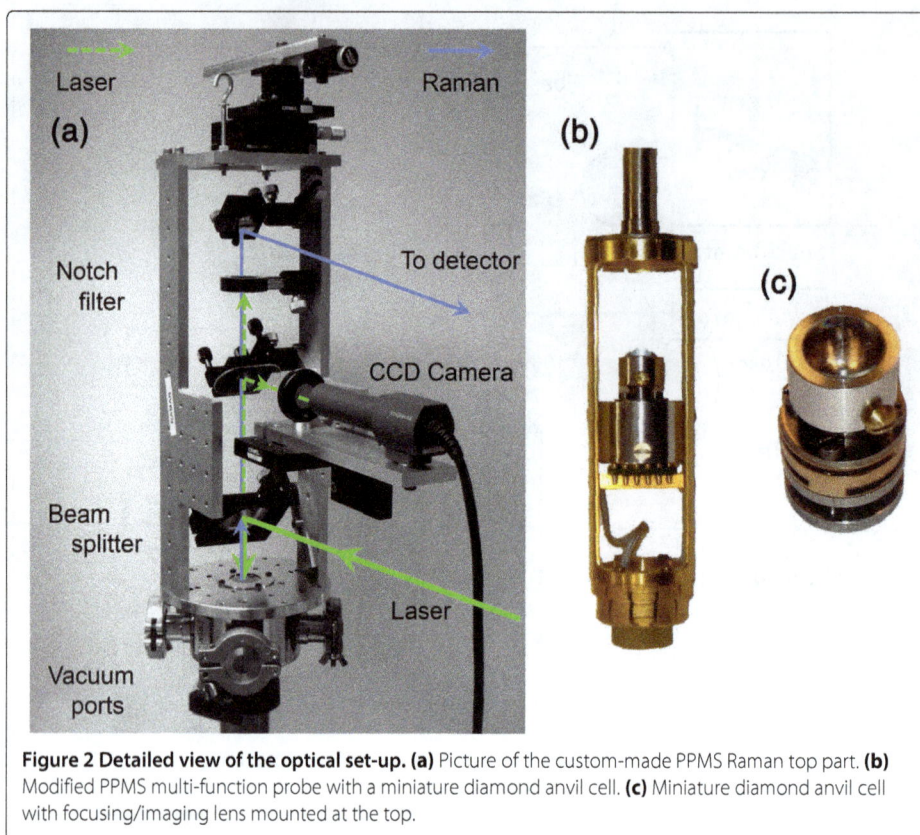

Figure 2 Detailed view of the optical set-up. (a) Picture of the custom-made PPMS Raman top part. **(b)** Modified PPMS multi-function probe with a miniature diamond anvil cell. **(c)** Miniature diamond anvil cell with focusing/imaging lens mounted at the top.

Alignment and measurement routine

Measurements are taken in the back scattering geometry. Due to space limitation inside the cryostat and low temperature conditions right angle scattering geometry is difficult to achieve in the described set-up. For sample illumination and focusing, white light is coupled via an optical fiber into the set-up and applied in transmission. Prior the measurements the optical system is pre-aligned and tested outside the cryostat using a room-temperature probe stand.

A standard (high-pressure) experiment is done in the following way. (i) The sample is mounted onto the Raman probe using ambient-pressure sample holder or DAC. For high-pressure experiments specific sample-, ruby- and pressure transmitting medium loading procedures have to be applied as described in Ref. [14]. (ii) The sample is aligned with the optical set-up and inserted into the PPMS cryostat. (iii) Single measurements are performed manually or an automatized measurement sequence is programmed using the PPMS third-party option together with the spectrometer software scripting option.

Exemplary results and discussion

Ruby fluorescence lines

Cr^{3+}-doped Al_2O_3, also known as ruby, is a well-established standard material with interesting properties that led to significant technological advances both in academia and industry [15]. For instance, ruby was the material that led to the invention of the first laser by Maiman in 1960 [16]. The characteristic red color of ruby is related to electronic transitions involving the Cr^{3+} ions manifested in the two fluorescence lines, R1 and R2 [17].

Ruby is furthermore a well-established pressure gauge in optical high-pressure experiments [18] and hence has been studied extensively. Zeeman splitting in single crystalline ruby has been observed in pulsed magnetic fields up to 60 T and hydrostatic pressure up to 10 GPa [19]. In the presented PPMS Raman set-up static magnetic fields up to 9 T (16 T with different cryostat) and hydrostatic pressure up to 60 GPa can be achieved. In Figure 3 the magnetic field dependence for the ruby R1 line at 5 K and 34 GPa is depicted. Ruby can hence serve as a local (*in-situ*) probe of the magnetic field ($\Delta\lambda/\Delta H \sim 0.046$ nm/T).

Hydrogen conversion

Hydrogen is the most common element in the known universe. It exists in two different spin isomers, ortho hydrogen with two parallel proton spins and para hydrogen with two antiparallel proton spins. At room temperature and ambient pressure the ortho- to para hydrogen ratio is approximately 1:3. This equilibrium ratio is temperature dependent and at low temperatures the equilibrium state is para hydrogen. H_2 conversion, from ortho- to para hydrogen can be traced by measuring H_2 rotational modes (rotons), i.e. the integrated intensities of the rotational Raman peaks. Figure 4(a) shows H_2 conversion data collected using the PPMS Raman extension. The H_2 conversion rate is strongly enhanced under hydrostatic pressure [20]. In the presented PPMS Raman set-up the combined effect of hydrostatic pressure and high magnetic fields can be studied. The evolution of the rotational Raman modes (from 0 h (black line) to 8 h) at $\sim$20 K, 9 T magnetic field, and 27 GPa hydrostatic pressure is depicted. Vibrational Raman modes (vibrons) of hydrogen can be observed in a factor of 10 higher frequency range. Figure 4(b) illustrates the hydrogen $Q_1(1)$ vibron at temperatures close to room temperature (300 K) and at low temperature (5 K) for both zero and 9 T magnetic fields. Hydrogen vibron modes are sensitive to structural changes and a clear difference between fluid and solid hydrogen is observed.

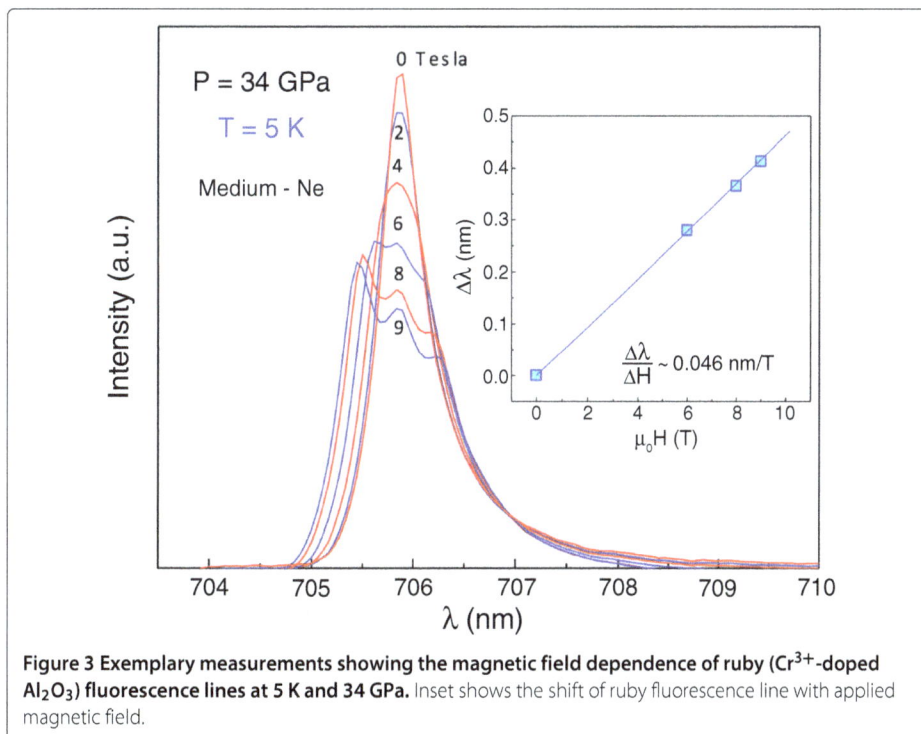

Figure 3 Exemplary measurements showing the magnetic field dependence of ruby (Cr^{3+}-doped Al$_2$O$_3$) fluorescence lines at 5 K and 34 GPa. Inset shows the shift of ruby fluorescence line with applied magnetic field.

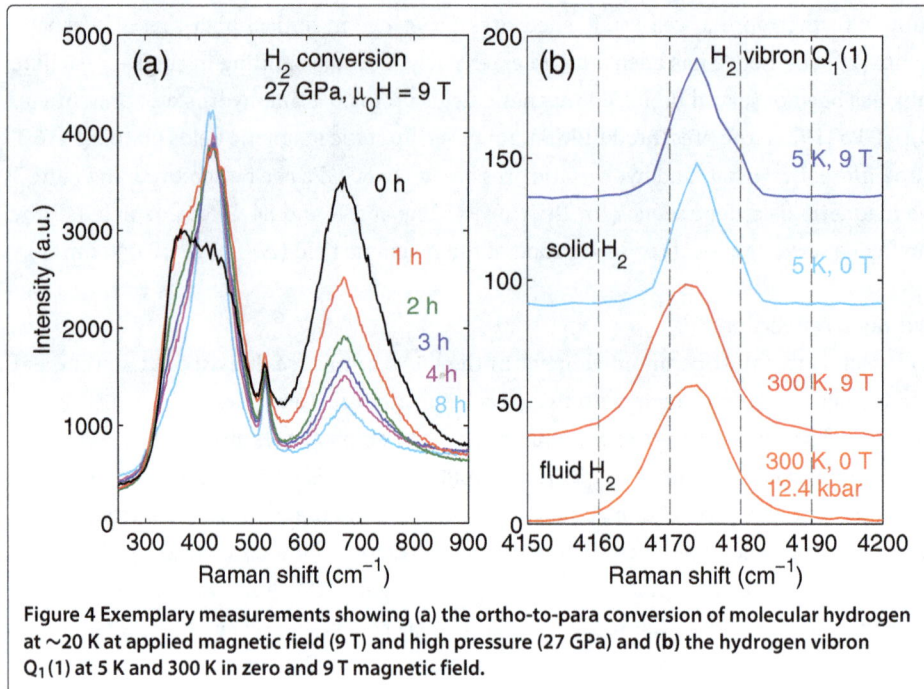

Figure 4 Exemplary measurements showing (a) the ortho-to-para conversion of molecular hydrogen at ~20 K at applied magnetic field (9 T) and high pressure (27 GPa) and (b) the hydrogen vibron $Q_1(1)$ at 5 K and 300 K in zero and 9 T magnetic field.

Spin-phonon coupling in a strongly correlated oxide

Langasite materials are well known for their piezoelectric and non-linear optical properties [21,22]. A recent interest in the magnetic and electrical properties of langasite materials is due to the discovery that they could crystalize in a Kagomé-like lattice structure consisting of rare-earth and transition metal cations [23,24]. While a detailed report of the study of the spin-phonon coupling in those materials is beyond the scope of this article, we present here Raman spectroscopy data collected in our set-up as a function of temperature and magnetic fields. Simple qualitative analysis are also included, in order to illustrate the typical studies which may be performed on such materials.

$Ba_3NbFe_3Si_2O_{14}$ crystallizes in the non-centrosymmetric trigonal P321 space group with 23 unit cell atoms. Magnetic Fe^{3+} ions are arranged in triangle units on a triangular lattice [23]. The structure is non-polar and hence no spontaneous polarization can be expected for point group 32 [25]. It has been reported that $Ba_3NbFe_3Si_2O_{14}$ undergoes an antiferromagnetic phase transition at approximately 27 K [23]. A much larger Curie-Weiss temperature of $\theta \approx -175$ K [26], reflects the large magnetic frustration in the system [27]. The magnetic transition is accompanied by a symmetry lowering from P321 to C2 [28]. Further details on the magnetic structure and Raman mode assignment can be found in literature [25,29,30].

Unpolarized Raman spectra on a single crystal of $Ba_3NbFe_3Si_2O_{14}$ recorded as a function of temperature and magnetic field are shown in Figure 5. All T and H dependent measurements were done with an the incident beam along the [1-10] direction. For better visualization spectra (50 - 300 K) have been shifted vertically with a linear offset. The phonon spectrum of $Ba_3NbFe_3Si_2O_{14}$ exhibits a frequency hardening and narrowing of the line-width on reducing temperature.

In a quasi-harmonic approximation, the frequency shift can be attributed to the volume change only and anharmonic effects due to phonon-phonon and spin-phonon

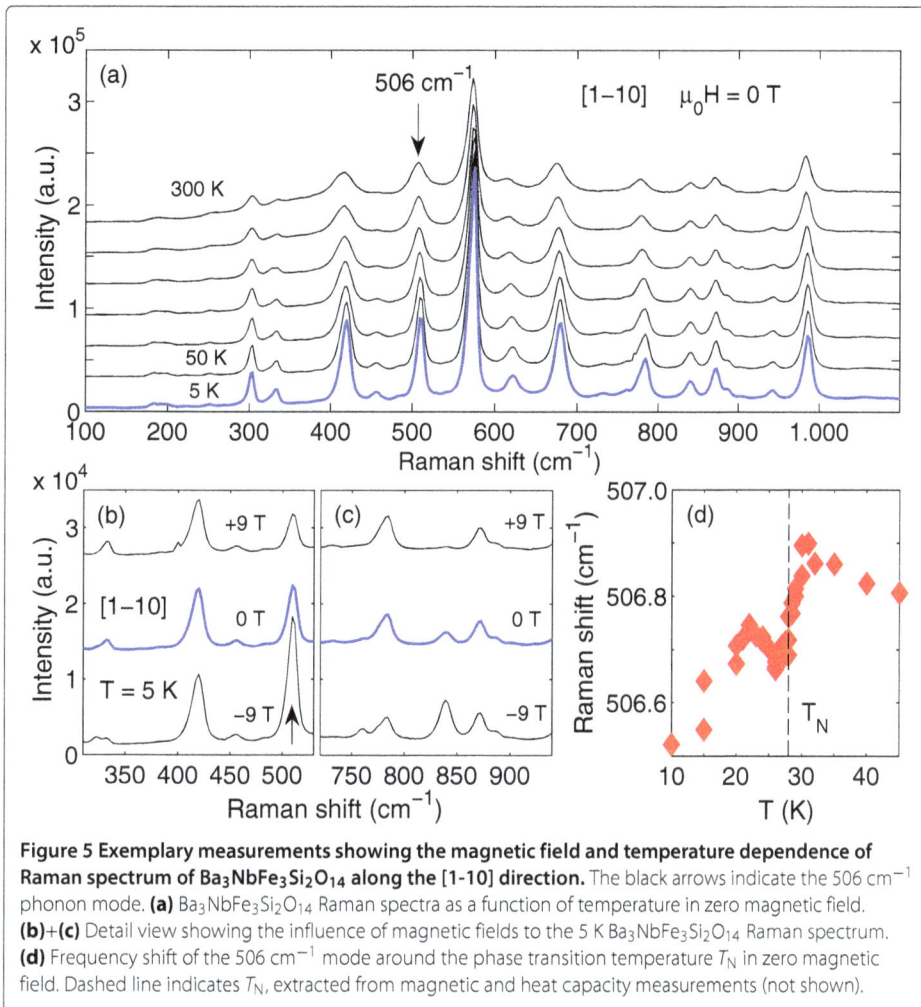

Figure 5 Exemplary measurements showing the magnetic field and temperature dependence of Raman spectrum of Ba₃NbFe₃Si₂O₁₄ along the [1-10] direction. The black arrows indicate the 506 cm⁻¹ phonon mode. **(a)** $Ba_3NbFe_3Si_2O_{14}$ Raman spectra as a function of temperature in zero magnetic field. **(b)+(c)** Detail view showing the influence of magnetic fields to the 5 K $Ba_3NbFe_3Si_2O_{14}$ Raman spectrum. **(d)** Frequency shift of the 506 cm⁻¹ mode around the phase transition temperature T_N in zero magnetic field. Dashed line indicates T_N, extracted from magnetic and heat capacity measurements (not shown).

interaction are ignored. Nevertheless, additional changes in the phonon frequencies when approaching T_N are evident, as shown in Figure 5(d). For instance, the 506 cm⁻¹ phonon mode reveals softening behavior with an onset above T_N and a subsequent hardening below 25 K. Similar behavior is observed for the other modes, suggesting spin-phonon coupling. Figure 5(d) also illustrates precision and stability of the Raman set-up, which allow detection of a magnetic transition involving a change of only a fraction of wavenumber.

The magnetic field dependence of the unpolarized Raman spectra of $Ba_3NbFe_3Si_2O_{14}$ at 5 K is shown in Figure 5(b)+(c). In this measurement the magnetic field was set to the highest value (9 T) and the first Raman spectrum was collected. Subsequently, the magnetic field was swept from 9 T to -9 T. At a first glance, a comparably 'strong' change in the integrated intensity of the 310, 416 and 506 cm⁻¹ modes and in the position of the 615 cm⁻¹ mode are found. Furthermore, several new modes appear at 250, 322, 760 and 840 cm⁻¹ when the field is changed from 9 T to -9 T. The influence of the magnetic field on phonon modes 506, 573, 675 and 982 cm⁻¹ is weak and only the 416 and 615 cm⁻¹ modes exhibit major changes in the frequency [31].

Recently, the magnetic (and structural) properties of related langasites were found to be greatly affected by application of pressure [32], suggesting the relevance of T,P,H-dependent Raman studies on $Ba_3NbFe_3Si_2O_{14}$ and other strongly correlated materials.

Conclusion

A Raman and luminescence spectroscopy system for the study of strongly correlated electron materials is presented. The system is designed as an extension of the Quantum Design, Inc. PPMS and allows for measurements in a large temperature range (3 - 350 K), high magnetic fields (0 - 9 T), high pressure (0 - 60 GPa) and electric field strengths up to 10 MV/m. Gaseous, liquid and solid materials, in amorphous, poly- and single-crystalline bulk form or as a thin film can be studied. Exemplary results for ruby and solid hydrogen at low temperature and high hydrostatic pressure and high magnetic field are shown. Finally results of the study of the spin-phonon coupling in the complex transition metal oxide $Ba_3NbFe_3Si_2O_{14}$ are presented and discussed.

Competing interests
The authors declare that they have no competing interests.

Authors' contributions
MH participated in building the experimental apparatus, performed experimental studies of ruby, H_2, and $Ba_3NbFe_3Si_2O_{14}$ and drafted the manuscript. PL initiated the study and participated in design and building the experimental apparatus and performed experimental studies of ruby and H_2. RM participated in design of the study on strongly correlated materials and helped to draft the manuscript. AG participated in experimental studies of ruby. VS participated experimental studies of ruby and H_2 and helped to draft the manuscript. All authors read and approved the final manuscript.

Acknowledgements
We are very grateful to Dr. Y. Tokunaga, and Profs Y. Taguchi and Y. Tokura for advices and support in the floating zone growth of $Ba_3NbFe_3Si_2O_{14}$ single-crystal. Financial support from the Carl Tryggers Stiftelse för Vetenskaplig Forskning (project CTS 09:217), the Swedish Foundation for International Cooperation in Research and Higher Education (STINT), the Swedish Research Council (VR) and the Göran Gustafsson Foundation is acknowledged. MH thanks the Swedish Research Council (VR, contract 2012-6562) for financial support and Profs M. Fiebig and O. Tjernberg for hosting him during his postdoc. AG thanks the Russian Foundation for Basic Researches (grant 14-02-00483-a), and the program of RAS "Physics of elementary particles, fundamental nuclear physics and nuclear technologies" for financial support.

Author details
[1] KTH Royal Institute of Technology, ICT Materials Physics, Electrum 229, SE-164 40 Kista, Sweden. [2] Department of Materials, ETH Zürich, Vladimir-Prelog-Weg 4, CH-8093 Zürich, Switzerland. [3] Department of Earth Sciences, Uppsala University, Villavägen 16, SE-752 36 Uppsala, Sweden. [4] Department of Engineering Sciences, Uppsala University, Box 534, SE-751 21 Uppsala, Sweden. [5] Institute of Crystallography, Russian Academy of Sciences, Leninsky pr. 59, 119333 Moscow, Russia. [6] Institute for Nuclear Research, Russian Academy of Sciences, 60-letiya Oktyabrya prospekt 7a, 117312 Moscow, Russia. [7] Geophysical Laboratory, Carnegie Institution of Washington, 5251 Broad Branch Road NW, 20015 Washington DC, USA.

References
1. Bibes M, Barthelemy A. Oxide Spintronics. IEEE Trans Elec Dev. 2007;54(5):1003.
2. Scott JF. Applications of magnetoelectrics. J Mater Chem. 2012;22:4567.
3. Choudhury D, Mandal P, Mathieu R, Hazarika A, Rajan S, Sundaresan A, et al. Near-Room-Temperature Colossal Magnetodielectricity and Multiglass Properties in Partially Disordered La_2NiMnO_6. Phys Rev Lett. 2012;108:127201.
4. Grinberg I, West DV, Torres M, Gou G, Stein DM, Wu L, et al. Perovskite oxides for visible-light-absorbing ferroelectric and photovoltaic materials. Nature. 2013;503(7477):509–12.
5. Hosono H. Exploring electro-active functionality of transparent oxide materials. Jpn J Appl Phys. 2013;52:090001.
6. Tokura Y, Nagaosa N. Orbital Physics in Transition-Metal Oxides. Science. 2000;288:462.
7. Devereaux TP, Hackl R. Inelastic light scattering from correlated electrons. Rev Mod Phys. 2007;79:175–233.
8. Guennou M, Bouvier P, Chen GS, Dkhil B, Haumont R, Garbarino G, et al. Multiple high-pressure phase transitions in $BiFeO_3$. Phys Rev B. 2011;84:174107.
9. Iliev MN, Gospodinov MM, Singh MP, Meen J, Truong KD, Fournier P, et al. Growth, magnetic properties, and Raman scattering of La_2NiMnO_6 single crystals. J Appl Phys. 2009;106(2):023515.
10. Quantum Design Inc. Physical Property Measurement System (PPMS). http://www.qdusa.com/products/ppms.html 2015 Accessed 6 Feb 2015.
11. Hudl M, Mathieu R, Ivanov SA, Weil M, Carolus V, Lottermoser T, et al. Complex magnetism and magnetic-field-driven electrical polarization of Co_3TeO_6. Phys Rev B. 2011;84:180404.

12. Isaacs ED, Heiman D. Fiber optics for Raman scattering at low temperatures and high-magnetic fields. Rev Sci Instrum. 1987;58(9):1672–4.
13. Matsuda Y, Kuroda N, Nishina Y. High-field, high-pressure, and low-temperature magneto-optical apparatus using a diamond anvil cell. Rev Sci Instrum. 1992;63(12):5764–6.
14. Gavriliuk AG, Mironovich AA, Struzhkin VV. Miniature diamond anvil cell for broad range of high pressure measurements. Rev. Sci. Instrum. 2009;80:043906.
15. Syassen K. Ruby under pressure. High Pressure Res. 2008;28(2):75–126.
16. Maiman TH. Stimulated Optical Radiation in Ruby. Nature. 1960;187:493–94.
17. García-Lastra JM, Barriuso MT, Aramburu JA, Moreno M. Origin of the different color of ruby and emerald. Phys. Rev. B. 2005;72:113104.
18. Mao HK, Xu J, Bell PM. Calibration of the Ruby pressure gauge to 800 kBar under quasi-hydrostatic conditions. J Geophys Res-Solid. 1986;91(B5):4673–6.
19. Millot M, Broto J-M, Gonzalez J. High-field Zeeman and Paschen-Back effects at high pressure in oriented ruby. Phys Rev B. 2008;78:155125.
20. Eggert JH, Karmon E, Hemley RJ, Mao HK, Goncharov AF. Pressure-enhanced ortho-para conversion in solid hydrogen up to 58 GPa. Proc Nat Acad Sci USA. 1999;96(22):12269–72.
21. Kaminskii AA, Mill BV, Khodzhabagyan GG, Konstantinova AF, Okorochkov AI, Silvestrova IM. Investigation of trigonal $(La_{1-x}Nd_x)_3Ga_5SiO_{14}$ crystals. Phys Stat Sol (A). 1983;80:387.
22. Sato J, Takeda H, Morikoshi H, Shimamura K, Rudolph P, Fukuda T. Czochralski growth of $RE_3Ga_5SiO_{14}$ (RE = La, Pr, Nd) single crystals for the analysis of the influence of rare earth substitution on piezoelectricity. J Cryst Growth. 1998;191:746.
23. Marty K, Simonet V, Ressouche E, Ballou R, Lejay P, Bordet P. Single domain magnetic helicity and triangular chirality in structurally enantiopure $Ba_3NbFe_3Si_2O_{14}$. Phys Rev Lett. 2008;101(24):247201.
24. Zhou HD, Lumata LL, Kuhns PL, Reyes AP, Choi ES, Dalal NS, et al. $Ba_3NbFe_3Si_2O_{14}$: A New Multiferroic with a 2D Triangular Fe^{3+} Motif. Chem Mater. 2009;21(1):156–9.
25. Birss RR. Symmetry and Magnetism: North-Holland Publishing Company - Amsterdam; 1966.
26. Marty K, Bordet P, Simonet V, Loire M, Ballou R, Darie C, et al. Magnetic and dielectric properties in the langasite-type compounds: A(3)BFe(3)D(2)O(14) (A = Ba, Sr, Ca; B = Ta, Nb, Sb; D = Ge, Si). Phys Rev B. 2010;81: 054416.
27. Greedan JE. Geometrically frustrated magnetic materials. J Mater Chem. 2001;11:37–53.
28. Lyubutin IS, Naumov PG, Mill' BV, Frolov KV, Demikhov EI. Structural and magnetic properties of the iron-containing langasite family $A_3MFe_3X_2O_{14}$ (A = Ba, Sr; M = Sb, Nb, Ta; X = Si, Ge) observed by Mössbauer spectroscopy. Phys Rev B. 2011;84:214425.
29. Li Y, Lu G, Yang H, Lan J, Zhang J, Huang W, et al. Lattice vibration of $Sr_3TaGa_3Si_2O_{14}$ single crystal. Phys Stat Solidi (b). 2007;244(2):518–28.
30. Kroumova E, Aroyo MI, Perez-Mato JM, Kirov A, Capillas C, Ivantchev S, et al. Bilbao crystallographic server: Useful databases and tools for phase-transition studies. Phase Transit. 2003;76(1-2):155–70.
31. Hudl M. Magnetic materials with tunable thermal, electrical, and dynamic properties: An experimental study of magnetocaloric, multiferroic, and spin-glass materials: Uppsala: Uppsala Dissertations from the Faculty of Science and Technology 98; 2012.
32. Gavriliuk AG, Lyubutin IS, Starchikov SS, Mironovich AA, Ovchinnikov SG, Trojan IA, et al. The magnetic P-T phase diagram of langasite $Ba_3TaFe_3Si_2O_{14}$ at high hydrostatic pressures up to 38 GPa. Appl Phys Lett. 2013;103(16):.

15

Quantum and classical nonlinear dynamics in a microwave cavity

Charles H Meaney[1], Hyunchul Nha[2], Timothy Duty[3] and Gerard J Milburn[1*]

*Correspondence:
milburn@physics.uq.edu.au
[1]Department of Physics, The
University of Queensland, St Lucia,
QLD 4072, Australia
Full list of author information is
available at the end of the article

Abstract

We consider a quarter wave coplanar microwave cavity terminated to ground via a superconducting quantum interference device. By modulating the flux through the loop, the cavity frequency is modulated. The flux is varied at twice the cavity frequency implementing a parametric driving of the cavity field. The cavity field also exhibits a large effective nonlinear susceptibility modelled as an effective Kerr nonlinearity, and is also driven by a detuned linear drive. We show that the semi-classical model corresponding to this system exhibits a fixed point bifurcation at a particular threshold of parametric pumping power. We show the quantum signature of this bifurcation in the dissipative quantum system. We further linearise about the below threshold classical steady state and consider it to act as a bifurcation amplifier, calculating gain and noise spectra for the corresponding small signal regime. Furthermore, we use a phase space technique to analytically solve for the exact quantum steady state. We use this solution to calculate the exact small signal gain of the amplifier.

Keywords: superconducting circuit; parametric amplifier; quantum noise

1 Introduction

Superconducting circuit quantum electrodynamics (circuit QED) [1] is increasingly being used to study systems in the quantum regime. This experimental context sees a superconducting coplanar waveguide act as a microwave cavity, in contrast to the optical frequency cavities of traditional cavity quantum electrodynamics (cavity QED). The microwave resonator is made from aluminium on a silicon substrate, and Josephson junctions are created by allowing the aluminium to oxidise before adding more aluminium. Such devices are placed in a dilution refrigerator, and experiments take place at cryogenic temperatures. Such low temperatures, close to the quantum ground state, allow quantum mechanical phenomena to become manifest. Recent engineering progress means that fabrication of these devices is possible [2].

In recent experiments at Chalmers [3], a quarter wave coplanar microwave cavity is terminated to ground via one or more superconducting quantum interference devices (SQUIDs), see Figure 1. By modulating the flux through the loop, the cavity frequency can be modulated. If the flux is varied at twice the cavity frequency this implements a parametric driving of the cavity field. The cavity field also exhibits a large effective nonlinear susceptibility that can be modelled as an intensity dependent phase shift [4].

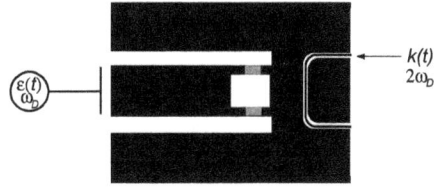

Figure 1 Schematic of the system under consideration in this paper. A schematic of the system under consideration. A superconducting microwave cavity of frequency ω_c has an effective Kerr nonlinearity χ, due primarily to the SQUID-loop connecting it to ground. It is driven both by a linear drive of amplitude ϵ at a detuning from the cavity of Δ, and also by a parametric drive at twice the cavity frequency at an amplitude κ.

This paper is structured as follows. In Section 2 we introduce the nonlinear microwave system considered in this paper. We establish a description in the form of a Markov master equation, one term of which being an effective Hamiltonian we derive. We also give the input-output formulation of the microwave system. In Section 3 we present a detailed analysis of the fixed point structure of the nonlinear microwave system in a semi-classical description, including bifurcation of the fixed points. We include dissipation of the microwave mode. In Section 4 we look at the steady state of the quantum system. This is done in a phase space representation based on the positive P-representation, both analytically, and numerically. We look for signatures of the semi-classical bifurcations. In Section 5 we analytically compute and plot the small signal gain. Then, in Section 6 we linearise the model and extract gain and noise spectra up to the threshold defined by the semi-classical bifurcation. Finally in Section 7 we summarise our results.

A similar model to that considered here has been given by Wustmann and Shumeiko [5]. Their discussion of the semiclassical steady states and fixed point structure parallels the discussion here but gives a more detailed description of the semiclassical dynamics. They also discuss the quantum noise features of the model using a linearised quantum Langevin approach. In addition to a linearised analysis of the gain and signal-to-noise ratio, we give an exact steady state solution for the quantum master equation using the positive P-representation. The steady state behaviour of the model we describe has been experimentally observed by Wilson et al. [6].

2 The dissipative Cassinian oscillator model

2.1 Master equation

We consider a superconducting microwave resonator connected through a superconducting quantum interference device (SQUID) loop to ground. The SQUID loop consists of two separated Josephson junctions; a magnetic flux can then be applied to loop to change the effective resonant frequency of the cavity [3]. The SQUID also induces a significant quartic nonlinearity. Following Wallquist et al. [4] we describe the integrated cavity + SQUID system in terms of an equivalent circuit composed of a capacitor and a nonlinear inductor. The Hamiltonian for one mode of the cavity field may be written in terms of this effective nonlinear oscillator as [4]

$$\mathcal{H} = E_C n^2 + E_L \phi^2 + \lambda \phi^4, \tag{1}$$

where $E_C = \frac{(2e)^2}{2C}$ represents the charging energy of the effective LC oscillator while n is the number of elementary charges on the capacitor, $E_L = \frac{\hbar^2}{2(2e)^2 L}$ represents the inductive energy of the effective oscillator, while ϕ represents the flux though the equivalent inductor and λ represents the inductive nonlinearity. This depends on the inductive energy scale and on the mode function of the cavity field $\lambda = E_L B$, where B is a geometric factor. Further details are given in Wallquist et al. give [4].

The system may be quantised by introducing the bosonic raising and lowering operators, defined in terms of the canonical variables of the cavity field as,

$$\hat{\phi} = \left(\frac{E_C}{4E_L}\right)^{\frac{1}{4}}(\hat{a} + \hat{a}^\dagger),$$

$$\hat{n} = -i\left(\frac{E_L}{4E_C}\right)^{\frac{1}{4}}(\hat{a} - \hat{a}^\dagger),$$

(2)

where $\hat{\psi}$ and $\hat{n}$ are the average phase across the junctions and charge on the junctions, respectively and C and L are the effective lumped capacitance and inductance, respectively, of the equivalent circuit for the cavity in terms of which the cavity resonant frequency is give by $\omega_c = \frac{1}{\sqrt{LC}}$. The total Hamiltonian, including the coherent driving and the parametric driving are then given by

$$\mathcal{H} = \hbar\omega_c\hat{a}^\dagger\hat{a} + \hbar\left(\epsilon^*\hat{a}e^{i\omega_D t} + \epsilon\hat{a}^\dagger e^{-i\omega_D t}\right) + \hbar\chi\hat{a}^{\dagger 2}\hat{a}^2 + \frac{\hbar}{2}\left(\kappa^*\hat{a}^2 e^{i2\omega_D t} + \kappa\hat{a}^{\dagger 2}e^{-i2\omega_D t}\right), \quad (3)$$

where ω_c is the cavity frequency, $\epsilon = |\epsilon|e^{i\upsilon}$ represents the coherent driving strength, κ represents the parametric driving strength, ω_D is the coherent driving frequency and we have assumed that the parametric driving is at twice the coherent driving frequency and υ is the phase difference between the coherent driving and the parametric driving as we have taken the phase of the parametric driving term as zero. In Section 6 we consider coherent homodyne detection of the cavity output. This means there is another phase in this problem; the phase choice for the local oscillator which may not be in phase with either the coherent or the parametric driving. The term proportional to χ represents a nonlinear (quartic) phase shift that arises from the nonlinear inductance of the SQUID loop. Quartic non-linearities in oscillators have been discussed in [7, 8]; parametric terms in the nano-electromechanical context have been discussed in [9–11].

For a realistic device we adopt a dissipative model. We model the microwave cavity resonator as being damped in a zero temperature heat bath. Such a model for the bath is certainly justified as the typical microwave cavity is at mK temperature and thus the mean excitation photon number is very close to zero [12]. The amplitude decay rate for the microwave cavity is γ. We then describe the dissipative dynamics with the master equation (with weak damping and the rotating wave approximation for the system-environment couplings). In an interaction picture at the coherent driving frequency this is

$$\frac{d\hat{\rho}}{dt} = -\frac{i}{\hbar}[\hat{\mathcal{H}}, \hat{\rho}] + \gamma\left(2\hat{a}\hat{\rho}\hat{a}^\dagger - \hat{a}^\dagger\hat{a}\hat{\rho} - \hat{\rho}\hat{a}^\dagger\hat{a}\right),$$

(4)

where $\hat{\rho}$ is the density matrix of the microwave cavity, and $\hat{\mathcal{H}}$ is the Hamiltonian in an interaction picture in a rotating frame with respect to the linear driving frequency. We have made the rotating wave approximation by ignoring terms with frequency $2\omega_D$ or above. We thus have

$$\hat{\mathcal{H}} = \hbar\Delta\hat{a}^\dagger\hat{a} + \hbar\left(\epsilon^*\hat{a} + \epsilon\hat{a}^\dagger\right) + \frac{\hbar}{2}\left(\kappa^*\hat{a}^2 + \kappa\hat{a}^{\dagger 2}\right) + \frac{\hbar\chi}{2}\hat{a}^{\dagger 2}\hat{a}^2,$$

(5)

where $\Delta = \omega_c - \omega_D$. In the absence of damping, classical trajectories arising from the parametric and nonlinear portion of this Hamiltonian are the ovals of Cassini, and that system is hence sometimes described as the 'Cassinian' oscillator; the quantum version of that part of the Hamiltonian system has been previously studied by Wielinga et al. [13] and more recently by Dykman and his collaborators [14].

The semi-classical dynamics and the exact quantum steady state can be found using the complex P-function of Drummond and Gardiner [15]. In this approach the density operator in terms of the off diagonal projectors onto oscillator coherent states

$$\hat{\rho}(t) = \int d\alpha \, d\beta \, P(\alpha, \beta, t) \frac{|\alpha\rangle\langle\beta^*|}{\langle\beta^*|\alpha\rangle}. \tag{6}$$

This function determines the normally ordered moments by

$$\langle \hat{a}^{\dagger m} \hat{a}^n \rangle = \int d\alpha \, d\beta \, P(\alpha, \beta, t) \alpha^n \beta^m. \tag{7}$$

It may seem surprising at first sight to notice that the Positive P-function has support in a phase space with twice as many canonical variables as the corresponding classical problem. There is a direct physical interpretation of the extra variables based on a measurement model in which there are twice as many readout channels for the canonical phase space variable [16]. This is required if the distributions are to give normally ordered moments directly via integration. In [17] a direct implementation using circuit QED of these additional channels is demonstrated and connection is made to the stationary normal ordered moments.

The master equation can then be converted into a Fokker-Planck like equation for the P-function,

$$\frac{\partial P(\alpha, \beta)}{\partial t} = \left\{ \partial_\alpha \big[(\gamma + i\Delta)\alpha + i\epsilon + i\chi\beta\alpha^2 + i\kappa\beta \big] + \partial_\beta \big[(\gamma - i\Delta)\beta - i\epsilon^* - i\chi\alpha\beta^2 - i\kappa^*\alpha \big] \right.$$
$$\left. + \partial_{\alpha\alpha}^2 \left[-i\frac{1}{2}(\kappa + \chi\alpha^2) \right] + \partial_{\beta\beta}^2 \left[i\frac{1}{2}(\kappa^* + \chi\beta^2) \right] \right\}, \tag{8}$$

where $\partial_\alpha = \frac{\partial}{\partial\alpha}$ and $\partial_{\alpha\alpha}^2 = \frac{\partial^2}{\partial\alpha\,\partial\alpha}$ etc. The corresponding stochastic differential equations are

$$d\alpha = -(\gamma + i\Delta)\alpha dt - i\epsilon dt - i(\chi\alpha^2 + \kappa)\beta dt + \left[-i(\kappa + \chi\alpha^2) \right]^{\frac{1}{2}} dz_1,$$
$$d\beta = -(\gamma - i\Delta)\beta dt + i\epsilon^* dt + i(\chi\beta^2 + \kappa^*)\alpha dt + \left[i(\kappa^* + \chi\beta^2) \right]^{\frac{1}{2}} dz_2. \tag{9}$$

The semi-classical equations are obtained by setting $\beta = \alpha^*$ in the drift term and ignoring the diffusion term, and are thus

$$\frac{d\alpha}{dt} = -(\gamma + i\Delta + i\chi|\alpha|^2)\alpha - i\epsilon - i\kappa\alpha^*, \tag{10}$$

from which it is apparent that the nonlinearity appears as a nonlinear detuning. This ensures that the instability in the $\chi = 0$ model when $\kappa = \gamma$ does not arise.

3 Semi-classical fixed point structure

3.1 No coherent driving, $\epsilon = 0$

We first consider the case of no driving field $\epsilon = 0$. The semi-classical equations of motion (10) are then given by

$$\frac{d\alpha}{dt} = -(\gamma + i\Delta)\alpha - i\chi\alpha|\alpha|^2 - i\kappa\alpha^*. \tag{11}$$

The semi-classical steady states, $\alpha_0 = \sqrt{n_0}e^{i\theta_0}$, are given by $\bar{n}_0 = 0$, and

$$\bar{n}_0 = -\Delta' \pm \sqrt{(\kappa')^2 - 1}, $$
$$\sin(2\theta_0) = -\frac{1}{\kappa'}, \tag{12}$$

where we have defined the scaled variables

$$\bar{n}_0 = \frac{\chi}{\gamma}n_0, $$
$$\kappa' = \frac{\kappa}{\gamma}, \tag{13}$$
$$\Delta' = \frac{\Delta}{\gamma}.$$

In order to determine the stability of the fixed points, we linearise the equations of motion around the fixed points. Thus we have the semi-classical linearised equation of motion for $\delta\alpha = \alpha - \alpha_0$ and $\delta\alpha^* = \alpha^* - \alpha_0^*$

$$\begin{bmatrix} \frac{d(\delta\alpha)}{dt} \\ \frac{d(\delta\alpha^*)}{dt} \end{bmatrix} \approx \mathbf{M}_{\alpha\alpha^*} \begin{bmatrix} \delta\alpha \\ \delta\alpha^* \end{bmatrix}, \tag{14}$$

where

$$\mathbf{M}_{\alpha\alpha^*} = \begin{bmatrix} -\gamma - i\Delta - i2\chi|\alpha_0|^2 & -i\chi\alpha_0^2 - i\kappa \\ i\chi(\alpha_0^*)^2 + i\kappa & -\gamma + i\Delta + i2\chi|\alpha_0|^2 \end{bmatrix}. \tag{15}$$

In the limit of no parametric pumping ($\kappa = 0$), this Jacobian matches the result obtained by Babourina-Brooks et al. in [8]. Stability of the fixed point requires all the eigenvalues of the Jacobian to have a real part less than or equal to zero [18]. A real part of exactly zero indicates marginal stability in that parameter direction, where the fixed point is neither attractive nor repulsive. Real parts strictly less than zero are attracting fixed points which draw in nearby regions in phase space. In general, stability may depend on more coupling parameter combinations than those which define the fixed points.

The origin is a fixed point for all parameter values. Indeed, the origin is the only fixed point for $\kappa < \gamma$, the 'below threshold' regime. This fixed point is stable for $\kappa^2 < \gamma^2 + \Delta^2$. Four additional fixed points occur as antipodal pairs for $\kappa > \gamma$; the 'above threshold' regime. The first additional pair of fixed points, which we will call the 'stable pair', and are given by

$$(\bar{n}_0, 2\theta_0) = \left(-\Delta' + \sqrt{(\kappa')^2 - 1}, -\pi + \text{arccsc}\,\kappa'\right), \tag{16}$$

exists for $\Delta < \sqrt{(\kappa)^2 - \gamma^2}$, and is always stable. The second additional pair of fixed points, which we will call the 'unstable pair'

$$(\bar{n}_0, 2\theta_0) = \left(-\Delta' - \sqrt{(\kappa')^2 - 1}, -\operatorname{arccsc} \kappa'\right), \tag{17}$$

exists for $\Delta < -\sqrt{(ka)^2 - \gamma^2}$, and is always unstable. We note that the unstable pair of fixed points can only exist for negative detuning, and that whenever the unstable pair of fixed points exists, the first pair of fixed points exists also. Also, we note that for $\epsilon = 0$, $\hat{a} \rightarrow -\hat{a}$ is a symmetry of the system, and thus pairs of antipodal fixed points are the expected semi-classical result. We plot the radial and angular components of the semi-classical fixed points in Figure 2. The colour in these plots shows the stability.

The bifurcating behaviour of the semi-classical steady state is depicted in the 'phase diagram' of Figure 3. There are two qualitatively different transitions that can take place in

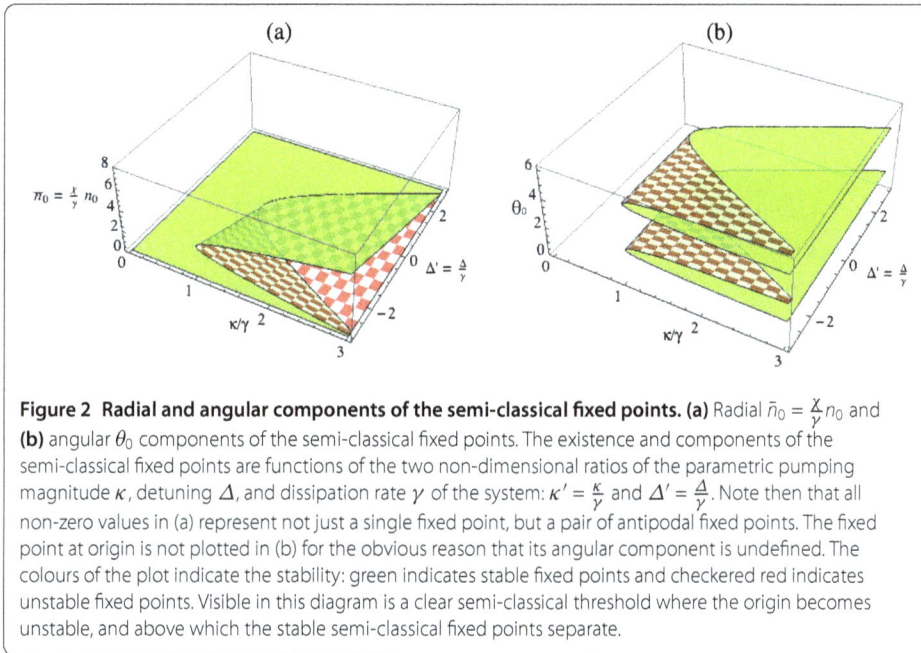

Figure 2 Radial and angular components of the semi-classical fixed points. (a) Radial $\bar{n}_0 = \frac{\chi}{\gamma} n_0$ and **(b)** angular θ_0 components of the semi-classical fixed points. The existence and components of the semi-classical fixed points are functions of the two non-dimensional ratios of the parametric pumping magnitude κ, detuning Δ, and dissipation rate γ of the system: $\kappa' = \frac{\kappa}{\gamma}$ and $\Delta' = \frac{\Delta}{\gamma}$. Note then that all non-zero values in (a) represent not just a single fixed point, but a pair of antipodal fixed points. The fixed point at origin is not plotted in (b) for the obvious reason that its angular component is undefined. The colours of the plot indicate the stability: green indicates stable fixed points and checkered red indicates unstable fixed points. Visible in this diagram is a clear semi-classical threshold where the origin becomes unstable, and above which the stable semi-classical fixed points separate.

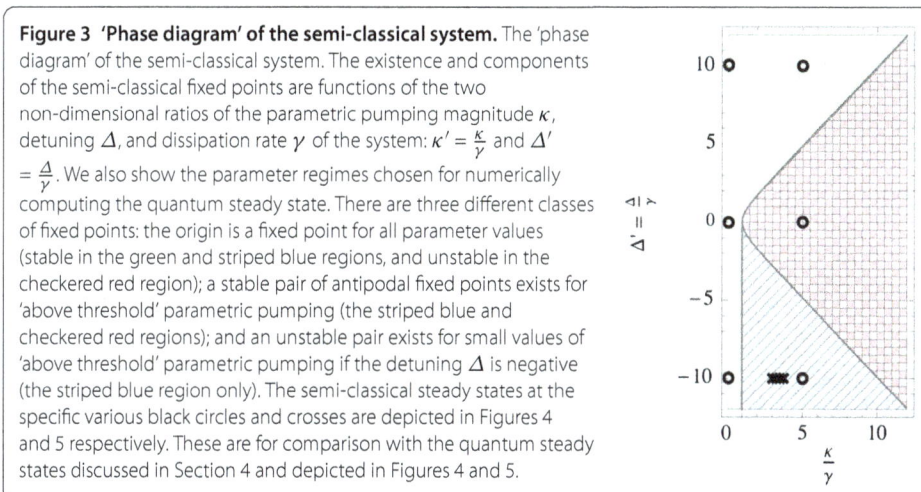

Figure 3 'Phase diagram' of the semi-classical system. The 'phase diagram' of the semi-classical system. The existence and components of the semi-classical fixed points are functions of the two non-dimensional ratios of the parametric pumping magnitude κ, detuning Δ, and dissipation rate γ of the system: $\kappa' = \frac{\kappa}{\gamma}$ and $\Delta' = \frac{\Delta}{\gamma}$. We also show the parameter regimes chosen for numerically computing the quantum steady state. There are three different classes of fixed points: the origin is a fixed point for all parameter values (stable in the green and striped blue regions, and unstable in the checkered red region); a stable pair of antipodal fixed points exists for 'above threshold' parametric pumping (the striped blue and checkered red regions); and an unstable pair exists for small values of 'above threshold' parametric pumping if the detuning Δ is negative (the striped blue region only). The semi-classical steady states at the specific various black circles and crosses are depicted in Figures 4 and 5 respectively. These are for comparison with the quantum steady states discussed in Section 4 and depicted in Figures 4 and 5.

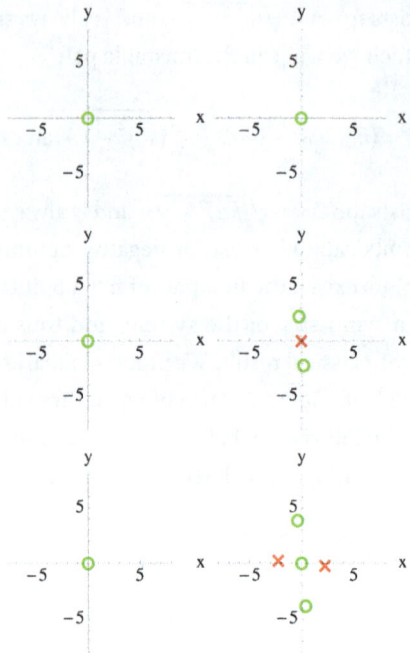

Figure 4 Semi-classical steady states of the microwave system for different regions of the phase diagram. Semi-classical steady states of the microwave system for different regions of the 'phase diagram' in Figure 3. These nine plots show the semi-classical fixed points that correspond to the parameters of the nine black circles in Figure 3. Specifically, from top-left to bottom-right, moving form left to right, then top to bottom, these parameter values are: $(\kappa', \Delta') = (0, 10), (5, 10), (0, 0), (5, 0), (0, -10),$ and $(5, -10)$. The colours indicate stability; the green circle markers are stable fixed points, and the red cross markers are unstable fixed points. Clearly visible are the qualitative differences in each region separated by the semi-classical bifurcations. These semi-classical fixed points can be compared with the quantum steady state Wigner functions for the same set of parameters in Figure 6.

the semi-classical system as it moves from being 'below threshold' to 'above threshold'. First, for positive detuning ($\Delta > 0$), the threshold parametric pumping $\kappa' = 1$ is effectively increased by the detuning to $\kappa' = \sqrt{1 + \Delta'^2}$ (the solid green/checkered red boundary in Figure 3. At this threshold, the semi-classical system undergoes a supercritical pitchfork bifurcation where the stable origin goes unstable, and the stable pair of fixed points emerges from the origin and grows in separation with increasing parametric pumping (the checkered red region in the upper half-plane in Figure 3). Alternatively, for all values of negative detuning ($\Delta < 0$), at the threshold parametric pumping $\kappa' = 1$ two saddle-node bifurcations produce both the stable and unstable fixed point pairs (the solid green/striped blue boundary in Figure 3). The origin remains stable, and the two newly created pairs of fixed points then exist for parametric pumping above threshold ($\kappa' > 1$) until pumping reaches the even higher value $\kappa' = \sqrt{1 + \Delta'^2}$. Between these two values (the striped blue region in Figure 3) with increasing parametric pumping, the stable pair increases in separation and the unstable pair moves to the origin. At the higher parametric pumping $\kappa' = \sqrt{1 + \Delta'^2}$ (the striped blue/checkered red boundary in Figure 3) the unstable pair annihilates in a subcritical pitchfork bifurcation at the origin, and the origin becomes unstable for all higher parametric pumping $\kappa' > \sqrt{1 + \Delta'^2}$. The stable pair of fixed points continues to grow in separation with further increased parametric pumping (the checkered red region in the lower half-plane in Figure 3).

We now illustrate the steady state behaviour of the semi-classical microwave system in different 'phases'. We choose a point in parameter space from each region of the semi-classical 'phase diagram' of Figure 3. These choices are marked with the black circles in that figure. These steady states are shown in Figure 4. Some other near identical steady states, corresponding to the black crosses in Figure 3 are depicted in Figure 5. These are plotted for comparison with the quantum version of the system in Section 4.

As well as the bifurcation structure, an important observation to make at this point is that the existence and bifurcations of the fixed points depend upon only three parame-

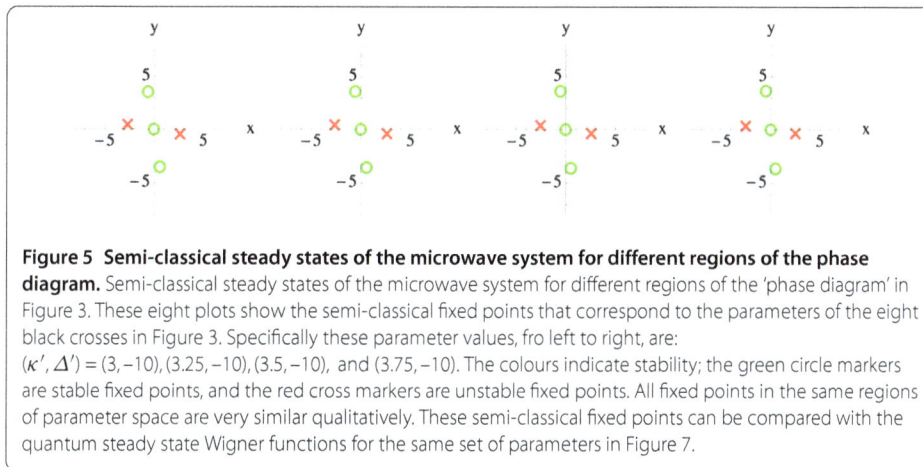

Figure 5 Semi-classical steady states of the microwave system for different regions of the phase diagram. Semi-classical steady states of the microwave system for different regions of the 'phase diagram' in Figure 3. These eight plots show the semi-classical fixed points that correspond to the parameters of the eight black crosses in Figure 3. Specifically these parameter values, fro left to right, are:
$(\kappa', \Delta') = (3, -10), (3.25, -10), (3.5, -10),$ and $(3.75, -10)$. The colours indicate stability; the green circle markers are stable fixed points, and the red cross markers are unstable fixed points. All fixed points in the same regions of parameter space are very similar qualitatively. These semi-classical fixed points can be compared with the quantum steady state Wigner functions for the same set of parameters in Figure 7.

ters: the magnitude of the parametric pumping rate κ; the detuning Δ; and the dissipation rate γ. And specifically, only the two non-dimensional ratios of them, here we chose $\kappa' = \frac{\kappa}{\gamma}$ and $\Delta' = \frac{\Delta}{\gamma}$. Thus, the below threshold to above threshold transition of the parametric oscillator is independent of the size of the induced Kerr nonlinearity χ. However, the separation of the semi-classical fixed points $\sqrt{\bar{n}_0} = \sqrt{\frac{\chi}{\gamma}n_0}$, and thus the degree and visibility of the above-threshold oscillations, depends on the scaling parameter $\frac{\chi}{\gamma}$. Thus, to see the semi-classical fixed points move significantly away from the origin, and thus to observe significant above-threshold behaviour we require a significantly large nonlinearity χ as well as parametric pumping κ.

It will also prove useful to consider the non-dissipative limit $(\gamma \to 0)$ of the semi-classical equations on resonance $(\Delta \to 0)$. The fixed points are no longer stable zero dimensional attractors but rather elliptical fixed points corresponding to stable small oscillations in the corresponding Hamiltonian model. One easily sees that the fixed points occur at

$$\alpha_0^2 = -\frac{\kappa}{\chi}. \tag{18}$$

3.2 Including coherent driving, $\epsilon \neq 0$

Similar to our definitions $\alpha_0 = \sqrt{\bar{n}_0}e^{i\theta_0}$ and $\bar{n}_0$, we introduce the scaled Cartesian coordinates $\bar{x}_0$ and $\bar{y}_0$ such that $\alpha_0 = x_0 + iy_0$ and

$$\bar{x}_0 = \sqrt{\frac{\chi}{\gamma}}x_0,$$
$$\bar{y}_0 = \sqrt{\frac{\chi}{\gamma}}y_0. \tag{19}$$

We also define the scaled linear driving

$$\bar{\epsilon}' = \sqrt{\frac{\chi}{\gamma}}\frac{\epsilon}{\gamma}. \tag{20}$$

In terms of these, the fixed points of the semi-classical equation of motion (10) satisfy the quintic equation

$$
\begin{aligned}
0 = {}& 4\kappa'^2 \bar{y}_0^5 - 4\Im\{\bar{\epsilon}'\}\kappa' \bar{y}_0^4 + \left[4\left(\kappa' + \Delta'\kappa'^2 - \kappa'^3\right) + |\bar{\epsilon}'|^2\right]\bar{y}_0^3 \\
& + 2\left[\Re\{\bar{\epsilon}'\}(\Delta' - 2\kappa') + \Im\{\bar{\epsilon}'\}\left(1 + 2\kappa'(\Delta' - 2\kappa')\right)\right]\bar{y}_0^2 \\
& + \left[1 + \Delta'^2 - \kappa'^2 - 3\Re\{\bar{\epsilon}'\}\Im\{\bar{\epsilon}'\} + \Im\{\bar{\epsilon}'\}^2(\Delta' - 5\kappa')\right]\bar{y}_0 \\
& + \left[\Re\{\bar{\epsilon}'\} + \Im\{\bar{\epsilon}'\}(\Delta' + \kappa') + \Im\{\bar{\epsilon}'\}^3\right].
\end{aligned}
\tag{21}
$$

We notice that were there to be only a small input signal and not a large linear drive with a small input signal on top ($\epsilon = \bar{\epsilon}' = 0$), then the quintic factorises into the quadratic

$$
0 = \bar{y}_0\left(4\kappa'^2\left(\bar{y}_0^2\right)^2 + 4\left(\kappa' + \Delta'\kappa'^2 - \kappa'^3\right)\bar{y}_0^2 + 1 + \Delta'^2 - \kappa'^2\right).
\tag{22}
$$

This of course defines the tractable analytic fixed points given in Section 3.1.

Unfortunately, solving a quintic equation analytically in terms of radicals can lead to unhelpful expressions, and is not even always possible. We can of course numerically solve for the fixed points for certain parameter values, but we leave non-perturbative exploration of the steady states of the $\epsilon \neq 0$ system for a later study. Instead, we will ultimately expand the Positive P function as a power series in ϵ in Section 5.

4 Quantum steady state

In the previous section we described the fixed point bifurcations of the semi-classical system. Here, we investigate whether there is a signature of those semi-classical bifurcations present in the full quantum system. This can be done exactly using the positive P function, or numerically by computing the quantum steady state density operator in a truncated number basis and then constructing a phase space quasi probability density (e.g. a Q function) in different regions of the semi-classical 'phase diagram' of Figure 3. As we will show, by changing the coupling parameters so as to be on different sides of a semi-classical bifurcation, there is a corresponding qualitative change in the quantum steady state. This kind of correspondence principle has proven to be the case for other dissipative nonlinear quantum systems [19–21].

4.1 Steady state via the positive P function

The steady state solution of (8) can be found as the potential conditions are satisfied [22]. The steady state solution can be written as

$$
P_s(\alpha, \beta) = \mathcal{N}e^{-V(\alpha,\beta)},
\tag{23}
$$

where the potential function is given by

$$
\begin{aligned}
V(\alpha, \beta) = {}& -2\alpha\beta - \lambda \ln\left(\chi\alpha^2 + \kappa\right) - \lambda^* \ln\left(\chi\beta^2 + \kappa\right) \\
& - \frac{2\epsilon}{\sqrt{\chi\kappa}}\arctan\frac{\alpha}{A_0} - \frac{2\epsilon^*}{\sqrt{\chi\kappa}}\arctan\frac{\beta}{A_0^*},
\end{aligned}
\tag{24}
$$

where

$$\lambda = -1 + \frac{\Delta}{\chi} - i\frac{\gamma}{\chi}, \tag{25}$$

and $A_0 = \sqrt{-\alpha_0^2} = \sqrt{\frac{\kappa}{\chi}}$ determines the semi-classical fixed points of the corresponding Hamiltonian (non-dissipative) model, see (18). This may be written in an alternate form by noting that $\arctan x = i\frac{1}{2} \ln \frac{1-ix}{1+ix}$,

$$P_s(\alpha, \beta) = \mathcal{N} \left(\frac{\alpha_0 - \alpha}{\alpha_0 + \alpha} \right)^\mu \left(\frac{\alpha_0^* - \beta}{\alpha_0^* + \beta} \right)^{\mu^*} \left(\chi \alpha^2 + \kappa \right)^\lambda \left(\chi \beta^2 + \kappa \right)^{\lambda^*} e^{2\alpha\beta}, \tag{26}$$

where $\mu = i\frac{\epsilon}{\sqrt{\chi\kappa}}$.

Before we can compare this distribution to the phase space structure of the semi-classical fixed points we must face the unusual feature that the Positive P function has support on a phase space with twice as many dimensions as the corresponding classical problem. The semi-classical subspace corresponds to $\beta = \alpha^*$. If it were not for the noise terms in the stochastic differential equations, (9), we could start on this subspace and never leave it. The noise however will drive the dynamics off the semi-classical subspace. Despite this we can find a very close correspondence between the semi-classical fixed points and the form of the steady state Positive P function.

We first discuss the correspondence for the case of no coherent driving, $\epsilon = 0$. The peaks of the steady state positive P function will be located at the minimum of the corresponding potential function, that is to say, the solutions of, $\partial_\alpha V = \partial_\beta V = 0$. This gives

$$\beta = -\frac{\lambda \alpha}{\alpha^2 - \alpha_0^2},$$
$$\alpha = -\frac{\lambda^* \beta}{\beta^2 - \alpha_0^{*2}}, \tag{27}$$

where we have used (18). A little algebra shows that these are equivalent to

$$\left(\beta^2 - \alpha_0^{*2} \right)\left(\alpha^2 - \alpha_0^2 \right) = |\lambda|^2, \tag{28}$$

$$\frac{1 - (\frac{\alpha_0}{\alpha})^2}{1 - (\frac{\alpha_0^*}{\beta})^2} = \frac{\lambda}{\lambda^*}. \tag{29}$$

There are two classes of solutions: $\beta = \alpha^*$ and $\beta = -\alpha^*$. We will refer to the first of these as the semi-classical subspace and the second as the nonclassical.

We first consider the semi-classical subspace. With $\beta = \alpha^*$, the first equation in (27) should be compared with the semi-classical steady state from (11), which may be written as

$$\alpha^* = -\frac{(\frac{\Delta}{\chi} - i\frac{\gamma}{\chi})\alpha}{\alpha^2 + \frac{\kappa}{\chi}}. \tag{30}$$

In the limit of small quantum noise, $\chi \to 0, \kappa \to 0$, such that $\frac{\kappa}{\chi}$ = constant we find that

$$\lambda \approx \frac{\Delta}{\chi} - i\frac{\gamma}{\chi}, \tag{31}$$

and, in the semi-classical subspace, the P-function is peaked on the semi-classical steady states.

In the model of Wolinsky and Carmichael [23] the nonlinear detuning χ becomes complex, thus describing nonlinear damping, and the dynamics of the positive P-function takes a very similar form to that considered here. In particular the additional fixed points of the non classical dimension are also present. As they describe, the non classical subspace allows the noise to drive a stochastic process that corresponds to the nonclassical features of the steady state solution. In the case of strong nonlinearity they show that the steady state positive P-function on the non classical subspace is localised on the non classical fixed points and that these peaks reflect the fact that the steady state is close to a superposition of two coherent states localised on the classical fixed points.

The explicit solutions to (27) are not straightforward; they are

$$(\alpha, \beta) = (0, 0), \tag{32}$$

$$\left(\pm_1 \frac{1}{\alpha_0^*} \sqrt{|\alpha_0|^4 + i\lambda \Im\{\lambda\} \pm_2 \lambda \sqrt{|\alpha_0|^4 - \Im\{\lambda\}^2}}, \right.$$

$$\left. \mp_2 \pm_1 \frac{1}{\alpha_0} \sqrt{|\alpha_0|^4 - i\lambda^* \Im\{\lambda\} \pm_2 \lambda^* \sqrt{|\alpha_0|^4 - \Im\{\lambda\}^2}} \right).$$

These are very close, though not exactly coincident, with the semi-classical fixed points derived in Section 3.

4.2 Numerical steady state

To perform the numerical computation of the quantum steady state we use the Quantum Optics MATLAB toolbox [24]. To do this we approximate the infinite basis of the microwave cavity oscillator; we choose to do this by truncating in the Fock (number) basis. This means that we must choose couplings such that the bifurcation takes place sufficiently close to the origin to be accurately approximated by the truncation. This is roughly because a coherent state of amplitude α has a mean occupation number of $|\alpha|^2$. Given the quantum steady state typically (as we shall see direct evidence of in this section) has support centred on the semi-classical steady state, fixed points far from the origin (high $|\alpha|$) will produce high occupations and thus inaccurate results if we truncate in the Fock (number) basis.

We choose a point in parameter space from each region of the semi-classical 'phase diagram' of Figure 3. These choices are marked with the black circle markers in that figure. Semi-classically, the corresponding steady states were shown in Figure 4. We now look at the quantum steady state through the Wigner function of the steady state density matrix. The Wigner function is defined as $W(x, y) = \frac{1}{\pi \hbar} \int_{-\infty}^{\infty} dz \langle x - z | \hat{\rho} | x + z \rangle e^{i \frac{2yz}{\hbar}}$; for more on the Wigner function see [15, 25]. These Wigner functions are shown in Figure 6. There are clear signature of the semi-classical bifurcations. The quantum steady state has support centred on the stable semi-classical fixed points, something which has been previously observed in [19–21].

However, in two of the Wigner functions of Figure 6 (those corresponding to the striped blue region of Figure 3) there are three semi-classical stable fixed points, yet only two main regions of quantum steady state density. To investigate this further, we consider the

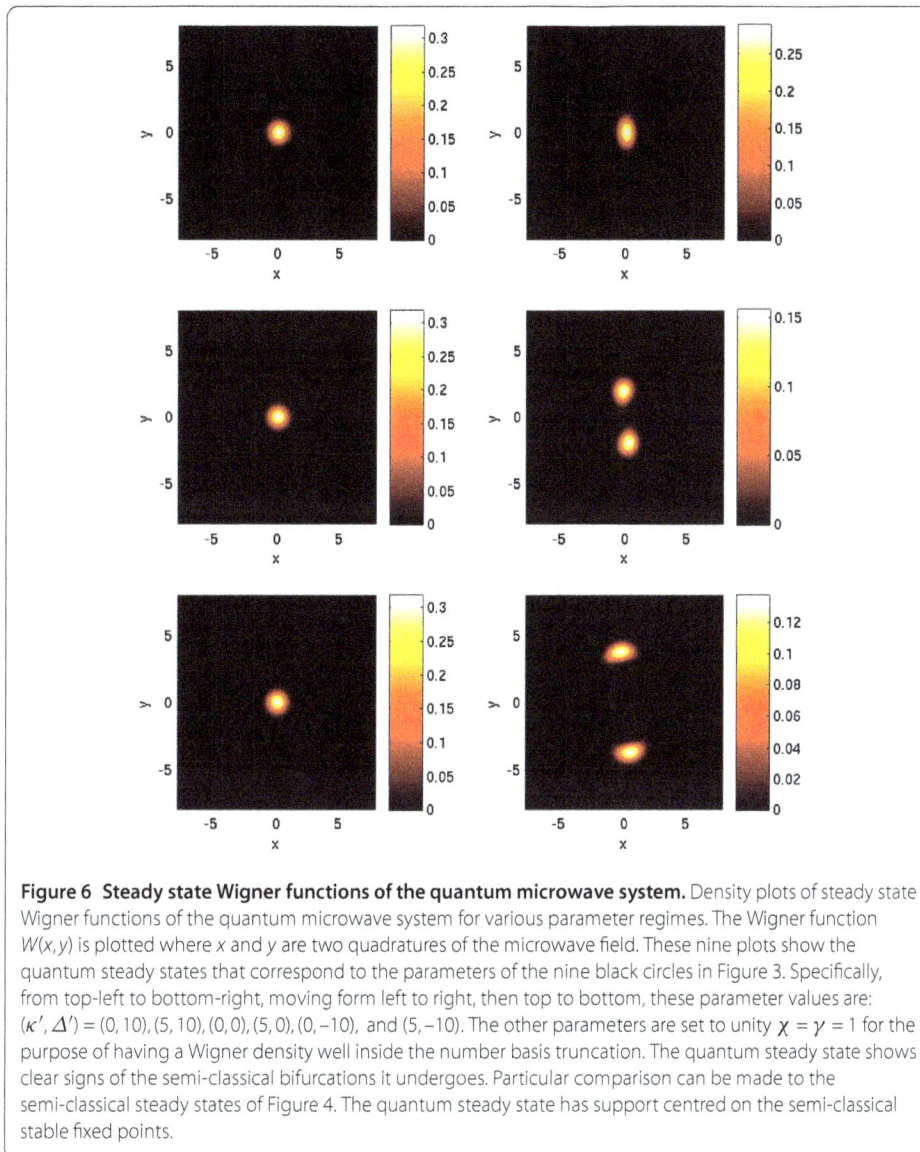

Figure 6 Steady state Wigner functions of the quantum microwave system. Density plots of steady state Wigner functions of the quantum microwave system for various parameter regimes. The Wigner function $W(x,y)$ is plotted where x and y are two quadratures of the microwave field. These nine plots show the quantum steady states that correspond to the parameters of the nine black circles in Figure 3. Specifically, from top-left to bottom-right, moving form left to right, then top to bottom, these parameter values are: $(\kappa', \Delta') = (0, 10), (5, 10), (0, 0), (5, 0), (0, -10),$ and $(5, -10).$ The other parameters are set to unity $\chi = \gamma = 1$ for the purpose of having a Wigner density well inside the number basis truncation. The quantum steady state shows clear signs of the semi-classical bifurcations it undergoes. Particular comparison can be made to the semi-classical steady states of Figure 4. The quantum steady state has support centred on the semi-classical stable fixed points.

quantum steady states corresponding to small parameter changes in this region. In particular, we look at the quantum steady states corresponding to the parameter space choices marked with black crosses in Figure 3. Semi-classically, the corresponding steady states were shown in Figure 5. The corresponding Wigner functions are shown in Figure 7. Interestingly, there is a gradual transition from quantum steady state support centred on the semi-classical stable fixed point at the origin, to support centred on the separated stable pair. This transition does not correspond to any semi-classical bifurcation, and at this stage is a quantum feature we cannot explain or predict semi-classically. We mention it here to suggest one direction for future investigation of this system.

5 The small signal gain

The positive P-function directly determines the normally ordered steady state moments of the intracavity field. We now need to chose the contour of integration so that the nor-

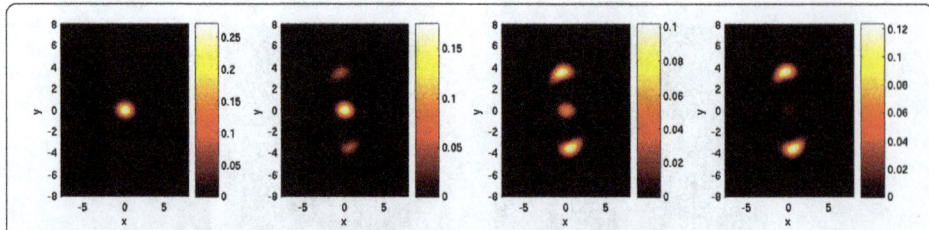

Figure 7 Steady state Wigner functions of the quantum microwave system. Density plots of steady state Wigner functions of the quantum microwave system for small parameter changes in the blue parameter region of Figure 2. The Wigner function $W(x, y)$ is plotted where x and y are two quadratures of the microwave field. These eight plots show the quantum steady states that correspond to the parameters of the eight black crosses in Figure 3. Specifically from left to right these parameter values are: $(\kappa', \Delta') = (3, -10), (3.25, -10), (3.5, -10),$ and $(3.75, -10)$. The other parameters are set to unity $\chi = \gamma = 1$ for the purpose of having a Wigner density well inside the number basis truncation. Comparison should be made with the semi-classical steady states of Figure 5. The quantum steady state shows support that shifts from being centred on the semi-classical stable fixed point at the origin, to being centred on the separated stable pair. This transition does not correspond to any semi-classical bifurcation. While the semi-classical steady states of Figure 5 are quite insensitive to small parameter shifts in regions bounded by semi-classical bifurcations, the corresponding quantum steady states have a marked qualitative change.

malization constant $\mathcal{N}$ is fixed. To this end we define the integrals

$$A_{mn} = \mathcal{N}^{-1} \int d\alpha \, d\beta \, \alpha^n \beta^m P_s(\alpha, \beta), \tag{33}$$

and express the normally ordered moments as

$$\langle \hat{a}^{\dagger m} \hat{a}^n \rangle = \frac{A_{mn}}{A_{00}}. \tag{34}$$

If we wish to regard this system as an amplifier, we need to calculate the mean cavity field amplitude $\langle \hat{a} \rangle$ as a function of ϵ for the case that $\epsilon \ll \kappa$. With this in mind we expand the solution in a Taylor series in ϵ

$$P_s(\alpha, \beta) = P_s^{(0)}(\alpha, \beta) - 2\mu \sum_{k=0}^{\infty} \frac{1}{2k+1} \left(\frac{\alpha}{\alpha_0} \right)^{2k+1} P_s^{(0)}(\alpha, \beta)$$

$$- 2\mu^* \sum_{k=0}^{\infty} \frac{1}{2k+1} \left(\frac{\beta}{\alpha_0^*} \right)^{2k+1} P_s^{(0)}(\alpha, \beta), \tag{35}$$

where $P_s^{(0)}(\alpha, \beta)$ is the exact steady state solution for $\epsilon = 0$. Then

$$A_{mn} = A_{mn}^{(0)} - 2\mu \sum_{k=0}^{\infty} \frac{1}{2k+1} \left(\frac{\alpha}{\alpha_0} \right)^{2k+1} A_{m,n+2k+1}^{(0)}$$

$$- 2\mu^* \sum_{k=0}^{\infty} \frac{1}{2k+1} \left(\frac{\beta}{\alpha_0^*} \right)^{2k+1} A_{m+2k+1,n}^{(0)}. \tag{36}$$

In this form we can see that the normalisation for $P_s(\alpha, \beta)$ is the same as that for $P_s^{(0)}(\alpha, \beta)$ as the integrals $A_{0,2k+1}^{(0)}$ vanish.

If we now substitute (26) (with $\epsilon = 0 \Rightarrow \mu = 0$) into (33), and use the Beta function identity

$$\left(1 - e^{i2\pi\alpha}\right)\left(1 - e^{i2\pi\beta}\right)B(\alpha, \beta) = \int_C t^{\alpha-1}(1 - t)^{\beta-1}\,dt, \tag{37}$$

then we obtain the moments for zero coherent driving

$$A_{mn}^{(0)} = \frac{\kappa^{\lambda+\lambda^*+1}\left(1 - e^{i2\pi(\lambda+1)}\right)\left(1 - e^{i2\pi(\lambda^*+1)}\right)}{-4\chi}\left(-\frac{\kappa}{\chi}\right)^{(n+m)/2}$$
$$\times \sum_{l=0}^{\infty}\left\{\frac{1}{l!}\left(\frac{2\kappa}{\chi}\right)^l[1 + (-1)^{l+n}][1 + (-1)^{l+m}]\right.$$
$$\left.\times B\left(\lambda + 1, \frac{l+n+1}{2}\right)B\left(\lambda^* + 1, \frac{l+m+1}{2}\right)\right\}. \tag{38}$$

Since we will always be interested in ratios of these, we can omit the leading constant; this then exactly matches the expression found by Kryuchkyan and Kheruntsyan [26]

$$A_{mn}^{(0)} = \left(-\frac{\kappa}{\chi}\right)^{(n+m)/2}\sum_{l=0}^{\infty}\left\{\frac{1}{l!}\left(\frac{2\kappa}{\chi}\right)^l[1 + (-1)^{l+n}][1 + (-1)^{l+m}]\right.$$
$$\left.\times B\left(\lambda + 1, \frac{l+n+1}{2}\right)B\left(\lambda^* + 1, \frac{l+m+1}{2}\right)\right\}. \tag{39}$$

We first consider the steady state mean intra-cavity photon number with no coherent signal,

$$\langle\hat{a}^\dagger\hat{a}\rangle^{(0)} = \frac{A_{11}^{(0)}}{A_{00}^{(0)}}. \tag{40}$$

In Figure 8 we plot this as a function of the parametric driving strength. Note that we do not see a bistable curve as in Figure 2. The reason for this is that the quantum steady state gives a long time average which averages over all possible switching events between the two semi-classical steady states in the bistable region. The quantum steady state is a double peaked distribution in the complex P representation with each peak localised near one or the other semi-classical fixed points in the bistable region.

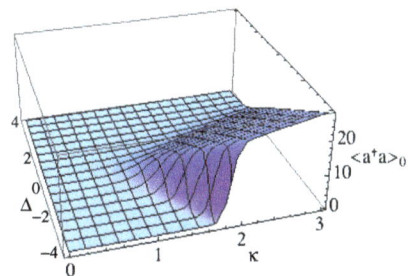

Figure 8 The steady state mean photon number in the cavity for no linear driving. The steady state mean photon number in the cavity for no coherent driving $\epsilon = 0$ as a function of the parametric pump magnitude κ and the detuning Δ. The corresponding semi-classical fixed points plotted in Figure 2 showed bi-stability for negative detuning $\Delta < 0$ which does not occur in the quantum steady state. Time units are chosen so that $\gamma = 1$ and $\chi = 0.25$.

We can now write the moments of the intra-cavity field as,

$$\langle \hat{a}^{\dagger m} \hat{a}^n \rangle = \langle \hat{a}^{\dagger m} \hat{a}^n \rangle^{(0)} - 2\mu \sum_{k=0}^{\infty} \frac{1}{2k+1} \left(\frac{1}{\alpha_0} \right)^{2k+1} \langle \hat{a}^{\dagger m} \hat{a}^{n+2k+1} \rangle^{(0)}$$

$$- 2\mu^* \sum_{k=0}^{\infty} \frac{1}{2k+1} \left(\frac{1}{\alpha_0^*} \right)^{2k+1} \langle \hat{a}^{\dagger m+2k+1} \hat{a}^n \rangle^{(0)}, \tag{41}$$

where $\langle \cdots \rangle_0$ denotes the steady state average for $\epsilon = 0$. In particular, the average amplitude in the cavity at steady state is

$$\langle \hat{a} \rangle = \langle \hat{a} \rangle^{(0)} - 2\mu \sum_{k=0}^{\infty} \frac{1}{2k+1} \left(\frac{1}{\alpha_0} \right)^{2k+1} \langle \hat{a}^{2k+2} \rangle^{(0)}$$

$$- 2\mu^* \sum_{k=0}^{\infty} \frac{1}{2k+1} \left(\frac{1}{\alpha_0^*} \right)^{2k+1} \langle \hat{a}^{\dagger 2k+1} \hat{a} \rangle^{(0)}, \tag{42}$$

where we have used $\langle \hat{a} \rangle_0 = 0$. Explicitly, this average amplitude is

$$\langle \hat{a} \rangle = -2 \frac{\alpha_0}{N} \sum_{k=0}^{\infty} \sum_{r=0}^{\infty} \frac{1}{2k+1} \frac{2^{2r} |\alpha_0|^{4r}}{(2r)!} \Gamma\left(\frac{1}{2} + r \right) \Gamma\left(\frac{3}{2} + r + k \right)$$

$$\times \left[\frac{\mu}{\Gamma(\frac{3}{2} + r + \lambda^*) \Gamma(\frac{5}{2} + r + k + \lambda)} + \frac{\mu^* |\alpha_0|^2}{\Gamma(\frac{5}{2} + r + \lambda) \Gamma(\frac{5}{2} + r + k + \lambda^*)} \right], \tag{43}$$

where

$$N = \sum_{s=0}^{\infty} \frac{2^{2s} |\alpha_0|^{4s}}{(2s)!} \frac{|\Gamma(\frac{1}{2} + s)|^2}{|\Gamma(\frac{3}{2} + s + \lambda)|^2} \tag{44}$$

we recall that $\alpha_0^2 = -\frac{\kappa}{\chi}$, $\lambda = -1 + \frac{\Delta}{\chi} - i\frac{\gamma}{\chi}$, and $\mu = i\frac{\epsilon}{\sqrt{\chi\kappa}} = i\frac{1}{\sqrt{\frac{\kappa}{\chi}}} \frac{\epsilon}{|\chi|}$. Writing $\epsilon = |\epsilon| e^{i\upsilon}$, we can obtain the magnitude of the cavity field at steady state $|\langle \hat{a} \rangle|$,

$$|\langle \hat{a} \rangle| = G\left(\frac{\kappa}{\chi}, \frac{\Delta}{\chi}, \frac{\gamma}{\chi}, \upsilon \right) \left| \frac{\epsilon}{\chi} \right|, \tag{45}$$

where the gain $G = G(\frac{\kappa}{\chi}, \frac{\Delta}{\chi}, \frac{\gamma}{\chi}, \upsilon) \geq 0$ is

$$G = \frac{2}{R} \left| \sum_{k=0}^{\infty} \sum_{r=0}^{\infty} \frac{1}{2k+1} \frac{2^{2r}}{(2r)!} \left(\frac{\kappa}{\chi} \right)^{2r} \Gamma\left(\frac{1}{2} + r \right) \Gamma\left(\frac{3}{2} + r + k \right) \right.$$

$$\left. \times \left[\frac{1}{S} - \frac{(\frac{\kappa}{\chi}) e^{-2i\upsilon}}{S^*} \right] \right|, \tag{46}$$

where

$$R = \sum_{s=0}^{\infty} \frac{2^{2s}}{(2s)!} \left(\frac{\kappa}{\chi} \right)^{2s} \frac{|\Gamma(\frac{1}{2} + s)|^2}{|\Gamma(\frac{1}{2} + s + \frac{\Delta}{\chi} - i\frac{\gamma}{\chi})|^2},$$

$$S = \Gamma\left(\frac{1}{2} + r + \frac{\Delta}{\chi} + i\frac{\gamma}{\chi} \right) \Gamma\left(\frac{3}{2} + r + k + \frac{\Delta}{\chi} - i\frac{\gamma}{\chi} \right).$$

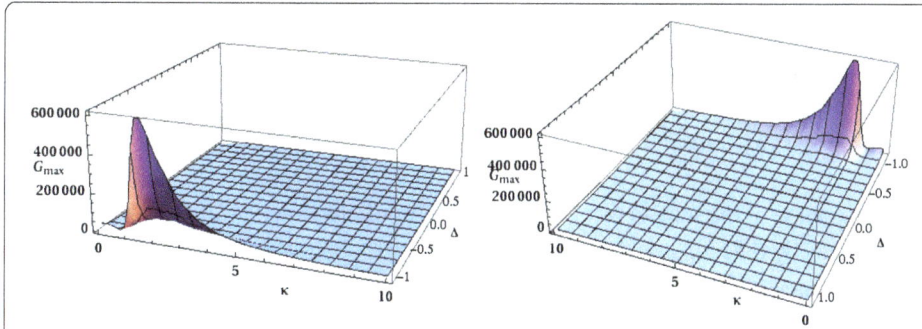

Figure 9 The maximum gain versus the pump magnitude and detuning. The maximum gain $G_{max} = \max\{G\}$ at the optimal signal phase v, versus the parametric pump magnitude κ and detuning Δ, with time units chosen so that $\gamma = 1$ and $\chi = 0.25$. The two plots show different camera perspectives of the same plotted data. The plot is made from summing 300 terms of the appropriate hypergeometric series. The summation is not normalised, and thus the gain values are only correct up to a scale; however, the shape of the plot is indicative.

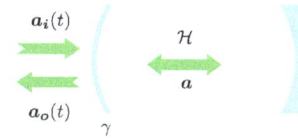

Figure 10 The superconducting microwave resonator modelled as a single-sided cavity for use as an input-output formulation. The superconducting microwave resonator modelled as a single-sided cavity for use as an input-output formulation. The incoming field mode operator is $\hat{a}_i(t)$ and the outgoing field mode operator is $\hat{a}_o(t)$. Loss from the microwave cavity occurs at a rate γ. Note here that γ is the coefficient of the amplitude decay, the coefficient for the photon number loss is 2γ. The Hamiltonian dynamics of the cavity mode $\hat{a}$ are governed by the Interaction picture Hamiltonian $\mathcal{H}$.

In Figure 9 we plot the maximum gain $G_{max} = \max\{G\}$ for a given parametric pump strength κ and detuning Δ. We have plotted the maximum gain by choosing the optimal signal phase v at each set of parameters. Comparing this to Figure 8 for the case when there is no coherent driving, we see that the gain is a maximum around the critical parametric driving strength in the bi-stable, negatively detuned region.

6 Linearised quantum system

6.1 Input-output formalism

We consider the microwave cavity with the input-output formulation of quantum optics, as originally described by Collett and Gardiner in [27]. To do this, we model the super-conducting microwave resonator as a single-sided cavity as depicted in Figure 10.

The input and output fields are treated explicitly with their mode annihilation operators $\hat{a}_i$ and $\hat{a}_o$ respectively. With this formulation, the quantum stochastic differential equation we obtain for the microwave resonator field mode operators $\hat{a}$ and $\hat{a}^{\dagger}$ are

$$\frac{d\hat{a}}{dt} = -\frac{i}{\hbar}[\hat{a}, \mathcal{H}] - \gamma\hat{a} + \sqrt{2\gamma}\hat{a}_i(t),$$

$$\frac{d\hat{a}^{\dagger}}{dt} = -\frac{i}{\hbar}[\hat{a}^{\dagger}, \mathcal{H}] - \gamma\hat{a}^{\dagger} + \sqrt{2\gamma}\hat{a}_i^{\dagger}(t),$$
(47)

where the input field is effectively white noise, uncorrelated in time,

$$\left[\hat{a}_i(t), \hat{a}_i^{\dagger}(t')\right] = \delta(t - t')\hat{I}.$$
(48)

The probability per unit time to detect a photon in the input field is $2\gamma \langle \hat{a}_i^\dagger(t)\hat{a}_i(t)\rangle$. Finally, the relationship between the input, output, and cavity fields is given by

$$\hat{a}_o(t) = \sqrt{2\gamma}\hat{a}(t) + e^{i\xi}\hat{a}_i(t), \tag{49}$$

where the phase of the second term, the reflected input, may vary with the system. For an almost perfectly reflecting mirror of an optical cavity we have $\xi = \pi$ and $e^{i\xi} = -1$, here we choose this phase as an appropriate approximation.

We also look at the various fields in the frequency domain, by defining the frequency-domain operators as the time-domain operators' Fourier transforms,

$$\hat{\tilde{a}}(\omega) = \mathcal{F}_{t\to\omega}\{\hat{a}(t)\},$$

$$\hat{\tilde{a}}_i(\omega) = \mathcal{F}_{t\to\omega}\{\hat{a}_i(t)\}, \tag{50}$$

$$\hat{\tilde{a}}_o(\omega) = \mathcal{F}_{t\to\omega}\{\hat{a}_o(t)\},$$

where we have used the Fourier Transform convention $\tilde{f}(\omega) = \mathcal{F}_{t\to\omega}\{f(t)\} = \frac{1}{\sqrt{2\pi}} \times \int_{-\infty}^{\infty} dx f(t)e^{-i\omega t}$. In the frequency domain, the input field is also uncorrelated in frequency,

$$\left[\hat{\tilde{a}}_i(\omega), \hat{\tilde{a}}_i^\dagger(\omega')\right] = \delta(\omega - \omega')\hat{I}, \tag{51}$$

and the relationship between the input, output, and cavity fields is then given by

$$\hat{\tilde{a}}_o(\omega) = \sqrt{2\gamma}\hat{\tilde{a}}(\omega) - \hat{\tilde{a}}_i(\omega). \tag{52}$$

6.2 Gain spectra

Recall our quantum equations of motion for the microwave cavity field (47). We linearise the system about a semi-classical fixed point α_0 as we did semi-classically in (14). This gives us the linearised equation of motion for the fluctuation

$$\begin{bmatrix} \frac{d(\widehat{\delta a}(t))}{dt} \\ \frac{d(\widehat{\delta a}^\dagger(t))}{dt} \end{bmatrix} = \mathbf{M}_{\alpha\alpha*} \begin{bmatrix} \widehat{\delta a}(t) \\ \widehat{\delta a}^\dagger(t) \end{bmatrix} + \sqrt{2\gamma} \begin{bmatrix} \hat{a}_i(t) \\ \hat{a}_i^\dagger(t) \end{bmatrix}, \tag{53}$$

where

$$\widehat{\delta a}(t) = \hat{a}(t) - \alpha_0. \tag{54}$$

In the frequency domain the linearised equation of motion (53) becomes

$$i\omega \begin{bmatrix} \widehat{\widetilde{\delta a}}(\omega) \\ \widehat{\widetilde{\delta a}}^\dagger(-\omega) \end{bmatrix} = \mathbf{M}_{\alpha\alpha*} \begin{bmatrix} \widehat{\widetilde{\delta a}}(\omega) \\ \widehat{\widetilde{\delta a}}^\dagger(-\omega) \end{bmatrix} + \sqrt{2\gamma} \begin{bmatrix} \hat{\tilde{a}}_i(\omega) \\ \hat{\tilde{a}}_i^\dagger(-\omega) \end{bmatrix}. \tag{55}$$

We rewrite this to obtain an expression for the microwave cavity field fluctuation in terms of the input radiation,

$$\begin{bmatrix} \widehat{\widetilde{\delta a}}(\omega) \\ \widehat{\widetilde{\delta a}}^\dagger(-\omega) \end{bmatrix} = \sqrt{2\gamma}(i\omega\mathbf{I} - \mathbf{M}_{\alpha\alpha*})^{-1} \begin{bmatrix} \hat{\tilde{a}}_i(\omega) \\ \hat{\tilde{a}}_i^\dagger(-\omega) \end{bmatrix}. \tag{56}$$

Using this expression, together with our input-output expression (52), allows us to obtain an expression for the output fluctuation of the microwave cavity in terms of the input,

$$
\begin{bmatrix} \hat{\bar{a}}_o(\omega) \\ \hat{\bar{a}}_o^\dagger(-\omega) \end{bmatrix} = \mathbf{G}(\omega) \begin{bmatrix} \hat{\bar{a}}_i(\omega) \\ \hat{\bar{a}}_i^\dagger(-\omega) \end{bmatrix},
\tag{57}
$$

where the gain matrix $\mathbf{G}(\omega)$ is

$$
\mathbf{G}(\omega) = \begin{bmatrix} G_{11}(\omega) & G_{12}(\omega) \\ G_{21}(\omega) & G_{22}(\omega) \end{bmatrix} = 2\gamma \left(i\omega \mathbf{I} - \mathbf{M}_{\alpha\alpha^*} \right)^{-1} - \mathbf{I}.
\tag{58}
$$

Note that $\hat{\bar{a}}_o(\omega)$ is the output fluctuation in the frequency domain. If we re-introduced the coherent term we would obtain the full output amplitude in the frequency domain, $\hat{\bar{a}}_o(\omega) = \hat{\bar{a}}_o(\omega) + \sqrt{4\pi\gamma}\alpha_0\delta(\omega)\hat{I}$.

Now, we can rewrite our Jacobian $\mathbf{M}_{\alpha\alpha^*}$ from (15) in terms of the parameters we defined in (13) for our semi-classical steady states,

$$
\mathbf{M}_{\alpha\alpha^*} = \gamma \begin{bmatrix} -1 - i(\Delta' + 2\bar{n}_0) & -i\bar{n}_0 e^{i2\theta_0} - i\kappa' \\ i\bar{n}_0 e^{-i2\theta_0} + i\kappa' & -1 + i(\Delta' + 2\bar{n}_0) \end{bmatrix}.
\tag{59}
$$

We now introduce two other useful parameters, $\Lambda_0 \in \mathbb{R}$, and $\omega' \in \mathbb{R}$,

$$
\begin{aligned}
\Lambda_0 &= \left(\kappa'\right)^2 - \Delta'^2 + \bar{n}_0\left(-4\Delta' + 2\kappa'\cos(2\theta_0) - 3\bar{n}_0\right), \\
\omega' &= \frac{\omega}{\gamma}.
\end{aligned}
\tag{60}
$$

In terms of these parameters, the gain matrix $\mathbf{G}(\omega)$ is

$$
\begin{aligned}
\mathbf{G}(\omega) = & \frac{1}{\Lambda_0 + (\omega' - i)^2} \\
& \times \begin{bmatrix} -1 - \Lambda_0 - \omega'^2 + 2i(\Delta' + 2\bar{n}_0) & 2i\bar{n}_0 e^{i2\theta_0} + 2i\kappa \\ -2i\bar{n}_0 e^{-2i\theta_0} - 2i\kappa & -1 - \Lambda_0 - \omega'^2 - 2i(\Delta' + 2\bar{n}_0) \end{bmatrix}.
\end{aligned}
\tag{61}
$$

To calculate the gain measured at an arbitrary phase, we first define the quadrature operator in the frequency domain $\hat{X}_\phi(\omega) = \hat{\bar{a}}_o(\omega)e^{i\phi} + \hat{\bar{a}}_o^\dagger(-\omega)e^{-i\phi}$, which can be written in terms of our gain matrix as

$$
\hat{X}_\phi(\omega) = \begin{bmatrix} e^{i\phi} & e^{-i\phi} \end{bmatrix} \mathbf{G}(\omega) \begin{bmatrix} \hat{\bar{a}}_i(\omega) \\ \hat{\bar{a}}_i^\dagger(-\omega) \end{bmatrix}.
\tag{62}
$$

This expression reduces to

$$
\hat{X}_\phi(\omega) = g_\phi(\omega)\left(\hat{\bar{a}}_i(\omega)e^{i(\phi+\zeta(\omega))} + \hat{\bar{a}}_i^\dagger(-\omega)e^{-i(\phi+\zeta(\omega))}\right),
\tag{63}
$$

where our signal gain $g_\phi(\omega)$ at phase ϕ is

$$
g_\phi(\omega) = \frac{|-1 - \Lambda_0 - \omega'^2 - 2ie^{-2i\phi}(\kappa' + \bar{n}_0 e^{-2i\theta_0}) + 2i(\Delta' + 2\bar{n}_0)|}{\Lambda_0 + (\omega' - i)^2},
\tag{64}
$$

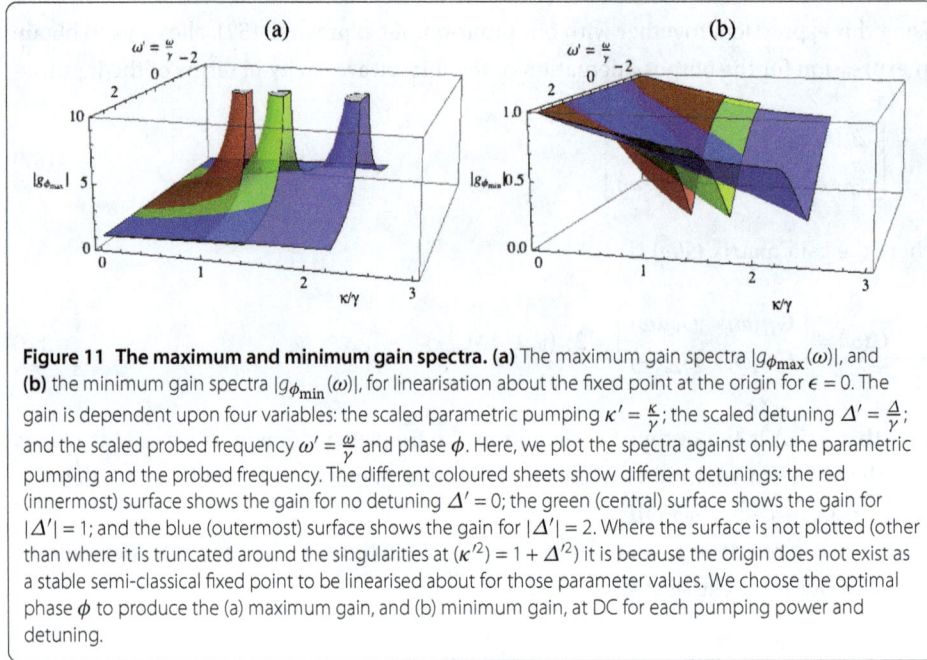

Figure 11 The maximum and minimum gain spectra. (a) The maximum gain spectra $|g_{\phi_{\max}}(\omega)|$, and **(b)** the minimum gain spectra $|g_{\phi_{\min}}(\omega)|$, for linearisation about the fixed point at the origin for $\epsilon = 0$. The gain is dependent upon four variables: the scaled parametric pumping $\kappa' = \frac{\kappa}{\gamma}$; the scaled detuning $\Delta' = \frac{\Delta}{\gamma}$; and the scaled probed frequency $\omega' = \frac{\omega}{\gamma}$ and phase ϕ. Here, we plot the spectra against only the parametric pumping and the probed frequency. The different coloured sheets show different detunings: the red (innermost) surface shows the gain for no detuning $\Delta' = 0$; the green (central) surface shows the gain for $|\Delta'| = 1$; and the blue (outermost) surface shows the gain for $|\Delta'| = 2$. Where the surface is not plotted (other than where it is truncated around the singularities at $(\kappa'^2) = 1 + \Delta'^2$) it is because the origin does not exist as a stable semi-classical fixed point to be linearised about for those parameter values. We choose the optimal phase ϕ to produce the (a) maximum gain, and (b) minimum gain, at DC for each pumping power and detuning.

and the frequency-dependent phase shift $\zeta(\omega)$ is

$$\zeta(\omega) = \arg\left(-1 - \Lambda_0 - \omega'^2 - 2i e^{-2i\phi}\left(\kappa' + \bar{n}_0 e^{-2i\theta_0}\right) + 2i\left(\Delta' + 2\bar{n}_0\right)\right). \tag{65}$$

Note that while our signal gain $g_\phi(\omega)$ is complex for non-zero frequency, it is real for the DC frequency in this frame.

If we consider our analytically solved case of no linear driving bias ($\epsilon = \bar{\epsilon}' = 0$), and linearise about the 'below threshold' stable fixed point at the origin, then $\Lambda_0 = \kappa'^2 - \Delta'^2 < 1$ and our gain $|g_\phi(\omega)|$ at a phase ϕ can be plotted against the scaled parametric pumping κ' and detuning Δ'. We plot the maximum gain $|g_{\phi_{\max}}(\omega)|$ for each value of parametric pumping and detuning (optimising ϕ to find the maximum gain at DC for each pair of these parameters) in Figure 11.

6.3 Squeezing spectra

Having derived the gain matrix (61) relating the input to the output, we are now also in a position to investigate the squeezing spectrum of the microwave system. Recall our quadrature operator in the frequency domain $\hat{X}_\phi(\omega) = \hat{\tilde{a}}_o(\omega)e^{i\phi} + \hat{\tilde{a}}_o^\dagger(-\omega)e^{-i\phi}$ written in terms of our gain matrix in (62). Our squeezing spectrum is the variance of this quadrature operator. We thus define this squeezing spectrum $S_\phi(\omega)$, again in the frequency domain, to be

$$S_\phi(\omega) = \int_{-\infty}^{\infty} \langle \hat{X}_\phi(\omega), \hat{X}_\phi(\omega') \rangle \, d\omega', \tag{66}$$

where the notation for the covariance bracket is $\langle \hat{A}, \hat{B} \rangle = \langle \hat{A}\hat{B} \rangle - \langle \hat{A} \rangle \langle \hat{B} \rangle$. We then use the linearity of both arguments of the covariance bracket to express the variance of the quadrature operator as

$$S_\phi(\omega) = \int_{-\infty}^{\infty} \begin{bmatrix} e^{i\phi} & e^{-i\phi} \end{bmatrix} \mathbf{G}(\omega) \begin{bmatrix} \langle \hat{\tilde{a}}_i(\omega), \hat{\tilde{a}}_i(\omega') \rangle & \langle \hat{\tilde{a}}_i(\omega), \hat{\tilde{a}}_i^\dagger(-\omega') \rangle \\ \langle \hat{\tilde{a}}_i^\dagger(-\omega), \hat{\tilde{a}}_i(\omega') \rangle & \langle \hat{\tilde{a}}_i^\dagger(-\omega), \hat{\tilde{a}}_i^\dagger(-\omega') \rangle \end{bmatrix}$$
$$\times \mathbf{G}(\omega')^T \begin{bmatrix} e^{i\phi} \\ e^{-i\phi} \end{bmatrix} d\omega'. \tag{67}$$

Then, using the commutation relation of (51) we can rewrite $S_\phi(\omega)$ in terms of normally-ordered variances of the input field as

$$S_\phi(\omega) = \int_{-\infty}^{\infty} \begin{bmatrix} e^{i\phi} & e^{-i\phi} \end{bmatrix} \mathbf{G}(\omega) \begin{bmatrix} \langle \hat{\tilde{a}}_i(\omega), \hat{\tilde{a}}_i(\omega') \rangle & \langle \hat{\tilde{a}}_i^\dagger(-\omega), \hat{\tilde{a}}_i(\omega') \rangle \\ & + \delta(\omega + \omega') \\ \langle \hat{\tilde{a}}_i^\dagger(-\omega), \hat{\tilde{a}}_i(\omega') \rangle & \langle \hat{\tilde{a}}_i^\dagger(-\omega), \hat{\tilde{a}}_i^\dagger(-\omega') \rangle \end{bmatrix}$$
$$\times \mathbf{G}(\omega')^T \begin{bmatrix} e^{i\phi} \\ e^{-i\phi} \end{bmatrix} d\omega'. \tag{68}$$

To proceed we now use the statistics of the input field. A coherent input field has zero normally-ordered variances ($\langle \hat{\tilde{a}}_i(\omega), \hat{\tilde{a}}_i(\omega') \rangle = \langle \hat{\tilde{a}}_i^\dagger(-\omega'), \hat{\tilde{a}}_i(\omega) \rangle = \langle \hat{\tilde{a}}_i^\dagger(-\omega), \hat{\tilde{a}}_i^\dagger(-\omega') \rangle = 0$). Thus, the only non-zero term in central matrix is the delta function term arising from the commutation relations. Using this, we can compute the integral of the matrix expression, and our squeezing spectrum reduces to

$$S_\phi(\omega) = G_{11}(\omega)G_{12}(-\omega)e^{i2\phi} + G_{11}(\omega)G_{22}(-\omega)$$
$$+ G_{21}(\omega)G_{12}(-\omega) + G_{21}(\omega)G_{22}(-\omega)e^{-i2\phi}, \tag{69}$$

or explicitly,

$$S_\phi(\omega) = \frac{|-1 - \Lambda_0 - \omega'^2 - i2e^{-i2\phi}(\kappa' + \bar{n}_0 e^{-i2\theta_0}) + i2(\Delta' + 2\bar{n}_0)|^2}{|\Lambda_0 + (\omega' - i)^2|^2}. \tag{70}$$

If we consider our analytically solved case of no linear driving bias ($\epsilon = \bar{\epsilon}' = 0$), and linearise about the 'below threshold' stable fixed point at the origin, then $\Lambda_0 = \kappa'^2 - \Delta'^2 < 1$ and our squeezing $S_\phi(\omega)$ at a phase ϕ can be plotted against the scaled parametric pumping κ' and detuning Δ'. We plot the squeezing spectrum for each value of parametric pumping and detuning (setting the phase ϕ to be that which gives the minimum noise searching over all frequencies for each pair of these parameters) in Figure 12.

6.4 Signal to noise ratio

For operation of the microwave system as a bifurcation amplifier, the parameters which result in maximum gain may not result in minimum noise. Instead, rather than optimising for maximum gain or minimum noise individually, the quantity which we wish to maximise is the signal to noise ratio. However, we see from our expressions for the gain (64) and noise (70), that $S_\phi(\omega) = |g_\phi(\omega)|^2$, and that our signal to noise ratio is thus unity,

$$\text{SNR}_\phi(\omega) = \frac{|g_\phi(\omega)|}{\sqrt{S_\phi(\omega)}} = 1. \tag{71}$$

Figure 12 The minimum possible noise spectra.
The minimum possible noise spectra $S_{\phi_{\min}}(\omega)$ for
linearisation about the fixed point at the origin for ϵ
= 0. The noise spectrum is dependent on four
parameters: the scaled parametric pumping $\kappa' = \frac{\kappa}{\gamma}$;
the scaled detuning $\Delta' = \frac{\Delta}{\gamma}$; and the scaled probed
frequency $\omega' = \frac{\omega}{\gamma}$ and phase ϕ. Here we plot the
noise spectrum against only the parametric
pumping and the probed frequency. The different
coloured sheets show different detunings: the red
(innermost) surface shows the noise for no detuning
$\Delta' = 0$; the green (central) surface shows the noise
for $|\Delta'| = 1$; and the blue (outermost) surface shows the noise for $|\Delta'| = 2$. Where the surface is not plotted it
is because the origin does not exist as a stable semi-classical fixed point to be linearised about for those
parameter values. We choose the optimal phase ϕ to produce the minimum noise $S_\phi(\omega)$ searching over all
frequencies for each pumping power and detuning.

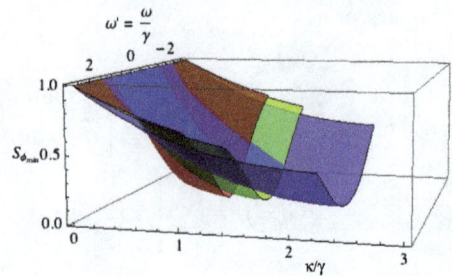

For the linearised system, this equality holds for all values of all parameters (parametric
pumping, detuning, and cavity dissipation), all probed frequencies and phases, and regardless of which semi-classical fixed point we choose to linearise about.

Physically the means that our system is acting as a parametric amplifier. The quadrature
of maximum gain is the same as the quadrature of maximum noise, and vice-versa for the
minimum gain and noise. We can thus use this microwave system to amplify a signal to a
measurable level without affecting its signal to noise ratio.

7 Conclusion

In this paper we detailed the quantum and semi-classical structure of a superconducting
microwave resonator connected through a SQUID loop to ground. In particular we observed that the semi-classical model contains a bifurcation structure, and that the remains
of this structure are still visible in the full quantum mechanical steady state. Furthermore,
we showed it can be used as a bifurcation amplifier. We did this analysis by: linearising
about the semi-classical steady state below the 'threshold' of the amplifier; by truncating
the oscillator basis in the Fock basis and numerically computing the quantum phase space
at steady state; and also by computing the exact quantum steady state by using an analytical
phase space technique.

First, we showed that the corresponding semi-classical model has its fixed points determined by a quintic polynomial. We showed that for the small linear signal regime $\epsilon = 0$,
that this quintic factors and is analytically solvable. This semi-classical system then undergoes a bifurcation of its semi-classical steady state with increased parametric pumping
power. This bifurcation gives a threshold for the amplifier and occurs when the parametric pumping power equals the cavity decay, with adjustment for a detuned drive,
$|\kappa|^2 = \gamma^2 + \Delta^2$. The sign of the detuning specifies the form of the bifurcations. For a positive
detuning $\Delta \geq 0$, the origin undergoes a supercritical pitchfork bifurcation at the thresh-
old. For negative detuning $\Delta < 0$, the origin instead loses its stability at $|\kappa|^2 = \gamma^2 + \Delta^2$ in
a subcritical pitchfork bifurcation with two intermediate pairs of fixed points created in
saddle-node bifurcations when the parametric pumping power reaches $|\kappa|^2 = \gamma^2$. The nu-
merically calculated quantum steady states were shown to have clear signatures of these
semi-classical steady state bifurcations. Specifically, the Wigner function representation
of the quantum phase space was seen to have support on the semi-classical fixed points.

In addition to numerically computing the quantum phase space at steady state by truncating the oscillator basis, we also calculated the exact quantum steady state. This was done following the work of Kruchkyan and Kheruntsyan [26] by using the Positive P representation. The method took advantage of the fact that the potential conditions were satisfied. The exact quantum phase space density at steady state was seen to be peaked in the vicinity of the corresponding semi-classical fixed points.

We showed that the quantum device functioned as a bifurcation amplifier until threshold. We calculated the small signal gain of the amplifier using the exact quantum steady state. We also approximated this by linearising the steady state about the semi-classical below-threshold fixed point using the input-output formalism of Collett and Gardiner [27]. With this procedure we also calculated noise spectra, and we showed that the signal to noise ratio at all frequencies and phases was equal to unity. We thus showed that the quarter-wave microwave resonator considered can be made to act as a parametric amplifier. This device can take a signal from a nano-electromechanical system and amplify it to a measurable level without affecting its signal to noise ratio.

Competing interests
The authors declare that they have no competing interests.

Authors' contributions
The paper was written by CPM who did the numerical simulations and, together with GJM, performed the analytic calculations. HN obtained the solution in Section 4. TD provided some of the experimental parameters.

Author details
[1]Department of Physics, The University of Queensland, St Lucia, QLD 4072, Australia. [2]Department of Physics, Texas A & M University at Qatar, PO Box 23874, Doha, Qatar. [3]Department of Physics, The University of New South Wales, Kensington, NSW 2052, Australia.

Acknowledgements
This work was supported by the Australian Research Council grants FF0776191 and CE110001014.

References
1. Devoret MH, Girvin S, Schoelkopf R: *Ann. Phys.* 2007, **16**:767-779. doi:10.1002/andp.200710261.
2. Schoelkopf RJ, Girvin SM: *Nature* 2008, **451**:664-669. doi:10.1038/451664a.
3. Sandberg M, Wilson CM, Persson F, Bauch T, Johansson G, Shumeiko V, Duty T, Delsing P: *Appl. Phys. Lett.* 2008, **42**:203501. doi:10.1063/1.2929367.
4. Wallquist M, Shumeiko VS, Wendin G: *Phys. Rev. B* 2006, **74**:224506.
5. Wustmann W, Shumeiko V: *Phys. Rev. B* 2013, **87**:184501.
6. Wilson CM, Duty T, Sandberg M, Persson F, Shumeiko V, Delsing P: *Phys. Rev. Lett.* 2010, **105**:233907.
7. Kozinsky I, Postma HWC, Kogan O, Husain A, Roukes ML: *Phys. Rev. Lett.* 2007, **99**:207201. doi:10.1103/PhysRevLett.99.207201.
8. Babourina-Brooks E, Doherty A, Milburn GJ: *New J. Phys.* 2008, **10**:105020. doi:10.1088/1367-2630/10/10/105020.
9. Woolley MJ, Doherty AC, Milburn GJ, Schwab KC: *Phys. Rev. A* 2008, **78**:062303. doi:10.1103/PhysRevA.78.062303.
10. Hertzberg JB, Rocheleau T, Ndukum T, Savva M, Clerk AA, Schwab KC: *Nat. Phys.* 2010, **6**:213.
11. Rocheleau T, Ndukum T, Macklin C, Hertzberg JB, Clerk AA, Schwab KC: *Nature* 2010, **463**:72.
12. Wallraff A, Schuster DI, Blais A, Frunzio L, Huang R-S, Majer J, Kumar S, Girvin SM, Schoelkopf RJ: *Nature* 2004, **431**:162.
13. Wielinga B, Milburn GJ: *Phys. Rev. A* 1993, **48**:2494. doi:10.1103/PhysRevA.48.2494.
14. Marthaler M, Dykman M: *Phys. Rev. A* 2006, **73**:042108.
15. Walls DF, Milburn GJ: *Quantum Optics*. 2nd edition. Berlin: Springer; 2008.
16. Braunstein SL, Caves CM, Milburn GJ: *Phys. Rev. A* 1991, **43**:1153.
17. Eichler C, Bozyigit D, Wallraff A: *Phys. Rev. A* 2012, **86**:032106.
18. Hilborn RC: *Chaos and Nonlinear Dynamics*. Oxford: Oxford University Press; 1994.
19. Hines AP, Dawson CM, McKenzie RH, Milburn GJ: *Phys. Rev. A* 2004, **70**:022303. doi:10.1103/PhysRevA.70.022303.
20. Meaney CP, Duty T, McKenzie RH, Milburn GJ: *Phys. Rev. A* 2010, **81**:043805. doi:10.1103/PhysRevA.81.043805.
21. Meaney CP, McKenzie RH, Milburn GJ: *Phys. Rev. E* 2011, **83**:056202.
22. Carmichael HJ: *Statistical Methods in Quantum Optics 1*. Berlin: Springer; 2008.
23. Wolinsky M, Carmichael HJ: *Phys. Rev. Lett.* 1988, **60**:1836.
24. Tan SM: **Quantum optics and computation toolbox for MATLAB**; 2002.
25. Zachos CK, Fairlie DB, Curtright TL (Eds): World Scientific.
26. Kruchkyan GY, Kheruntsyan KV: *Opt. Commun.* 1996, **127**:230-236.
27. Collett MJ, Gardiner CW: *Phys. Rev. A* 1984, **30**:1386.

Permissions

The contributors of this book come from diverse backgrounds, making this book a truly international effort. This book will bring forth new frontiers with its revolutionizing research information and detailed analysis of the nascent developments around the world.

We would like to thank all the contributing authors for lending their expertise to make the book truly unique. They have played a crucial role in the development of this book. Without their invaluable contributions this book wouldn't have been possible. They have made vital efforts to compile up to date information on the varied aspects of this subject to make this book a valuable addition to the collection of many professionals and students.

This book was conceptualized with the vision of imparting up-to-date information and advanced data in this field. To ensure the same, a matchless editorial board was set up. Every individual on the board went through rigorous rounds of assessment to prove their worth. After which they invested a large part of their time researching and compiling the most relevant data for our readers.

The editorial board has been involved in producing this book since its inception. They have spent rigorous hours researching and exploring the diverse topics which have resulted in the successful publishing of this book. They have passed on their knowledge of decades through this book. To expedite this challenging task, the publisher supported the team at every step. A small team of assistant editors was also appointed to further simplify the editing procedure and attain best results for the readers.

Apart from the editorial board, the designing team has also invested a significant amount of their time in understanding the subject and creating the most relevant covers. They scrutinized every image to scout for the most suitable representation of the subject and create an appropriate cover for the book.

The publishing team has been an ardent support to the editorial, designing and production team. Their endless efforts to recruit the best for this project, has resulted in the accomplishment of this book. They are a veteran in the field of academics and their pool of knowledge is as vast as their experience in printing. Their expertise and guidance has proved useful at every step. Their uncompromising quality standards have made this book an exceptional effort. Their encouragement from time to time has been an inspiration for everyone.

The publisher and the editorial board hope that this book will prove to be a valuable piece of knowledge for researchers, students, practitioners and scholars across the globe.

List of Contributors

Kyle M Sundqvist
Electrical and Computer Engineering, Texas A&M University, College Station, TX 77843, USA

Per Delsing
Microtechnology and Nanoscience, Chalmers University of Technology, Göteborg, SE-412 96, Sweden

Muriel Brengues
Center for Applied NanoBioscience and Medicine, The University of Arizona College of Medicine, 425 N. 5th Street, Phoenix, AZ 85004, USA

David Liu
Scottsdale Clinical Research Institute, Scottsdale Healthcare, 10510 N. 92nd Street, Scottsdale, AZ 85258, USA

Ronald Korn
Scottsdale Clinical Research Institute, Scottsdale Healthcare, 10510 N. 92nd Street, Scottsdale, AZ 85258, USA

Frederic Zenhausern
Center for Applied NanoBioscience and Medicine, The University of Arizona College of Medicine, 425 N. 5th Street, Phoenix, AZ 85004, USA
Scottsdale Clinical Research Institute, Scottsdale Healthcare, 10510 N. 92nd Street, Scottsdale, AZ 85258, USA

Bradley D Hauer
Department of Physics, University of Alberta, T6G 2E1 Edmonton, AB, Canada

Paul H Kim
Department of Physics, University of Alberta, T6G 2E1 Edmonton, AB, Canada

Callum Doolin
Department of Physics, University of Alberta, T6G 2E1 Edmonton, AB, Canada

Allison JR MacDonald
Department of Physics, University of Alberta, T6G 2E1 Edmonton, AB, Canada

Hugh Ramp
Department of Physics, University of Alberta, T6G 2E1 Edmonton, AB, Canada

John P Davis
Department of Physics, University of Alberta, T6G 2E1 Edmonton, AB, Canada

Benjamin A Bircher
Center for Cellular Imaging and NanoAnalytics, Biozentrum, University of Basel, Mattenstrasse 26, CH-4058 Basel, Switzerland

Roger Krenger
Center for Cellular Imaging and NanoAnalytics, Biozentrum, University of Basel, Mattenstrasse 26, CH-4058 Basel, Switzerland

Thomas Braun
Center for Cellular Imaging and NanoAnalytics, Biozentrum, University of Basel, Mattenstrasse 26, CH-4058 Basel, Switzerland

Seonghwan Kim
Department of Mechanical and Manufacturing Engineering, University of Calgary, T2N 1N4 Calgary, AB, Canada

Dongkyu Lee
Department of Chemical and Materials Engineering, University of Alberta, T6G 2V4 Edmonton, AB, Canada

Thomas Thundat
Department of Chemical and Materials Engineering, University of Alberta, T6G 2V4 Edmonton, AB, Canada

Marvin Weyland
Physikalisch-Technische Bundesanstalt, Bundesallee 100, 38116 Braunschweig, Germany
Max-Planck-Institut für Kernphysik, Saupfercheckweg 1, 69117 Heidelberg, Germany

Xueguang Ren
Physikalisch-Technische Bundesanstalt, Bundesallee 100, 38116 Braunschweig, Germany
Max-Planck-Institut für Kernphysik, Saupfercheckweg 1, 69117 Heidelberg, Germany

Thomas Pflüger
Physikalisch-Technische Bundesanstalt, Bundesallee 100, 38116 Braunschweig, Germany
Max-Planck-Institut für Kernphysik, Saupfercheckweg 1, 69117 Heidelberg, Germany

Woon Yong Baek
Physikalisch-Technische Bundesanstalt, Bundesallee 100, 38116 Braunschweig, Germany

Klaus Bartschat
Department of Physics and Astronomy, Drake University, Des Moines, IA 50311, USA

Oleg Zatsarinny
Department of Physics and Astronomy, Drake University, Des Moines, IA 50311, USA

Dmitry V Fursa
ARC Centre for Antimatter-Matter Studies, Curtin University, Perth, WA 6845, Australia

Igor Bray
ARC Centre for Antimatter-Matter Studies, Curtin University, Perth, WA 6845, Australia

Hans Rabus
Physikalisch-Technische Bundesanstalt, Bundesallee 100, 38116 Braunschweig, Germany

Alexander Dorn
Max-Planck-Institut für Kernphysik, Saupfercheckweg 1, 69117 Heidelberg, Germany

Surabhi Joshi
Department of Mechanical Engineering, McGill University, 817 Sherbrooke Street West, Montreal, Quebec H3A 0C3, Canada

Sherman Hung
Department of Mechanical Engineering, McGill University, 817 Sherbrooke Street West, Montreal, Quebec H3A 0C3, Canada

Srikar Vengallatore
Department of Mechanical Engineering, McGill University, 817 Sherbrooke Street West, Montreal, Quebec H3A 0C3, Canada

Ann-Lauriene Haag
Department of Physics, McGill University, 3600 Rue University, Montreal, QC H3A 2T8, Canada

Yoshihiko Nagai
Research Institute ofthe McGill University Health Centre, 2155 Guy Street, Montreal, QC H3H 2R9, Canada

R Bruce Lennox
Department of Chemistry and FQRNT Centre for Self Assembled Chemical Structures, McGill University, 801 Sherbrooke Street West, Montreal QC H3A 2K6, Canada

Peter Grütter
Department of Physics, McGill University, 3600 Rue University, Montreal, QC H3A 2T8, Canada

Ricardo Jose S Guerrero
World Premier International (WPI) Research Center Initiative, International Center for Materials Nanoarchitectonics (MANA), National Institute for Materials Science (NIMS), 1-1 Namiki, 305-0044 Tsukuba, Japan
University of Waterloo, 200 University Ave W, N2L 3G1 Waterloo, Canada

Francis Nguyen
World Premier International (WPI) Research Center Initiative, International Center for Materials Nanoarchitectonics (MANA), National Institute for Materials Science (NIMS), 1-1 Namiki, 305-0044 Tsukuba, Japan
University of Waterloo, 200 University Ave W, N2L 3G1 Waterloo, Canada

Genki Yoshikawa
World Premier International (WPI) Research Center Initiative, International Center for Materials Nanoarchitectonics (MANA), National Institute for Materials Science (NIMS), 1-1 Namiki, 305-0044 Tsukuba, Japan

Rico Otto
Department of Chemistry and Biochemistry, University of California, San Diego, 9500 Gilman Drive, La Jolla, CA, 92093–0340, USA

Amelia W Ray
Department of Chemistry and Biochemistry, University of California, San Diego, 9500 Gilman Drive, La Jolla, CA, 92093–0340, USA

Jennifer S Daluz
Department of Chemistry and Biochemistry, University of California, San Diego, 9500 Gilman Drive, La Jolla, CA, 92093–0340, USA

Robert E Continetti
Department of Chemistry and Biochemistry, University of California, San Diego, 9500 Gilman Drive, La Jolla, CA, 92093–0340, USA

Adrian Ghita
School of Physics and Astronomy, University of Nottingham, Nottingham NG7 2RD, UK

Flavius C Pascut
School of Physics and Astronomy, University of Nottingham, Nottingham NG7 2RD, UK

Virginie Sottile
School of Medicine, University of Nottingham, Nottingham NG7 2RD, UK

Chris Denning
School of Medicine, University of Nottingham, Nottingham NG7 2RD, UK

Ioan Notingher
School of Physics and Astronomy, University of Nottingham, Nottingham NG7 2RD, UK

Ruben Wiese
Institut für Experimentelle und Angewandte Physik, Leibnizstr. 19, Kiel D-24098, Germany

Holger Kersten
Institut für Experimentelle und Angewandte Physik, Leibnizstr. 19, Kiel D-24098, Germany

Georg Wiese
Formerly Institut für Plasmaforschung und Technologie, Felix-Hausdorff-Str. 2, Greifswald D-17489, Germany

René Bartsch
Formerly Institut für Plasmaforschung und Technologie, Felix-Hausdorff-Str. 2, Greifswald D-17489, Germany

Ralph van Oorschot
MA3 Solutions, Eindhoven, The Netherlands

Hector Hugo Perez Garza
Department of Precision and Microsystems Engineering, Delft University of Technology, Delft, The Netherlands

Roy J S Derks
MA3 Solutions, Eindhoven, The Netherlands

Urs Staufer
Department of Precision and Microsystems Engineering, Delft University of Technology, Delft, The Netherlands

Murali Krishna Ghatkesar
Department of Precision and Microsystems Engineering, Delft University of Technology, Delft, The Netherlands

Matthias Hudl
KTH Royal Institute of Technology, ICT Materials Physics, Electrum 229, SE-164 40 Kista, Sweden
Department of Materials, ETH Zürich, Vladimir-Prelog-Weg 4, CH-8093 Zürich, Switzerland

Peter Lazor
Department of Earth Sciences, Uppsala University, Villavägen 16, SE-752 36 Uppsala, Sweden

Roland Mathieu
Department of Engineering Sciences, Uppsala University, Box 534, SE-751 21 Uppsala, Sweden

Alexander G Gavriliuk
Institute of Crystallography, Russian Academy of Sciences, Leninsky pr. 59, 119333 Moscow, Russia
Institute for Nuclear Research, Russian Academy of Sciences, 60-letiya Oktyabrya prospekt 7a, 117312 Moscow, Russia

Viktor V Struzhkin
Geophysical Laboratory, Carnegie Institution of Washington, 5251 Broad Branch Road NW, 20015 Washington DC, USA

Charles H Meaney
Department of Physics, The University of Queensland, St Lucia, QLD 4072, Australia

Hyunchul Nha
Department of Physics, Texas A & M University at Qatar, PO Box 23874, Doha, Qatar

Timothy Duty
Department of Physics, The University of New South Wales, Kensington, NSW 2052, Australia

Gerard J Milburn
Department of Physics, The University of Queensland, St Lucia, QLD 4072, Australia